Topics in Parallel and Distributed Computing

Sushil K. Prasad • Anshul Gupta • Arnold Rosenberg
Alan Sussman • Charles Weems
Editors

Topics in Parallel and Distributed Computing

Enhancing the Undergraduate Curriculum:
Performance, Concurrency, and Programming
on Modern Platforms

 Springer

Editors
Sushil K. Prasad
Georgia State University
Atlanta, GA, USA

Anshul Gupta
IBM Research AI
Yorktown Heights, NY, USA

Arnold Rosenberg
University of Massachusetts Amherst
Amherst, MA, USA

Alan Sussman
University of Maryland
College Park, MD, USA

Charles Weems
University of Massachusetts Amherst
Amherst, MA, USA

ISBN 978-3-030-06591-1   ISBN 978-3-319-93109-8   (eBook)
https://doi.org/10.1007/978-3-319-93109-8

This Springer imprint is published by the registered company Springer Nature Switzerland AG
The registered company address is: Gewerbestrasse 11, 6330 Cham, Switzerland

Contents

Editors' Introduction and Roadmap

Sushil K. Prasad, Anshul Gupta, Arnold L. Rosenberg, Alan Sussman, and Charles Weems

The premise of the NSF-supported Center for Parallel and Distributed Computing Curriculum Development and Educational Resources (CDER) is that every computer science (CS) and computer engineering (CE) undergraduate student should achieve a basic skill level in parallel and distributed computing (PDC). This book is a companion to our 2015 book, the first product of the CDER Book Project.[1] The book series we have embarked on addresses the lack of adequate textbook support for integrating PDC-related topics into undergraduate courses, especially in the early curriculum.[2]

---

[1]Prasad, Gupta, Rosenberg, Sussman, and Weems. 2015. Topics in Parallel and Distributed Computing: Introducing Concurrency in Undergraduate Courses, 1st Edition, Morgan Kaufmann, ISBN : 9780128038994, Pages: 360. http://grid.cs.gsu.edu/~tcpp/curriculum/?q=cedr_book

[2]This material is based upon work partially supported by the National Science Foundation under Grants IIS 1143533, CCF 1135124, CCF 1048711 and CNS 0950432. Any opinions, findings, and conclusions or recommendations expressed in this material are those of the author(s) and do not necessarily reflect the views of the National Science Foundation.

S. K. Prasad (✉)
Georgia State University, Atlanta, GA, USA
e-mail: sprasad@gsu.edu

A. Gupta
IBM Research AI, Yorktown Heights, NY, USA

A. L. Rosenberg · C. Weems
University of Massachusetts Amherst, Amherst, MA, USA

A. Sussman
University of Maryland, College Park, MD, USA

© Springer International Publishing AG, part of Springer Nature 2018
S. K. Prasad et al. (eds.), *Topics in Parallel and Distributed Computing*,
https://doi.org/10.1007/978-3-319-93109-8_1

1

Why the CDER Book Project?

A curriculum working group drawn from the IEEE Technical Committee on Parallel Processing (TCPP), the National Science Foundation (NSF), and sibling communities such as the ACM and industry, has taken up the challenge of proposing and refining curricular guidelines for blending PDC-related concepts into early-stage undergraduate curricula in computational areas. A first version of the group's guidelines for a core curriculum that includes PDC was released in December 2012.[3] This curriculum and related activities – see Appendix for a brief description of the NSF/TCPP Curriculum Initiative – have spawned a vibrant international community of educators who are committed to PDC education. It is from this community that the desirability of the *CDER Book Project*, a series of books to support both instructors and students of PDC, became evident.

Curricular guidelines such as those promulgated by both us and the CS2013 ACM/IEEE Computer Science Curriculum Joint Task Force are an essential first step in propelling the teaching of PDC-related material into the twenty-first century. But such guidelines are only a first step: both instructors and students will benefit from suitable textual material to effectively translate guidelines into the curriculum. Moreover, experience to this point has made it clear that the members of the PDC community have much to share with each other and with aspiring new members, in terms of creativity in forging new directions and experience in evaluating existing ones. The Book Project's goal is to engage the community to address the need for suitable textbooks and related textual material to integrate PDC topics into the lower level core courses (which we affectionately, and hopefully transparently, refer to as CS1, CS2, Systems, Data Structures and Algorithms, Logic Design, etc.), and, as appropriate, into upper level courses. The current edited book series intends, over time, to cover all of these proposed topics.

In 2016, we invited proposals for chapters on teaching parallel and distributed computing topics, suitable for either instructors or students, specifically on topics from the current NSF/TCPP curriculum guidelines for introductory courses that have not been addressed by the chapters in the earlier book. Subsequently, we saw good community interest in authoring chapters for higher level elective courses as well. To address this, we extended the scope of this book to both lower-level core courses and more advanced, specialized topics in parallel and distributed computing that are targeted at students in upper level classes. The book has evolved organically based on contributions received in response to calls for book chapters. All contributions have been rigorously reviewed internally by the editors and externally by experts.

---

[3]Prasad, S. K., Chtchelkanova, A., Dehne, F., Gouda, M., Gupta, A., Jaja, J., Kant, K., La Salle, A., LeBlanc, R., Lumsdaine, A., Padua, D., Parashar, M., Prasanna, V., Robert, Y., Rosenberg, A., Sahni, S., Shirazi, B., Sussman, A., Weems, C., and Wu, J. 2012. NSF/IEEE-TCPP Curriculum Initiative on Parallel and Distributed Computing – Core Topics for Undergraduates, Version I, Online: https://grid.cs.gsu.edu/~tcpp/curriculum/, 55 pages.

Book Organization

This book has two parts.

Part I – For instructors: These chapters are aimed at instructors to provide background, scholarly materials, insights into pedagogical strategies, and descriptions of experience with both strategies and materials. The emphasis is on the basic concepts and references on what and how to teach PDC topics in the context of the existing topics in various core courses.

Part 2 – For students: These chapters provide supplemental textual material for core courses that students can rely on for learning and exercises. These are envisioned as being at the quality of a textbook presentation, with many illustrations and examples, following a sequence of smaller steps to build larger concepts. We envision the student materials as supplemental sections that could be inserted into existing texts by instructors.

Print and Free Web Publication: While a print version through a renowned commercial publisher will foster our dissemination efforts in a professional format, the preprint versions of all the chapters of this book series will be freely available on the CDER Book Project website.[4]

Chapter Organization: This introductory chapter is organized as follows. Section "Chapter Introductions" gives brief outlines of each of the ten subsequent chapters. Section "How to Find a Topic or Material for a Course?" provides a roadmap for the readers to find suitable chapters and sections within these which are relevant for specific courses or PDC topics from the NSF/TCPP Curriculum. Section "Editor and Author Biographical Sketches" contains biographical sketches of the editors and authors. Appendix gives a brief history of the NSF/TCPP Curriculum Initiative.

Chapter Introductions

Part I: For Instructors

Chapter 2, *What Do We Need to Know about Parallel Algorithms and Their Efficient Implementation?*, by Vladimir Voevodin, Alexander Antonov, and Vadim Voevodin, explores a two-phase paradigm for teaching parallel algorithmics. A student is first taught to investigate an algorithmic problem in a machine-independent manner, learning to recognize opportunities for exploiting concurrency and to identify inherent sequentiality that will preclude such exploitation. After having mastered

---

[4]CDER Book Project – Free Preprint Version: http://cs.gsu.edu/~tcpp/curriculum/?q= CDER_Book_Project

this inherent aspect of the problem, the student is taught to investigate the problem in the context of a variety of computing platforms. This two-phase paradigm allows a student to approach computing with an understanding of the opportunities and challenges provided by the structure of the problem to be solved, as well as the opportunities and challenges provided by the structure of the computing platform one has access to. This chapter is intended for the instructors of introductory courses on parallel algorithms.

In chapter 3, titled *Modules for Teaching Parallel Performance Concepts*, Apan Qasem discusses three teaching modules targeting parallel performance concepts. The first module discusses fundamental concepts in parallel computing performance and mainly targets a CS1 course, highlighting parallel programming tools and performance metrics, and provides several sample exercises. The second module targets an upper-level operating systems class and focuses on communication and synchronization and how they affect performance for parallel applications, introducing the concepts of data dependences, synchronization, race conditions, and load balancing, again providing several sample exercises. The third module focuses on performance measurement and estimation of parallel systems, targeting compiler and computer architecture classes. This module reviews basic performance concepts and discusses advanced concepts such as strong and weak scaling, linear and super-linear speedup, and latency vs. bandwidth measurements in the context of OpenMP, and provides two sample exercises.

Chapter 4, *Scalability in Parallel Processing*, by Yanik Ngoko and Denis Trystram, provides a broad exposure to the notion of scalability, which is so important in modern (and future) large-scale parallel computing environments. The chapter discusses how scalability manifests itself and the many ways in which the "degree" of scalability is measured. The classical laws of Amdahl and Gustafson provide a central focus. The original arguments leading to those laws are described, accompanied by a reexamination of the laws' applicability in today's machines and computational problems. The notion of scalability is then further examined in the light of the evolution of the field of computing, with explorations of modern resource-sharing techniques and the more specific issue of reducing energy consumption. The chapter ends with a statistical approach to the design of scalable algorithms, specifically by organizing teams of parallel algorithms that "cooperate" in solving a single problem. The technically sophisticated aspects of organizing such cooperations is illustrated using the classical Satisfiability Problem. This chapter is intended for intermediate and advanced courses on the design and analysis of parallel algorithms.

In chapter 5, titled *Energy Efficiency Issues in Computing Systems*, Krishna Kant introduces energy efficiency issues in computer systems. Traditionally, computing has focused only on performance at all levels including circuits, architecture, algorithms, and systems. With power consumption and power density playing a central role at all these levels, it is crucial to teach students about power and performance tradeoffs. Power and energy issues are gaining importance in the context of mobile and embedded systems as well as server farms and data centers, although for different reasons. This chapter introduces topics like the basics of energy, power and thermal issues in computing, importance of and technology trends in power consumption, power-performance tradeoffs, power states, power adaptation, and energy efficiency of parallel programs.

Chapter 6, *Scheduling for fault-tolerance*, by Guillaume Aupy and Yves Robert, addresses a problem that has plagued large-scale parallel computing since its development in the 1970s and 1980s – fault tolerance. The electronically "aggressive" circuitry that enables high-performance large-scale parallel computing is vulnerable to both (permanent) failures and (transient) faults. Achieving high performance in practice, even for perfectly parallel applications, therefore demands the use of techniques that cope with these vulnerabilities. This chapter discusses the challenges of coping with faults and failures and introduces three simple strategies to achieve this: checkpointing, prediction, and work replication. Scheduling techniques are developed to optimize these three strategies. This chapter is intended for intermediate and advanced courses on the design and analysis of parallel algorithms. An operational understanding of elementary probability theory is necessary for true mastery of this material.

Part 2: For Students

In chapter 7, titled *MapReduce – The Scalable Distributed Data Processing Solution*, Bushra Anjum provides students with an overview of how to process large datasets using the MapReduce programming model. Along with multiple examples of MapReduce applications, the chapter provides an outline of the basic functions that must be written to build a MapReduce application, and also discusses how the map and reduce steps in a distributed MapReduce system (i.e., Hadoop) execute a MapReduce application with scalable performance. The chapter also discusses the strengths and limitations of the MapReduce model, addressing scalability, flexibility, and fault tolerance. Finally, the chapter discusses higher level services built on top of the basic Hadoop MapReduce system.

In chapter 8, titled *The Realm of Graphical Processing Unit (GPU) Computing*, Vivek Pallipuram and Jinzhu Gao provide an introduction to general-purpose graphical processing unit (GPGPU) computing using the Compute Unified Device Architecture (CUDA) programming model. The chapter extensively covers the GPGPU architecture as viewed by a CUDA programmer and CUDA concepts including CUDA thread management, memory management, and performance optimization strategies. The chapter pedagogically reinforces the CUDA concepts using parallel patterns such as matrix-matrix multiplication and convolution. The chapter includes several active-learning exercises that engage students with the text. Throughout this chapter, students will develop an ability to write effective CUDA codes for maximum application performance. The chapter is intended for an upper-level undergraduate course with object-oriented programming and data structures using C++ as prerequisites. The chapter can also be used in a sophomore- or junior-level software engineering course, or in an undergraduate elective course dedicated to high-performance computing using a specialized architecture. Because the chapter covers the GPGPU architecture and programming in detail, a prior exposure to CS1/CS2 level programming with basic computer organization is desirable.

In chapter 9, titled *Managing Concurrency in Mobile User Interfaces with Examples in Android*, Konstantin Läufer and George K. Thiruvathukal discuss parallel and distributed computing from a mobile application development perspective, specifically addressing concurrency in interactive, GUI-based applications on the Android platform. The chapter gives an overview of GUI-based applications and frameworks, then looks at implementing simple interactive application behavior in the Android mobile application development framework using a running example. More complex use cases are introduced that enable discussing event handling and timers, to further show how GUI applications display all the benefits and costs of concurrent execution. Finally, the chapter closes with a deeper exploration of long-running compute-bound applications, where the problem is to maintain responsiveness to user requests.

In chapter 10, titled *Parallel Programming for Interactive GUI Applications*, Nasser Giacaman and Oliver Sinnen show students how to use Java threads to implement a graphical user interface (GUI) that is responsive even while computation is being done. Because this example of concurrency is concrete and visual, it can be introduced fairly early in the curriculum. If the first course in programming makes active use of the Java GUI framework, then this will be a modest extension of that coverage. By at least the second course in programming (again if GUI programming is already included), and certainly in a sophomore software engineering class, this material can be used as a means to introduce many ideas that are basic to PDC, and get students thinking in terms of using explicit concurrency to take advantage of the capabilities of modern systems. The foundation that is laid by this material could easily be extended, for example, in a programming with data structures course, to introduce thread-safe processing of larger structures, including algorithms such as parallel merge sort.

Chapter 11, titled *Scheduling in Parallel and Distributed Computing Systems* by Srishti Srivastava and Ioana Banicescu addresses the important topic of mapping tasks onto computational resources for parallel execution. The chapter provides an introduction to scheduling in PDC systems such that it can be understood by undergraduate students who are exposed to this topic for the first time. It contains detailed taxonomy of scheduling methods and comparisons between different methods from the point of view of applicability as well performance metrics such as runtime, speedup, efficiency, etc.

How to Find a Topic or Material for a Course?

Table 1 lists the remaining chapters in the book, core/elective undergraduate courses they can be used for (see list below), and their prerequisites, if any. More detailed chapter-wise tables which follow list the topics covered in each chapter.

Relevant Courses and Prerequisites

CORE COURSES:
CS0:   Computer Literacy for Non-majors
CS1:   Introduction to Computer Programming (First Course)
CS2:   Second Programming Course in the Introductory Sequence
Systems: Introductory Systems/Architecture Course
DS/A: Data Structures and Algorithms
CE1:   Digital Logic (First Course)

ADVANCED/ELECTIVE COURSES:
Arch2:   Advanced Elective Course on Architecture
Algo2:   Elective/Advanced Algorithm Design and Analysis (CS7)
Lang: Programming Language/Principles (after introductory sequence)
SwEngg: Software Engineering
ParAlgo: Parallel Algorithms
ParProg: Parallel Programming
Compilers: Compiler Design
Networking: Communication Networks
DistSystems: Distributed Systems
OS: Operating Systems

Chapters and Topics

The following tables list the topics covered in each chapter. The depth of coverage of each topic is indicated by the intended outcome of teaching that topic, expressed using Bloom's taxonomy of educational objectives:

K   = Know the term
C   = Comprehend so as to paraphrase/illustrate
A   = Apply it in some way

Table 1 Relevant Courses and Prerequisites

Chap #	Short title	Primary core course	Other courses	Prerequisites
		Part I		
2	Parallel algorithms and implementation	CS0	CS1, DS/A, ParAlgo	–
3	Parallel performance concepts	CS1, Systems	OS, DS/A	DS/A for advanced modules
4	Scalability	Systems	CS2, ParAlgo	Math maturity
5	Energy efficiency	CE1	CS2, DS/A	Math maturity
6	Scheduling for fault tolerance	Systems	CS2, DS/A	CS1, Probabilities
		Part II		
7	MapReduce	DS/A	CS2, ParAlgo, DistSystems	CS0, CS1
8	GPU computing	Systems	Arch 2, ParProg	CS1, CS2
9	Mobile user interfaces	DS/A	Lang, ParProg	CS1, CS2
10	Interactive GUI applications	CS2 DS/A	SwEng, ParProg	CS1, CS2
11	Scheduling	DS/A	ParAlgo, DistSystems	Basic PDC terms and concepts

Editor and Author Biographical Sketches

Editors

Anshul Gupta is a Principal Research Staff Member in IBM Research AI at IBM T.J. Watson Research Center. His research interests include sparse matrix computations and their applications in optimization and computational sciences, parallel algorithms, and graph/combinatorial algorithms for scientific computing. He has coauthored several journal articles and conference papers on these topics and a textbook titled "Introduction to Parallel Computing." He is the primary author of Watson Sparse Matrix Package (WSMP), one of the most robust and scalable parallel direct solvers for large sparse systems of linear equations.

Sushil K. Prasad (BTech'85 IIT Kharagpur, MS'86 Washington State, Pullman; PhD'90 Central Florida, Orlando – all in Computer Science/Engineering) is a Professor of Computer Science at Georgia State University and Director of Distributed and Mobile Systems (DiMoS) Lab. Sushil has been honored as an ACM Distinguished Scientist in Fall 2013 for his research on parallel data structures and applications. He was the elected chair of IEEE Technical Committee on Parallel Processing for two terms (2007–2011), and received its

Chapter 2: What Do We Need to Know About Parallel Algorithms and Their Efficient Implementation?

PDC concept	Chapter section		
	2.1	2.2	2.3
Performance issues	C	C	C
Information structure	C	C	C
Data locality	C	C	
Computational intensity	K		
Resource of parallelism	C	C	C
Computational kernel		K	
Serial complexity		K	C
Parallel complexity		K	C
Load balancing		C	A
Determinacy of an algorithm		C	
Scalability		C	A
Efficiency		C	C
Race conditions			A

Chapter 3: Modules for Teaching Parallel Performance Concepts

PDC concept	Chapter section		
	3.2	3.3	3.4
Speedup	K		C
Efficiency	K		C
Linear and super linear speedup	K		C
Strong and weak scaling	K		C
Amdahl's law	C		A
Power vs. time trade-offs	K		A
Task granularity		A	
Load balancing		A	
Communication and synchronization		C	
Scheduling and thread mapping		A	
SMP and NUMA			C
Data locality			C

highest honors in 2012 – IEEE TCPP Outstanding Service Award. Currently, he is leading the NSF-supported IEEE-TCPP curriculum initiative on parallel and distributed computing with a vision to ensure that all computer science and engineering graduates are well-prepared in parallelism through their core courses in this era of multi- and many-cores desktops and handhelds. His current research interests are in Parallel Data Structures and Algorithms, and Computation over Geo-Spatiotemporal Datasets over Cloud, GPU and Multicore Platforms. Sushil is currently a Program Director leading the Office of Advanced Cyberinfrastructure (OAC) Learning and Workforce Development crosscutting

Chapter 4: Scalability in Parallel Processing

PDC concept	Chapter section				
	4.1	4.2	4.3	4.4	4.5
Scalability	K	C			
Speedup		C	C	C	A
Efficiency		C	C	C	A
Data parallelism			K		
Isoefficiency		C			
Amdahl' law		K	C	C	A
Gustafson' law		K	C	C	A
Strong scaling		C			C
Weak scaling		C			C
Resource sharing				C	A
Energy efficiency		K		K	
P-completeness			K		
Algorithm portfolio					A

Chapter 5: Energy Efficiency Issues in Computing Systems

PDC concept	Chapter section						
	5.1	5.2	5.3	5.4	5.5	5.6	5.7
Energy efficiency in computing	C	C					
Power states and their Management			K	K			
Software energy efficiency					K		
Energy efficiency vs. parallelism						C	
Energy adaptation							N

Chapter 6: Scheduling for Fault-Tolerance

PDC concept	Chapter section					
	6.1	6.2	6.3	6.4	6.5	6.6
Why/What is par/Dist computing	K	K	K	K	K	K
Performance issues, Computation	K	A	C	A	A	K
Cluster	K	K	K	K	K	K
Performance measures		A	A	A	A	
Basic probabilities		C	A	C	C	
Programming SPMD		C				
Load balancing		K		K	A	
Scheduling		C		C	C	
Dynamic programming					A	

programs at U.S. National Science Foundation. His homepage is www.cs.gsu. edu/prasad.

Arnold L. Rosenberg holds the rank of Distinguished University Professor Emeritus in the School of Computer Science at the University of Massachusetts Amherst.

Chapter 7: MapReduce: The Scalable Distributed Data Processing Solution

PDC concept	Chapter section				
	7.1	7.2	7.3	7.4	7.5
Why/What is par/Dist computing	A	A	C	K	A
Concurrency	K	K	C		A
Cluster computing	A	C	A	K	K
Scalability	A	C	A	K	A
Speedup	K	C	A		
Divide & Conquer (parallel aspects)	C	A	K		A
Recursion (parallel aspects)			K		A
Scan (parallel-prefix)	K	A			C
Reduction (map-reduce)	K	A		K	A
Time	C	A			A
Sorting	K	A			A

Chapter 8: The Realm of Graphical Processing Unit (GPU) Computing

PDC concept	Chapter section								
	8.1	8.2	8.3	8.4	8.5	8.6	8.7	8.8	8.9
Data parallelism	C								
GPGPU devices		C	A	A	A	A	A	A	
nvcc compiler		A							
Thread management				A	A				
Parallel patterns							A	A	
Performance evaluation							A		
Performance optimization							A	A	
CUDA		A		A	A	A	A	A	
Advancements in GPU computing									K

Prior to joining UMass, Rosenberg was a Professor of Computer Science at Duke University from 1981 to 1986, and a Research Staff Member at the IBM Watson Research Center from 1965 to 1981. He has held visiting positions at Yale University and the University of Toronto, as well as research professorships at Colorado State University and Northeastern University. He was a Lady Davis Visiting Professor at the Technion (Israel Institute of Technology) in 1994, and a Fulbright Senior Research Scholar at the University of Paris-South in 2000. Rosenberg's research focuses on developing algorithmic models and techniques to exploit the new modalities of "collaborative computing" (wherein multiple computers cooperate to solve a computational problem) that result from emerging computing technologies. Rosenberg is the author or coauthor of more than 190 technical papers on these and other topics in theoretical computer science and discrete mathematics. He is the coauthor of the research book *Graph Separators, with Applications* and the author of the textbook *The Pillars of Computation*

Chapter 9: Managing Concurrency in Mobile User Interfaces with Examples in
Android

PDC concept	Chapter section						
	9.1	9.2	9.3	9.4	9.5	9.6	9.7
Why and what is PDC	K						
Tasks and threads	K		C			A	A
Thread safety		K	C	C	A	A	A
Race conditions		K	C	C	C	A	A
Thread/Task spawning			K			A	A
Synchronization			C	C	A	A	A
Nondeterminism			C			A	
Deadlocks				K			

Chapter 10: Parallel Programming for Interactive GUI Applications

PDC concept	Chapter section					
	10.1	10.2	10.3	10.4	10.5	10.6
Concurrency	C	A	A	C	A	A
Race conditions		C	A			
Thread safety			C	C		
GUI concurrency				C	A	A

Theory: State, Encoding, Nondeterminism; additionally, he has served as coeditor
of several books. Rosenberg is a Life Fellow of the ACM, a Life Fellow of the
IEEE, a Golden Core member of the IEEE Computer Society, and a member
of the Sigma Xi Research Society. Rosenberg received an A.B. in mathematics
at Harvard College and an A.M. and Ph.D. in applied mathematics at Harvard
University.

Alan Sussman is a Professor in the Department of Computer Science and Institute
for Advanced Computer Studies at the University of Maryland. Working with
students and other researchers at Maryland and other institutions he has published
numerous conference and journal papers and received several best paper awards
in various topics related to software tools for high performance parallel and
distributed computing, and has contributed chapters to six books. His research
interests include peer-to-peer distributed systems, software engineering for high
performance computing, and large scale data intensive computing. Software tools
he has built with his graduate students have been widely distributed and used in
many computational science applications, in areas such as earth science, space
science, and medical informatics. He is a subject area editor for the Parallel
Computing journal and an associate editor for IEEE Transactions on Services
Computing, and edited a previous book on teaching parallel and distributed
computing. He is a founding member of the Center for Parallel and Distributed
Computing Curriculum Development and Educational Resources (CDER). He
received his Ph.D. in computer science from Carnegie Mellon University.

Chapter 11: Scheduling in Parallel and Distributed Computing Systems

PDC concept	Chapter section			
	11.1	11.2	11.3	11.4
MIMD architecture	K			
Multicore	C			
SMP	N			C
Topologies		N	N	
Latency		K		
Heterogeneous		K		K
Data Parallel		C		
Computation	C	C	C	C
Load balancing		C	C	C
Distributed memory			C	C
Client server			C	
Static		C	C	
Dynamic		C	C	C
Asymptotic	C			
Communication		C	C	
Synchronization		C	C	
Speedup			A	
Efficiency		A	A	
Makespan		C		C
Concurrency		C		
Performance modeling				K
Fault tolerance	K			

Charles Weems is co-director of the Architecture and Language Implementation lab at the University of Massachusetts. His current research interests include architectures for media and embedded applications, GPU computing, and high precision arithmetic, and he has over 100 conference and journal publications. Previously he led development of two generations of a heterogeneous parallel processor for machine vision, called the Image Understanding Architecture, and co-directed initial work on the Scale compiler that was eventually used for the TRIPS architecture. He is the author of numerous articles, has served on many program committees, chaired the 1997 IEEE CAMP Workshop, the 1999 IEEE Frontiers Symposium, co-chaired IEEE IPDPS in 1999, 2000, and 2013, was general vice-chair for IPDPS from 2001 through 2005, is on the steering committees of EduPar and EduHPC. He has co-authored 28 introductory CS texts. He is a member of ACM, Senior Member of IEEE, a member of the Executive Committee of the IEEE TC on Parallel Processing, has been an editor for IEEE TPDS, Elsevier JPDC, and is an editor with Parallel Computing.

Authors

Bushra Anjum has a PhD in Computer Science from North Carolina State University and is currently serving as a Tech Lead for the Amazon Prime Program at Amazon, Inc. Alongside, she is also a visiting professor at the Computer Science Department of the California Polytechnic Institute, San Luis Obispo. Anjum has been extensively using Elastic MapReduce platform provided by Amazon Web Services for a few years now for job related tasks. Before joining industry, she served in academia full time both in the USA and in Pakistan for 5+ years. With unconventional career choices and international exposure, she brings the expertise of being an academician, a researcher and a practitioner at the same time.

Alexander Antonov is a leading researcher in Research Computing Center of Lomonosov Moscow State University (RCC MSU). His main research interests are related to research in such fields as parallel and distributed computing, performance and efficiency of computers, parallel programming, informational structure of algorithms and programs, application optimization and fine tuning, architecture of computers, benchmarks, etc. In 1999 Alexander Antonov received PhD degree on the subject of interprocedural analysis of programs. Alexander took part in a number of projects supported by the Ministry of Education and Sciences of the Russian Federation, Russian Foundation for Basic Research and Russian Science Foundation. Alexander is editor of Parallel.Ru Information analytical center for parallel computing. Alexander Antonov is one of the main developers of the AlgoWiki Open encyclopedia of parallel algorithmic features. At the present time Alexander Antonov takes part in different researches being conducted in RCC MSU that are devoted to efficiency analysis of parallel applications and supercomputer systems in general. He has published over 50 scientific papers with 4 books among them.

Guillaume Aupy received his PhD from ENS Lyon. He is currently a tenured researcher at Inria Bordeaux Sud-Ouest. His research interests include data-aware scheduling, reliability, energy efficiency in high-performance computing. He is the author of numerous articles, has served on many program committees. He was the technical program vice-chair of SC17.

Ioana Banicescu is a professor in the Department of Computer Science and Engineering at Mississippi State University (MSU). Between 2009 and 2017, she was also a Director of the NSF Center for Cloud and Autonomic Computing at MSU. Professor Banicescu received the Diploma in Electronics and Telecommunications from Polytechnic University – Bucharest, and the M.S. and the Ph.D. degrees in Computer Science from New York University – Polytechnic Institute. Her research interests include parallel algorithms, scientific computing, scheduling and load balancing algorithms, performance modeling, analysis and prediction, and autonomic computing. Between 2004–2017, she was an Associate Editor of the Cluster Computing journal and the International Journal on Computational Science and Engineering. Professor Banicescu, served and continues to serve on numerous research review panels for advanced research grants in the US and Europe, on steering and program committees of a number of international conferences, symposia and workshops, on the Executive Board and Advisory Board of the IEEE Technical Committee on Parallel Processing (TCPP). She has given many invited talks at universities, government laboratories, and at various national and international forums in the United States and overseas.

Jinzhu Gao received the Ph.D. degree in Computer Science from The Ohio State University in 2004. From June 2004 to August 2008, she worked at the Oak Ridge National Laboratory as a research associate and then the University of Minnesota, Morris, as an Assistant Professor of Computer Science. She joined the University of the Pacific (Pacific) in 2008 and is currently a Professor of Computer Science at Pacific. Her main research focus is on intelligent data visual analytics. Over the past 15 years, Dr. Gao has been working closely with application scientists and Silicon Valley technology companies to develop online data visual analytics and deep learning platforms to support collaborative science, mobile health, IoT data analytics, business operational visibility, and visual predicative analysis for industries. Her work has been published in top journals such as IEEE Transactions on Visualization and Computer Graphics, IEEE Transactions on Computers, and IEEE Computer Graphics and Applications.

Nasser Giacaman is a Senior Lecturer in the Department of Electrical and Computer Engineering at the University of Auckland, New Zealand. His disciplinary research interest includes parallel programming, particularly focusing on high-level languages in the context of desktop and mobile applications running on multi-core systems. He also researches Software Engineering Education by driving the development of tools and apps to help students learn difficult programming concepts.

Krishna Kant is a full professor in the department of computer and information science at Temple University. He has 37 years of combined experience in academia, industry, and government and has published in a wide variety of areas in computer science, telecommunications, and logistics systems. His current research interests span energy management, data centers, wireless networks, resilience in high performance computing, critical infrastructure security, storage systems, database systems, configuration management, and logistics networks. He is a fellow of IEEE.

Konstantin Läufer is a full professor of computer science at Loyola University Chicago. He received a PhD in computer science from the Courant Institute at New York University in 1992. His research interests include programming languages, software architecture, and distributed and pervasive computing systems. His recent focus in research and teaching has been on the impact of programming languages, methodologies, frameworks, and tools on software quality. Konstantin has repeatedly served as a consultant in academia and industry and is a co-inventor on two patents owned by Lucent Technologies.

Yanik Ngoko received his B.Sc. in Computer Science from University of Yaoundé I (UYI), Cameroon, his M.Sc. in parallel and numerical computing also from UYI, and his doctorate in Computer Science from the Institut National Polytechnique de Grenoble, France (2010). From 2011 to 2014, he was a postdoctoral researcher, first at the university of São Paulo and then at the university of Paris 13. Since October 2014, he is a research scientist at Qarnot computing and an associate member of the Laboratoire d'Informatique de Paris Nord (University of Paris 13). His research interests include parallel and distributed computing, web services, cloud computing, applications of edge computing to IoT.

Vivek Pallipuram (B.Tech.2008 NIT Trichy, MS2010 Clemson University, Ph.D.2013 Clemson University) is an Assistant Professor of Computer Engineering at University of the Pacific, Stockton, California. His research interests include high-performance computing (HPC), heterogeneous architectures such as general-purpose graphical processing units (GPGPUs) and Xeon Phi co-processors, Cloud computing, image processing, and random signal processing. His interests also include promoting HPC education and scientific computing in primarily-undergraduate universities. His work has been published in journals such as the Journal of Supercomputing, and Concurrency and Computation: Practice and Experience; and in top conferences such as IEEE Cluster and eScience. He is also a peer-reviewer for a number of international journals and conference proceedings. In the classroom, he strives to be a facilitator by engaging students using active-learning techniques. In addition to receiving information from the instructor, students interact with their peers via in-class group activities and gain valuable perspective. This process increases the influx of knowledge per student, promoting well-rounded and comprehensive learning. He enjoys teaching high-performance computing, computer systems and networks, random signals, and image processing.

Apan Qasem is an Associate Professor in the Computer Science Department at Texas State University. He received his PhD in 2008 from Rice University. Qasem directs the Compilers Research Group at Texas State where he and his students are working on a number of projects in the area of high-performance computing including developing intelligent software for improving programmer productivity and using GPUs for general-purpose computation. Qasem's research has received funding from the National Science Foundation, Department of Energy, Semiconductor Research Consortium (SRC), IBM, Nvidia and the Research Enhancement Program at Texas State. In 2012, he received an NSF CAREER award to pursue research in autotuning of exascale systems. Qasem has co-authored over 50 peer-reviewed publications including two that won best paper awards. He regularly teaches the undergraduate and graduate Compilers and Computer Architecture courses.

Yves Robert received the PhD degree from Institut National Polytechnique de Grenoble. He is currently a full professor in the Computer Science Laboratory LIP at ENS Lyon. He is the author of 7 books, 150 papers published in international journals, and 240 papers published in international conferences. He is the editor of 11 book proceedings and 13 journal special issues. He is the advisor of 30 PhD theses. His main research interests are scheduling techniques and resilient algorithms for large-scale platforms. He is a Fellow of the IEEE. He has been elected a Senior Member of Institut Universitaire de France in 2007 and renewed in 2012. He has been awarded the 2014 IEEE TCSC Award for Excellence in Scalable Computing, and the 2016 IEEE TCPP Outstanding Service Award. He holds a Visiting Scientist position at the University of Tennessee Knoxville since 2011.

Oliver Sinnen graduated in Electrical and Computer Engineering at RWTH Aachen University, Germany. Subsequently, he moved to Portugal, where he received his PhD from Instituto Superior Técnico (IST), University of Lisbon, Portugal in 2003. Since 2004 he is a (Senior) Lecturer in the Department of Electrical and Computer Engineering at the University of Auckland, New Zealand, where he leads the Parallel and Reconfigurable Computing Lab. His research interests include parallel computing and programming, scheduling and reconfigurable computing. Oliver authored the book "Task Scheduling for Parallel Systems", published by Wiley.

Srishti Srivastava is an Assistant Professor of Computer Science at the University of Southern Indiana. She received her Ph.D. in Computer Science at Mississippi State University in May 2015. Her research interests include dynamic load balancing in parallel and distributed computing, performance modeling, optimization, and prediction, robustness analysis of resource allocations, and autonomic computing. Srishti has authored and co-authored a number of articles published in renowned IEEE and ACM conferences, journals, and book chapters. Srishti has served on the program committees of international conference workshops such as, EduHPC, and EduPar. She has also been a peer reviewer for a number of international journals, and conference proceedings. She is a professional member of the IEEE computer society, ACM, Society for Industrial and Applied Mathematics (SIAM), Computing Research Association (CRA, CRA-W), Anita Borg Institute Grace Hopper Celebration (ABI-GHC), and an honor society of Upsilon Pi Epsilon (UPE). She is also a 2014 young researcher alumna of the Heidelberg Laureate Forum, Germany.

George K. Thiruvathukal received his PhD from the Illinois Institute of Technology in 1995. He is a full professor of computer science at Loyola University Chicago and visiting faculty at Argonne National Laboratory in the Mathematics and Computer Science Division, where he collaborates in high-performance distributed systems and data science. He is the author of three books, co-editor of a peer-reviewed collection, and author of various peer-reviewed journal and conference papers. His early research involved object-oriented approaches to parallel programming and the development of object models, languages, libraries, and tools (messaging middleware) for parallel programming, mostly based on C/C++ on Unix platforms. His subsequent work in Java resulted in the book *High-Performance Java Platform Computing*, Prentice Hall and Sun Microsystems Press, 2000. He also co-authored the book *Codename Revolution: The Nintendo Wii Platform* in the MIT Press Platform Studies Series, 2012. Recently, he co-edited *Software Engineering for Science*, Taylor and Francis/CRC Press, October 2016.

Denis Trystram is a Professor in Computer Science at Grenoble Institute of technology since 1991 and is now distinguished professor there. He was a senior member of Institut Universitaire de France from 2010 to 2014. He obtained in 2011 a Google research award in Optimization for his contributions in the field of multi-objective Optimisation. Denis is leading a research group on optimization of resource management for parallel and distributed computing platforms in a joint team with Inria. Since 2010, he is director of the international Master program in Computer Science at university Grenoble-Alpes. He has been elected recently as the director of the research pole in Maths and Computer Science in this university.

Vadim Voevodin is a senior research fellow in Research computing center of Lomonosov Moscow state university (RCC MSU). His main research interests are related to different aspects of high-performance computing: analysis of parallel program efficiency, development of system software, parallel programming, etc. Vadim Voevodin got his PhD in memory locality analysis in parallel computing. Also he was a main developer in a research devoted to the study of memory hierarchy usage. At the present time Vadim Voevodin is actively involved in different researches being conducted in RCC MSU that are devoted to efficiency analysis of parallel applications and supercomputer systems in general. One research is dedicated to detecting abnormal inefficient job behavior based on constant monitoring of supercomputer job flow. The other newly started research is aimed to develop a universal software tool suite that will help common users to conduct both large-scale efficiency analysis of the entire set of applications and a professional in-depth analysis of individual parallel applications, based on many researches previously done in RCC MSU. Another major research area concerns the analysis of supercomputer resource utilization and efficiency of using application packages installed on a supercomputer.

Vladimir Voevodin is Deputy Director of the Research Computing Center at Lomonosov Moscow State University. He is Head of the Department "Supercomputers and Quantum Informatics" at the Computational Mathematics and Cybernetics Faculty of MSU, professor, corresponding member of Russian academy of sciences. Vl. Voevodin specializes in parallel computing, super-computing, extreme computing, program tuning and optimization, fine structure of algorithms and programs, parallel programming technologies, scalability and efficiency of supercomputers and applications, supercomputing co-design technologies, software tools for parallel computers, and supercomputing education. His research, experience and knowledge became a basis for the supercomputing center of Moscow State University, which was founded in 1999 and is currently the largest supercomputing center in Russia. He has contributed to the design and implementation of the following tools, software packages, systems and online resources: V-Ray, X-Com, AGORA, Parallel.ru, hpc-education.ru, hpc-russia.ru, LINEAL, Sigma, Top50, OctoShell, Octotron, AlgoWiki. He has published over 100 scientific papers with 4 books among them. Voevodin is one of the founders of Supercomputing Consortium of Russian Universities established in 2008, which currently comprises more than 60 members. He is a leader of the major national activities on Supercomputing Education in Russia and General Chair of the two largest Russian supercomputing conferences.

Appendix: A Brief History of The NSF/TCPP Curriculum Initiative

The pervasiveness of computing devices containing multicore CPUs and GPUs, including PCs, laptops, tablets, and mobile devices, is making even casual users of computing technology beneficiaries of parallel processing. Certainly, technology has developed to the point where it is no longer sufficient for even basic programmers to acquire only the sequential programming skills that are the staple in computing curricula. The trends in technology point to the need for imparting a broad-based skill set in PDC technology at various levels in the educational fabric woven by Computer Science and Computer Engineering programs as well as their allied computational disciplines. To address this need, a curriculum working group drawn from the IEEE Technical Committee on Parallel Processing (TCPP), the National Science Foundation (NSF), and sibling communities such as the ACM and industry, has taken up the challenge of proposing and refining a curricular guidleines for blending PDC-related concepts into even early-stage undergraduate curricula in computational areas. This working group is built around a constant core of members and typically includes members from all segments of the computing world and the geographical world. A first version of the group's guidelines for a core curriculum that includes PDC was released informally in December, 2010, with a formal version[3] following in December 2012. The CS2013 ACM/IEEE Computer Science Curriculum Joint Task Force has recognized the need to integrate parallel and distributed computing topics in the early core courses in the computer science and computer engineering curriculum, and has collaborated with our working group in leveraging our curricular guidelines. The CS2013 curriculum[5] explicitly refers to the NSF/TCPP curricular guideines for comprehensive coverage of parallelism (and provides a direct hyperlink to the guidelines).

The enthusiastic reception of the CDER guidelines has led to a commitment within the working group to continue to develop the guidelines and to foster their adoption at an even broader range of academic institutions. Toward these ends, the Center for Curriculum Development and Educational Resources (CDER) was founded, with the five editors of this volume comprising the initial Board of Directors. An expanded version of the working group has taken up the task of

[5]The ACM/IEEE Computer Science Curricula 2013: (https://www.acm.org/binaries/content/assets/education/cs2013_web_final.pdf)

revising and expanding the 2012 NSF/TCPP curriculum during the 2016–2018 timeframe. One avenue for expansion has been to add special foci on a select set of important aspects of computing that are of particular interest today – Big Data, Energy-Aware Computing, Distributed Computing – and to develop Exemplars that will assist instructors in assimilating the guidelines' suggested topics into their curricula. CDER has initiated several activities toward the goal of fostering PDC education.

1. A *courseware repository*[6] has been established for pedagogical materials – sample lectures, recommended problem sets, experiential anecdotes, evaluations, papers, etc. This is a living repository. CDER invites the community to contribute existing and new material to it. The Exemplars aspect group is working to provide extensive set of exemplars for various topics and courses.
2. An *Early Adopter Program* has been established to foster the adoption and evaluation of the guidelines. This activity has fostered educational work on PDC at more than 100 educational institutions in North and South America, Europe, and Asia. The Program has thereby played a major role in establishing a worldwide community of people interested in developing and implementing PDC curricula. Additional early adopter training workshops and competitions are planned.
3. The *EduPar workshop series* has been established. The original instantiation of EduPar was as a satellite of the International Parallel and Distributed Processing Symposium (IPDPS). EduPar was – and continues to be – the first education-oriented workshop at a major research conference. The success of EduPar led to the development of a sibling workshop, EduHPC, at the Supercomputing Conference (SC) in 2013. In 2015 EduPar and EduHPC was joined by a third sibling workshop, Euro-EduPar, a satellite of the International Conference on Parallel Computing (EuroPar). CDER has also sponsored panels, and BOF and special sessions at the ACM Conference on Computer Science Education (SIGCSE).
4. A *CDER Compute Cluster* has been setup for free accesses by the early adopters and other educators and their students. The CDER cluster is a heterogeneous 14-node cluster featuring 280 cores, 1 TB of RAM, and GPUs that are able to sustain a mixed user workload.[7]

---

[6]CDER Courseware Repository: https://grid.cs.gsu.edu/~tcpp/curriculum/?q=courseware_management
[7]CDER Cluster free access: https://grid.cs.gsu.edu/~tcpp/curriculum/?q=node/21615

Part I
For Instructors

What Do We Need to Know About Parallel Algorithms and Their Efficient Implementation?

Vladimir Voevodin, Alexander Antonov, and Vadim Voevodin

Abstract The computing world is changing and all devices—from mobile phones and personal computers to high-performance supercomputers—are becoming parallel. At the same time, the efficient usage of all the opportunities offered by modern computing systems represents a global challenge. Using full potential of parallel computing systems and distributed computing resources requires new knowledge, skills and abilities, where one of the main roles belongs to understanding key properties of parallel algorithms. What are these properties? What should be discovered and expressed explicitly in existing algorithms when a new parallel architecture appears? How to ensure efficient implementation of an algorithm on a particular parallel computing platform? All these as well as many other issues are addressed in this chapter. The idea that we use in our educational practice is to split a description of an algorithm into two parts. The first part describes algorithms and their properties. The second part is dedicated to describing particular aspects of their implementation on various computing platforms. This division is made intentionally to highlight the machine-independent properties of algorithms and to describe them separately from a number of issues related to the subsequent stages of programming and executing the resulting programs.

Relevant core courses: Data Structures and Algorithms, Second Programming
  Course in the Introductory Sequence.
Relevant PDC topics: Parallel algorithms, computer architectures, parallel pro-
  gramming paradigms and notations, performance, efficiency, scalability, locality.
Learning outcomes: Faculty staff mastering the material in this chapter should
  be able to:

- Understand basic concepts of parallelism in algorithms and programs.
- Detect parallel (information) structure of algorithms.

V. Voevodin (✉) · A. Antonov · V. Voevodin
Lomonosov Moscow State University, Moscow, Russia
e-mail: voevodin@parallel.ru; asa@parallel.ru; vadim@parallel.ru

© Springer International Publishing AG, part of Springer Nature 2018   23
S. K. Prasad et al. (eds.), *Topics in Parallel and Distributed Computing*,
https://doi.org/10.1007/978-3-319-93109-8_2

- Understand deep relationship between properties of algorithms and features of computer architectures.
- Identify main features and properties of algorithms and programs affecting performance and scalability of applications.
- Use proper algorithms for different types of computer architectures.

Context for use: This chapter has to touch all the main areas of computer science and engineering: Architecture, Programming, Algorithms and Crosscutting topics. The primary area is Algorithms but these materials should be taught after learning the fundamentals of computer architecture and programming technologies. Materials of the chapter can be easily adapted for use in core, advanced or elective courses within bachelor's or master's curricula.

Introduction

Parallelism has been the "big thing" in the computing world in recent years. All devices run in parallel: supercomputers, clusters, servers, notebooks, tablets, smartphones...Even individual components are parallel: computing nodes can consist of several processors, processors have numerous cores, each core has several independent functional units that can be pipelined as well. All this hardware can work in parallel, provided that special software and the corresponding parallel algorithms are available.

After more than 60 years of development, a huge pool of software and algorithms has been accumulated for computers. The training process has been refined with the goal of learning programming technologies, and developing software, algorithms and methods to address various tasks. Now all of this is changing as the word "parallel" has literally found its way into everything: parallel programming technologies, parallel methods, parallel computing systems architecture, etc. Adding parallelism to existing training curricula definitely implies preserving the current serial programming methods, methodologies, technologies and algorithms, but many new things that never existed before need to be added [6, 12, 15]. How does one organize the parallel execution of a program to get a job completed faster? The question sounds simple, but answering it requires learning new ideas that have not been studied before.

In this chapter,[1] we present our experience in studying and teaching parallel methods of problem solving. This experience is based on using a large number of very different parallel computing systems: vector-pipeline, with shared and distributed memory, multi-core, computing systems with accelerators, and many

---
[1]The results were obtained in Lomonosov Moscow State University with the financial support of the Russian Science Foundation (agreement N 14-11-00190). The research is carried out using the equipment of the shared research facilities of HPC computing resources at Lomonosov Moscow State University.

others. Various forums for teaching parallel computing, parallel programming technologies, program and algorithm structures have been piloted at Lomonosov Moscow State University including general and special courses, seminars, practical computing exercises as part of the educational curricula at the Faculty of Computational Mathematics and Cybernetics, as well as at the annual MSU Summer Supercomputing Academy [1]. Many of the ideas described in this chapter were implemented in the national project "Supercomputing education" [11, 21].

The chapter consists of three sections. In the first section, we want to show, using numerous real-life examples, how many different properties of parallel algorithms and programs need to be taken into account to create efficient parallel applications. In the second section, these properties are described in a more systematic way, building on a structure that can be used to describe any algorithm. This helps to identify the most important properties for creating an efficient implementation. The description structure itself is universal, and not limited to any specific class of algorithms or methods. In the third section, we would like to show that the described materials can easily be incorporated into the educational process.

Before proceeding to the chapter we would like to make a special remark. This chapter is not a ready-to-use packaged lesson or a set of lessons, but rather ideas that should be presented throughout courses devoted to modern computational sciences and technologies. There is a high degree of freedom in choosing the methods for incorporating parallelism concepts into educational curricula, which do not require revolutionary changes and can be performed by existing academic staff within a current set of educational courses. We intentionally did not explain all the notions used in the chapter in a classical pedagogical way trying to concentrate on the main goal—to show a universal nature and wide use of parallel computing. From this point of view, our main target audience can be described as instructors who are already familiar with the subject and want to introduce parallel computing concepts into their courses, and parallel computing experts that teach related classes. At the same time it is really necessary to extend this audience involving a wide range of faculty staff into parallel computing as one of the most significant trends in computer science. The idea behind the chapter is to outline possible directions and ways how a teacher or instructor can incorporate parallel computing notions into any course.

What Knowledge of Algorithm Properties Is Needed in Practice?

In this section, we will consider several examples, focusing in each case on one property or another that determines how efficiently an algorithm is implemented. While reading the section, it may seem that we are conflating parallel algorithms, parallel computing for different platforms, and performance issues. In a sense, this is true but this is necessary. If we are discussing high performance computing, we have to consider parallel algorithms, programming technologies, and architectures all together to ensure high efficiency of the resulting code.

By giving examples, we are not trying to explain every minute detail, give definitions, or explain newly introduced concepts, especially since many of them are quite intuitive. Our goals here are different. On the one hand, we want to show the great diversity of questions that arise in practice, the answers to which are determined by knowledge of the fundamental properties of algorithms and programs. On the other hand, by analyzing examples, we will gradually identify the set of properties that must be included in an algorithm description, and which teachers need to point out to their students.

Even in Simple Cases, It Is Important to Understand the Algorithm Structure

Let's look at the classical algorithm for multiplying dense square matrices of size $n \times n$. Based on the formula

$$A_{ij} = \sum_k B_{ik} C_{kj},$$

it is quite natural to write the following version of the program (hereinafter in this paragraph matrix A is initialized with zeros):

```
for(i=0; i<n; ++i)
    for(j=0; j<n; ++j)
        for(k=0; k<n; ++k)
            A[i][j] += B[i][k] * C[k][j];
```

It has three nested loops and one assignment statement which calculates the element A_{ij}. The sequence of the loops in this example (i, j, k are the control parameters) is absolutely clear as it reflects the essence of the algorithm: for each element in matrix A (loop by i, loop by j), calculate the element A_{ij} (loop by k).

Let's perform a seemingly strange procedure: shuffle the three loops. We'll get a new fragment in which the loops can be organized in any of the six possible orders, for example (k, i, j) or (j, k, i). Will the new fragment provide the same results as the original program? A more general question also needs to be answered: "What loop order will provide the same result for the new program as the original version?" Below we show the two fragments mentioned above, with a loop order of (k, i, j) and (j, k, i); will the results of their execution be the same as those of the original fragment?

```
for(k=0; k<n; ++k)
    for(i=0; i<n; ++i)
        for(j=0; j<n; ++j)
```

```
              A[i][j] += B[i][k] * C[k][j];

for(j=0; j<n; ++j)
   for(k=0; k<n; ++k)
      for(i=0; i<n; ++i)
         A[i][j] += B[i][k] * C[k][j];
```

Questions like this may sound surprising, as there doesn't seem to be a reason why a formal loop interchange can result in a fragment equivalent to the original program. But the answer is even more surprising: in this example **any loop order** provides a result that is equal to the original fragment's results, accurate up to the rounding error. This begs two questions. Why does any loop order result in an equivalent fragment in this example? And the second question is why would we do something so strange as interchanging loops?

The first question can be answered by looking at the information structure of the matrix multiplication algorithm, shown in Fig. 1. The information structure is presented in a graph, where each vertex corresponds to one iteration of the three nested loops, and the vertices are connected with a directed edge [20, 22] if one vertex calculates the data used in another one. We see n^2 independent computational branches, where each branch corresponds to the innermost k-loop for certain values of i and j, i.e. the calculation of the element A_{ij}. The picture is worth a thousand words. First, we see at once that all n^2 elements in the resulting matrix A_{ij} can be calculated independently from one another: the algorithm has a tremendous resource of parallelism, offering good prerequisites for writing a parallel program. Second,

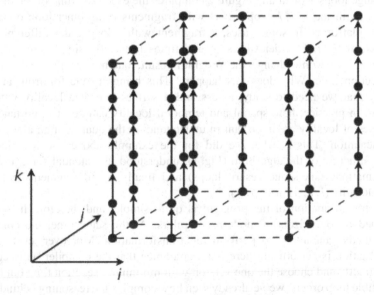

Fig. 1 Information structure of the matrix multiplication algorithm

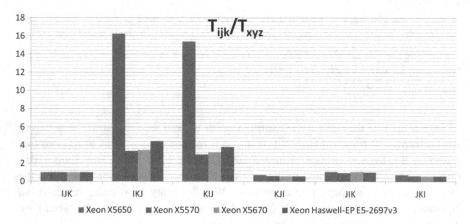

Fig. 2 Comparison of execution times for matrix multiplication programs with various loop orders against the execution time for the classical order (i, j, k), the higher bars, the faster execution of the order (x, y, z)

whatever the loop order in this fragment of the program: (k, i, j), (j, k, i) or any other, going through the vertices never violates the information relationship between the vertices; thus a fragment with any loop order is guaranteed to produce the same result as the original fragment. This feature of the information structure (information graph) for this algorithm explains the equivalence of the original and transformed fragments.

Now we only have to answer the second question: why did the need to interchange loops arise at all? Figure 2 compares the execution time of the original program (with the i, j, k loop order) with fragments using other loop orders on different platforms. In some cases, a fragment with a loop order different from the classical (i, j, k) order works several times faster! By simply changing the loop order, we won't change the program result, but may significantly reduce its execution time. Why does this happen? This brings to the forefront another property that we need to study, assess and describe—the data locality within a program. In practice, both spatial and temporal locality can be of importance, so both types of locality are important in understanding the quality of an algorithm's implementation. This is what we did for the example above: we revealed the parallel structure of the algorithm (Fig. 1), understood its potential for equivalent transformations (six sequences of loops) and finally found fragments with the highest locality (i, k, j) or (k, i, j).

This transformation of the program is quite simple, and therefore it is often performed by optimizers in modern compilers. At the same time, the compiler isn't actually guaranteed to perform the transformation. Moreover, even if the transformation is performed, there is no guarantee that the compiler will actually do it correctly and choose the one version with minimum execution time out of the six possible loop orders: we've already seen how complex the reasoning behind such a "simple" text change can be.

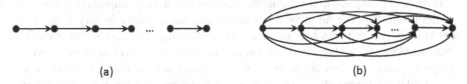

Fig. 3 Linear graph (**a**) and its transitive closure (**b**): quadratic increase in the volume of output information

Simple Properties Can Be Very Important, Too

No detail is an afterthought in an algorithm description, and even seemingly obvious properties and parameters need to be properly observed. Let's look at the volume of input and output data for an algorithm. These figures almost invariably follow the algorithm formulation and are therefore considered obvious and shrugged off as being of little relevance. At the same time, these figures aren't just important— they can completely determine the structure of the resulting program. Suppose we need to develop an algorithm to find the transitive closure of a directed graph. Let the graph consist of n vertices and m edges—these are the input parameters of the algorithm, which determine the input data volume. However, the output data volume for this task is strongly dependent not just on n and m, but also on the graph structure. In particular, if the input is a linear graph consisting of n vertices and $n - 1$ edges, its transitive closure will contain $n(n - 1)/2$ edges (see Fig. 3). This fact is no problem for processing relatively small graphs. However, it will be crucial for processing graphs representing social networks, which are obviously nonlinear but contain hundreds of millions of vertices and hundreds of billions of edges: due to immense volumes of output information (a quadratic dependence on the input data volume), the results will be impossible to store anywhere! The only way out of this situation is to restate the task so that it doesn't require listing every pair of vertices connected by edges. Input and output data volumes seem to be quite obvious parameters, but they do have to be thoroughly reviewed and described to understand the algorithm properties.

A New Look at Traditional Concepts

There are other arguments in favor of considering every detail when describing algorithm properties. Let's look at an array of input data V and the total number of operations N in an algorithm. Both values are well known, each one is of interest in and of itself and is frequently used in practice. But it is equally important to pay attention to the ratio $P = N/V$. P stands for "computational intensity" and represents the number of operations per unit of input data required to execute the algorithm.

Despite its simplicity, computational intensity is a very important feature of an algorithm. Suppose an algorithm's computational intensity is very high. This means input data requires a lot of processing before the algorithm's results can be obtained. As a result, this algorithm can be executed on an accelerator or a remote computing node, as the data transmission overhead will be low compared to the time it takes to process that data. This fact in particular explains why the Linpack test, with a computational intensity of n is so efficient on computers with distributed memory (about n^2 data elements are transferred to each computing node and about n^3 operations are performed on them). The same fact explains the low efficiency of an element-wise addition of two vectors using graphic accelerators: the time it takes to transfer $2n$ vector elements to a GPU completely offsets the rapid execution of n operations by the GPU.

In many cases determining the computational intensity requires taking output data into account, and not just input. In the vector addition example above, correctly evaluating the efficiency of the algorithm requires taking into account not just the time it takes to send input data to the GPU, but also the time it takes to get the results back. This will definitely reduce efficiency and decrease the computational intensity of this algorithm from $1/2$ to $1/3$, but that's the nature of the algorithm, and it must be considered. In practice, it is sometimes possible to increase the computational intensity by combining consecutive processing steps. For example, you might not want to return the results of a vector addition but instead continue processing at the accelerator, thereby eliminating unneeded data transfers.

Mathematics and Parallelism

The information structure of an algorithm is an important concept, but it should not be used alone to evaluate the parallelism potential of an algorithmic approach. The math behind the algorithm plays an equally important role. Let's look at the classical vector elements addition algorithm:

```
s = 0;
for(i=0; i<n; ++i)
    s += A[i];
```

The information graph for this algorithm is a linear graph (see Fig. 4a), where all vertices are connected with data dependency, which means only serial execution can be performed. Does this mean that neither the addition of vector elements, nor any other algorithms based on this operation can be used on parallel computers? Not exactly. The summation operation obeys the associative law, which allows us to tweak the original algorithm to achieve the appropriate degree of parallelism. Let's break all of the vector elements into non-overlapping groups, then find the subtotals for each group, and finally sum up the subtotals to get the total of all elements in a vector (see Fig. 4b). As the subtotals can be calculated independently (i.e. simultaneously), we now have a parallel version of the vector addition algorithm.

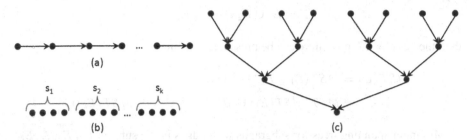

Fig. 4 The information structure of several vector element summation algorithms: (a) classical algorithm, (b) parallel calculation of partial sums and (c) pairwise summation method

The associative law allows the elements to be added in any order with the same result, so other methods of implementation are also possible. Figure 4c shows the information structure of the pairwise summation method, which is also a parallel modification of the original algorithm. All subtotals located at the same level can be calculated in parallel, moving between levels from top to bottom, until we get the desired result.

There are several important remarks regarding the serial-to-parallel process described above. First, it is math that enables us to make the key step: we would never get a parallel algorithm without using the associative law, based on just the knowledge of the information structure (Fig. 4a). One needs to pay attention to possibilities like this when explaining or describing algorithm's properties, otherwise their potential will not be fully revealed. Second, the associative law implies that all operations are executed precisely, but computers operate using approximations of the original numbers. When we change the summation order, we may get a different result. Often the difference falls within the rounding error accuracy, which is negligible in most cases, but the fact that the associative law may not always work in computer arithmetic (just like the commutative and distributive laws) is something to be kept in mind. This explains, to a certain extent, the lack of reproducible results for parallel applications executed on supercomputers with a high degree of parallelism: literally every global MPI operation is based on the associative law, which results in various rounding errors and ultimately in different results when executing the same application. This is a serious issue that complicates the transition to Exaflop systems with an enormous degree of parallelism [5, 10]. Third, one needs to understand clearly that Fig. 4a, b, c represent **different algorithms**. The original task is the same: summing up the vector elements, but the algorithms are different. There are many differences between these algorithms: different information structures, different parallel complexity, different complexity of the respective programs, different rounding errors...

A situation like this, where knowing the mathematical basics of an approach can increase the degree of parallelism, is important in practice, and should be taken into account. Let's look at the task of finding the minimum spanning tree in a weighted graph G with E edges and V vertices. Suppose $MST(E)$ is the procedure for finding the minimum spanning tree. If we break the set of edges E into k non-overlapping subsets $E_1, E_2, \ldots E_k$:

$$E = E_1 \cup E_2 \cup \cdots \cup E_k,$$

then the basic MST procedure can be presented as follows:

$$MST(E) = MST(E_1 \cup E_2 \cup \cdots \cup E_k)$$
$$= MST(MST(E_1) \cup MST(E_2) \cup \cdots \cup MST(E_k))$$

Minimum spanning trees for subgraphs with edges in the subsets E_1, E_2, ... E_k can be found independently from one another; therefore, $MST(E_1)$, $MST(E_2)$, ..., $MST(E_k)$ procedures can also be performed in parallel. If we leave this mathematical fact aside, the algorithm's potential will not be utilized in full, as the available degree of parallelism grows with an increasing number of subsets E_i. After finding $MST(E_1)$, $MST(E_2)$, ..., $MST(E_k)$, it is necessary to join the minimum spanning trees found and perform the MST operation once again; however, the advantage of using parallel computing will still be substantial for $|E| >> |V|$.

This correlation does more than just increase resource of the parallelism. Its variations are exceptionally useful when processing very large graphs, as individual subsets E_i can be entirely stored and efficiently processed within RAM.

Parallelism Can Be Inconvenient

If an algorithm has internal parallelism, this information is very important, but knowing this fact alone is not enough to make an efficient parallel program. Let's look at the example in Fig. 5a.

All iterations of the outer loop by parameter i are independent and can be executed in parallel (see Fig. 5b). To use the parallelism available in this fragment all we need to do is, for example, to place an OpenMP directive similar to "#pragma omp parallel for"—this makes the program parallel without any other modifications to its text. Moreover, all parallel branches will be perfectly balanced, as each one is used to execute n operations of the same kind.

The algorithm above has a very convenient structure and is easy to work with, but that's not always the case. Let's look at the example in Fig. 6a. The source code here looks very similar to the example we just looked at in Fig. 5a, but its information structure is completely different (see Fig. 6b).

None of the loops in the fragment Fig. 6a can be marked as parallel, since there are data dependencies in each dimension. But the fragment still has a great resource of parallelism; its serial complexity equals n^2 while the critical path of the information graph, reflecting the algorithm's parallel complexity, equals $2n - 2$. To show the possibility of parallel execution of the algorithm, we can draw diagonals, as shown in Fig. 7a: all diagonals must be accessed serially, one after the other in ascending order, while all vertices located on the same diagonal can be computed in parallel (this type of parallelism is called skewed parallelism). To describe this

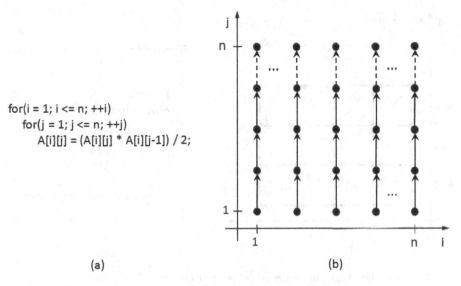

```
for(i = 1; i <= n; ++i)
   for(j = 1; j <= n; ++j)
      A[i][j] = (A[i][j] * A[i][j-1]) / 2;
```

(a)   (b)

Fig. 5 A fragment with "convenient" parallelism and its information graph

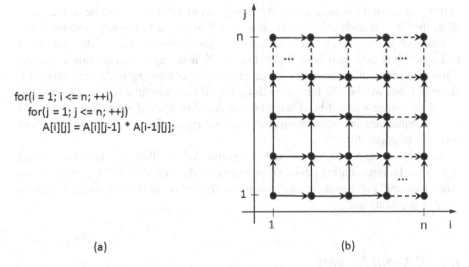

```
for(i = 1; i <= n; ++i)
   for(j = 1; j <= n; ++j)
      A[i][j] = A[i][j-1] * A[i-1][j];
```

(a)   (b)

Fig. 6 A fragment with "inconvenient" parallelism

execution method for the program, its text needs to be transformed; one possible transformation is shown in Fig. 7b. Moving from fragment Fig. 6a to fragment Fig. 7b is not a trivial task, as parallelism must be explicitly declared in most existing programming technologies.

The first reason why the parallelism in an algorithm can be called "inconvenient" is the need to transform the original code. Suppose the transformation has been completed; let's go back to example Fig. 7. The first diagonal contains just one

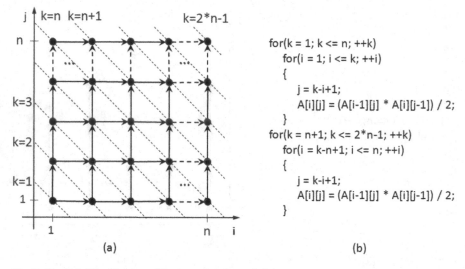

Fig. 7 Explicit identification of "inconvenient" parallelism

vertex, the second has two, the third—three; this number will increase to the value of n, then it will reverse its course, reaching 1 by the last diagonal. The available resource of parallelism change between steps, going from 1 to n and then returning to 1 again. It is extremely hard to develop an efficient way to execute this fragment: if too much computing resources (cores, processors, computing nodes) are allocated, some of them will be idle for a long time; but allocate too little resources—and the speed advantage will not be substantial compared to a serial implementation. This serious imbalance in computation is the second reason for the "inconvenience" of this type of parallelism.

How often do we focus on such properties of parallelism when we explain algorithm features during classes? In practice, the criterion of "convenient" or "inconvenient" parallelism in an algorithm is frequently the key factor in designing parallel applications.

It's All About Locality

Simple operations are efficiently implemented by a computer. This seems like an intuitively clear and correct thesis. It's much more complex in practice. The simplicity of an operation is primarily understood as a simple algorithm structure, but when we talk about implementation efficiency, it is important to take into account not just the algorithm but also the program implementing it. Niklaus Wirth called one of his books "Algorithms + Data Structures = Programs" [23], and "Data Structures" are often what determines the efficiency of an algorithm's implementation, even for very simple algorithms.

```
(a)    A[i] = B[i]*x + c;
(b)    A[i] = B[i]*x + C[i];
(c)    A[i] = B[i]*X[i] + C[i];
(d)    A[ind[i]] = B[ind[i]]*x+c;
(e)    A[ind[i]] = B[ind[i]]*x+C[ind[i]];
(f)    A[ind[i]] = B[ind[i]]*X[ind[i]]+C[ind[i]];
```

Fig. 8 Versions of the "triad" operation

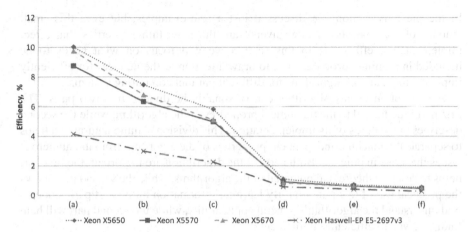

Fig. 9 Efficiency (the ratio of sustained performance to peak performance) of different options for "triad" operation

Let's look at several versions of a "triad", a basic operation used in many algorithms (see Fig. 8). Everything seems very clear and should not cause any efficiency implementation issues on modern processors: the structure is trivial, regular, no data dependencies, and the addition and multiplication operations are perfectly balanced. At the same time, the efficiency (the ratio of sustained performance to peak performance) for the simplest operation in Fig. 8a never exceeds 10% (see Fig. 9)! As we increase the number of input arrays (operations Fig. 8b, c), the efficiency falls even further.

What is the reason behind such low efficiency for a seemingly "perfect" operation? The main reason is the poor data locality in the resulting programs. The main data structure in every version Fig. 8a–c is the arrays, and the elements in each operation are accessed serially, but each element is used only once. This means that spatial data locality is relatively low, and temporal locality does not exist at all. As the number of input arrays increases, the situation only gets worse. How often do we explain to the students what data locality is and how it affects program efficiency?

Let's go further and look at sparse data structures, instead of dense ones (operations Fig. 8d–f). In this case we must use indirect addressing arrays (ind[i] array in Fig. 8), which further degrade the already low locality values and reduce efficiency to less than one percent! How often do we pay attention to this aspect of algorithms that operate with sparse arrays? One should not skip over a locality analysis when describing algorithm properties, otherwise the resulting program efficiency can be an unpleasant surprise.

Parallel Algorithms: What to Pay Attention to?

In the previous section, we discussed just a few examples, but even this small amount of material shows how diverse are the algorithm properties that affect implementation efficiency. In this section we will focus on what needs to be included in training curricula, so as to draw attention to the nuances for efficiently implementing parallel algorithms for different parallel computing systems.

A description of any algorithm can reasonably be divided into two parts. The first part is dedicated to the theoretical properties of the algorithm, while the second describes the features of its implementation. This division is quite natural and helps to separate the machine-independent properties of the algorithm from the numerous issues that arise in practice. Both parts of the description are important: the first part helps to describe theoretical potential of the algorithms, while the second part shows the practical use of this potential. By learning information in the first part, students will understand the algorithm's general applicability, while the second part will help finding a way to efficiently implement it.

Parallel Algorithms: Theoretical Potential

Let's look at an **algorithm's computational kernel**. It is the part of the algorithm that takes up most of the processing time. The computational kernel determines the quality of an algorithm's implementation in general, therefore it has to be an area of focus in the algorithm implementation process. If no acceptable implementation exists for an algorithm's computational kernel, it won't exist for the entire algorithm as well. Remember the analysis of the "triad" operation in the previous section (see Fig. 8): if we suppose it is the computational kernel of an algorithm, and the application uses sparse data structures, you can't expect high efficiency from the application in general (see Fig. 9).

The computational kernel does not have to be determined by operations on real numbers: addition, multiplication, division, square root, $\sin(x)$, $\cos(x)$, etc. For many algorithms, data load/store operations, boolean or integer operations are the biggest bottlenecks. This doesn't change anything, and a computational kernel consisting of such operations has to be singled out and described just as well. If most of the execution time for an algorithm is spent on matrix transposition, special attention must be paid to carefully storing and copying the data.

Another fact must be noted. Even though the full description of an algorithm can be quite large, the computational kernel is usually very compact, which allows the computational structure to be quickly understood and thus simplifies and speeds up the algorithm analysis.

Serial complexity, i.e. the number of operations that need to be executed in a serial implementation of the algorithm, is a highly important feature. Complexity is always expressed through parameters that determine the task size, and helps to quickly assess the viability of an algorithm's practical implementation. The operation type is not specified in any way, so whatever operations contribute the most to an algorithm's execution time shall be included in the formula for serial complexity. This can include operations on real numbers or integers, bitwise or memory loads/stores operations, array element updates, elementary functions, macro operations, etc. Arithmetic operations on real numbers prevail in LU decomposition, while large number factorization relies heavily on bitwise and boolean operations; this has to be reflected in the complexity evaluation.

If an algorithm has high complexity, then it must be used for large task sizes with extreme care. Moreover, if the algorithm is a component of another algorithm, then overall complexity can be even higher, and one should be even more careful. The computational complexity of a fast Fourier transform (Cooley–Tukey algorithm) for vectors with a length equal to powers of two equals $n \log_2 n$ complex addition operations and $(n \log_2 n)/2$ complex multiplication operations. At the same time, when looking at this algorithm, one should remember that a fast Fourier transform is often a basic component of other algorithms, being part of some large loops, which increases overall complexity.

All modern computers are parallel, so it is important to not just explain an algorithm, it is vital to simultaneously **show the algorithm's parallel structure**. This can be done, for example, with the help of an information graph, sometimes called an algorithm graph, data dependency graph or data-flow graph. Determining the parallel structure of an algorithm is always the first step in creating a parallel program, regardless of what specific parallel computing system the program is being written for. This step is very important, and if the algorithm's parallel structure is known (see Figs. 1, 5b, 7a), many subsequent decisions become obvious.

There are many possible options for representing the information structure of an algorithm. For some algorithms, the information structure must be shown in every detail; for others a macro structure is more important. A lot of information is available in various forms of information graph projections, which clarify the algorithm's regular components while hiding insignificant details. Sometimes it may be useful to show a pattern in the graph that changes with the values of external variables (e.g., matrix sizes): we often expect "similar" behavior in the information graph, but it isn't always obvious in practice.

Visualization of the information graph can be very useful for studying various algorithm properties. But the task of displaying an information graph is not trivial. To begin with, the information graph can potentially be endless, as the number of vertices and edges is determined by external variables which can be very large. In this situation it helps to look at likenesses, as described above, which consider graphs for different values of external variables as "similar": it is almost always enough to present one small graph, stating that graphs for other values will look "exactly the same." Not everything is so simple in practice, however; and one should be very careful here.

Next, an information graph is potentially a multi-dimensional object. The most natural coordinate system for placing vertices and edges in an information graph relies on the nested loops in an algorithm's implementation. If nested loops have not more than three levels (see Fig. 1), the graph can be placed in the traditional three-dimensional space, but complex loop constructs with nesting levels of four or more require special methods for presenting and displaying the graph.

There are many difficulties here, but also many ways to deliver the information to the students. The main task is to show the information structure of an algorithm so as to demonstrate all its key features, its parallel structure features, edge sets features, regularity areas and, vice versa, areas with an indeterministic structure dependent on the input data, etc. Teachers very rarely talk about parallel algorithm structure, while this is required in practice more and more often.

After telling about the parallel algorithm structure, one should proceed with **describing its resource of parallelism**. The main characteristic is **parallel complexity**, which is understood as the number of steps needed to execute this algorithm given an infinite number of processors (functional units, computing nodes, cores...). The concept of infinite parallelism is somewhat idealistic, but it helps to understand the advantages offered by the parallel execution of an algorithm. Parallel complexity of the fast Fourier transform (Cooley-Tukey algorithm) mentioned above for vectors with lengths equal to a power of two is $\log_2 n$, which means this algorithm can potentially be executed $1.5n$ times faster.

The concepts of a canonical parallel form or the critical path of an information graph are often used to evaluate and describe the resource of parallelism. The height of the parallel form, or the length of the critical path, determine the algorithm's parallel complexity, while the level width is determined by the number of processors needed at a specific level; all of this can be used to describe algorithm's properties. The complexity of an algorithm featuring the summation of n vector elements is reduced from $n - 1$ in a serial implementation to $\log_2 n$ in the parallel version, with the number of operations executed in parallel at each level (i.e. at each step of the pairwise parallel algorithm) falling from $n/2$ to 1.

Parallelism in an algorithm often has a natural hierarchical structure. This fact is very useful in practice and should be reflected in a description of algorithm's properties. This hierarchical parallelism structure is well reflected through a loop profile of the resulting program (including, in general, the call graph). Figure 10 shows a loop profile of a program, where each square bracket corresponds to a one loop of the program and nesting structure of the brackets repeats nesting structure of

Fig. 10 Resource of parallelism of a program: parallel loops are marked by '1' or '2' in the loop profile of the program

loops. Information that the outer loop (marked by '1') and all inner loops (marked by '2') are parallel (their iterations are independent) substantially improves the perception of the original algorithm's structure.

When explaining algorithms, it is important to pay attention to the **algorithm properties** which can prove important during implementation. We mentioned some of them in the previous section, namely **computational complexity** or **input/output data volume**: these properties are simple, but they often determine the quality of the future implementation.

Application efficiency and the **balance of the computation process** are two closely related concepts. The main challenge is that the balance can appear in different ways. This can include balancing between different types of operations, particularly between arithmetic operations (addition and multiplication) or between arithmetic operations and memory access operations. This can also include computational balance between different parallel branches of the algorithm. On the one hand, load balancing is a necessary condition for efficiency of a parallel algorithm. At the same time, this is a very challenging task, and one must explicitly show how many of these features the algorithm has. If ensuring of balance is not obvious, it is recommended to describe possible ways for solving a task.

An important aspect in practice is the **determinacy of an algorithm**, which can be understood as the consistency of the computational process structure. From this viewpoint, the classical multiplication of dense matrices is a highly deterministic algorithm, as its structure, given a fixed matrix size, does not depend on the input matrix elements. Multiplying a sparse matrix by a vector, when the matrix is stored in a special format, is no longer deterministic: data locality depends on the structure of the input matrices. An iterative algorithm with precision-based exit is also not a highly deterministic one: the number of iterations, and therefore the number of operations, changes depending on the input data.

The reason for pointing out determinacy as a property is clear: working with a deterministic algorithm is easier, since a structure, once found, will determine its implementation quality at all times. If determinacy is missing, this should be specially pointed out, along with a description of how indeterminacy affects the structure of the computational process.

A serious reason for the indeterminacy of a parallel program is a change in the execution order of associative operations. A typical example is the use of collective MPI operations by a group of parallel processes, such as summing elements of a distributed array. The MPI runtime system chooses the operation execution order assuming compliance with the associative law; rounding errors change for each program run, introducing changes in the program output. This is a serious issue often encountered on systems with massive parallelism, and it affects the reproducibility of results of parallel programs. If analysis of an algorithm's structure shows that a parallel application cannot work without collective operations, this property must also be kept in mind.

Interestingly, in some cases, determinacy can be "enforced" in an algorithm by introducing macro operations, which makes the structure not only deterministic but also more clearly understandable.

An important aspect is a **description of bit capacity needed to execute the algorithm's operations** (precision). In practice, executing all arithmetic operations on real numbers with double precision is rarely required, as this doesn't affect the algorithm's stability or the accuracy of the output. If most operations can be performed using a float type, and just a few fragments need to be changed to double, this fact must also be explicitly mentioned, as it can substantially improve implementation efficiency.

Parallel Algorithms: Implementation Features

In the beginning of this section we discussed two parts of the description for any algorithm: a description of its theoretical potential and its implementation features. The properties considered above are related to the first part of the description. This information is important and relevant, it has to be explained, but it is just as important to look ahead and point out some possible stumbling blocks that can be encountered in the process of implementation. This is what we will address below.

The issues of data locality and computation locality are rarely included in any training courses, but locality has a very high impact on the efficiency of program execution on modern computing platforms. To get the whole picture of an algorithm's implementation features, it is important to analyze both temporal and spatial locality, noting positive and negative factors related to locality, and under which conditions and situations they are caused. It is important to mention how locality changes when moving from a serial to a parallel implementation, and to highlight typical memory access patterns for a program implementing the given algorithm. We should also mention the potential correlation between the available programming language constructs and the locality exhibited by the resulting programs.

It is useful to show memory access profiles for computational kernels, which often explain the efficiency of the entire application. Figure 11 shows memory access profiles for programs that implement FFT, LU decomposition, dense matrix multiplication and random memory access. Each small red dot corresponds to one memory access operation. The X-axis shows the serial numbers of memory access operations arranged in the order they were performed during the program execution. It is similar to the timeline chart, but in this case we analyze only the order of memory accesses, not the particular time when they were performed. The Y-axis indicates the memory address used in each particular memory access operation.

For example, Fig. 11d, e show the memory access profiles for two versions of matrix multiplication algorithms, which makes it clear why the (IKJ) fragment is executed much faster than the (JKI) fragment: the profile in Fig. 11d has a much higher spatial and temporal locality than that in Fig. 11e.

Knowing the algorithm's resource of parallelism, it is important to show the **opportunities for equivalent transformation of the programs**: the students will see the program features for computers with certain architectures, and will feel

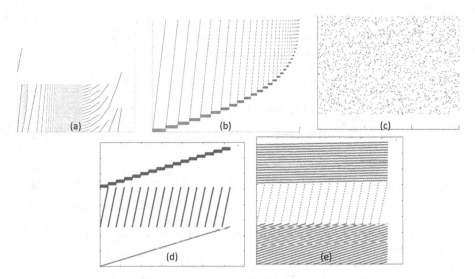

Fig. 11 Memory access profiles for programs implementing FFT (**a**), LU decomposition (**b**), random memory access (**c**) and dense matrix multiplication with different loop orders (**d, e**)

freedom in transforming the programs and obtaining program optimization skills. Let's go back to the program with loop profile shown in Fig. 10. It was mentioned that the outer loop (marked by '1') has many iterations, while the inner loops were very short. In this form, the parallel structure is suitable for SMP computers, allowing parallel execution of the outer loop iterations by processor cores. However, if the target system is a vector-pipeline computer, the efficiency will be low for sure: only the short innermost loops (marked by '2') can be vectorized. As a way out of this situation, we can perform a series of elementary transformations on the loops (see Fig. 12), moving the long outside loop (marked by '1') inside. The transformation is certainly not trivial, and the program must effectively be "turned inside out." During some transformations we obtain loops which are not perfectly nested ("dots" in the fourth and fifth loop profiles denote additional statements "between" loops), but all the transformations are fully equivalent (program's information structure remains intact, and we operate strictly within the available resource of parallelism). If you know about this freedom in code transformations, you can easily compose a variant of code which matches well any target architecture.

It is worth noting that such non-trivial transformations can't always be automatically identified and performed by the compiler, which means that the programmer himself must be aware of such features.

Scalability is one of the central notions in parallel computing, which shows how efficiently the algorithm and the program implementing it can use the available processing cores, processors and computing nodes. This is an important idea since all computers are parallel today, yet it is a highly complex one. Application's scalability potential is originally determined by the algorithm, but can be reduced

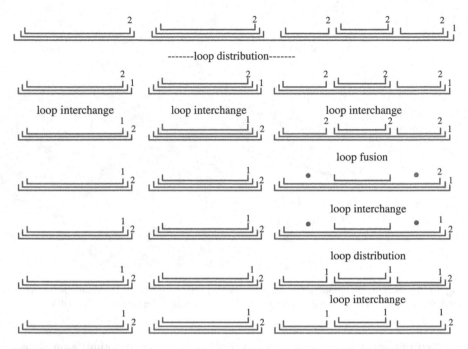

Fig. 12 Series of loop transformations that convert the outermost loop '1' in the original fragment into the innermost loop

substantially depending on the programming technology, bad data distribution, inadequate composition of the parallel program, and many other reasons.

Many things can be discussed with students here: strong scalability, weak scalability, wide scaling, possible reasons for low scalability. It is interesting to compare scalability of different algorithms that address the same task. It is important to show the connection between algorithm properties, program structure and computer architecture features, which lays the groundwork for co-design technologies and determines the scalability of parallel applications.

When explaining this idea, the most efficient argument is the behavior of the actual scalability of a given algorithm's various implementations, depending on the number of processors and the problem size. An important thing here is to find the right correlation between the number of processors used and the problem size to highlight all points of interest in the behavior of a parallel program, such as achieving maximum performance, and the more subtle issues that arise, for example, out of the algorithm's block structure or memory hierarchy.

Figure 13a shows scalability for an MPI implementation of the classical dense matrix multiplication algorithm depending on the number of processes and the size of the dataset. The chart clearly displays areas with higher performance corresponding to different levels of cache memory. Figure 13b shows good scalability for the Linpack benchmark: as the number of processors grows, so does the performance for

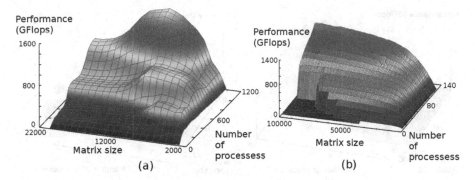

Fig. 13 Scalability of the MPI implementation of the dense matrix multiplication algorithm (**a**) and Linpack benchmark (**b**), MSU "Lomonosov" supercomputer

any matrix size shown in the chart. Some values are missing from the front part of Fig. 13b, as large tasks cannot fit into the memory of a small number of processors.

In addition to scalability, **performance and efficiency** are other concepts of particular interest in understanding the quality of a parallel algorithm implementation. These two notions are closely related and are often viewed not in abstract terms but in combination with a specific computing system. Efficiency can be understood differently: as parallel efficiency, or as efficiency compared to peak performance indicators for the computing system.

Algorithms or their implementations can possess specific features that prevent performance and efficiency from exceeding certain limits. These features need to be singled out and discussed with students, as they will likely run into something similar in practice. Figure 14 shows the performance and efficiency for Poisson's equation solution using the discrete Fourier transform depending on the number of processors and the problem size. Despite using serious computing resources (the experiments were conducted on the "Lomonosov" supercomputer [16] at Lomonosov Moscow State University), the program's performance was very low, with efficiency never exceeding 1.5% and falling quickly as the number of processors grew (the main reason of poor efficiency is low data locality). These facts need to be mentioned if we want students to have a realistic perception of what modern parallel computing systems are capable of, to understand peak performance figures correctly and to clearly understand the reasons why these situations arise.

Another subject for a more professional and detailed discussion is the search for the root causes for low performance and efficiency in parallel applications. The task is not a simple one. There isn't currently a single, universally accepted methodology or the respective tools to conduct such analysis. In fact, this requires conducting a comprehensive analysis adequate for modern supercomputer co-design technologies. One would need to assess and describe the efficiency of memory access operations, the efficiency of using the resource of parallelism inherent in an algorithm, the efficiency of using communication networks and particular features of the communication profile, the efficiency of input/output operations, and

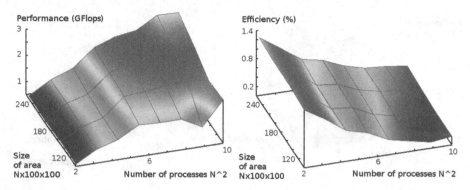

Fig. 14 Performance and efficiency of Poisson's equation solution using the discrete Fourier transform

many other aspects. Sometimes average aggregated characteristics of a program are generally sufficient; in some cases, it is necessary to show lower-level monitoring data such as CPU load, cache misses, InfiniBand network usage intensity, etc. Tools like TAU [18], Scalasca [17], Vampir [19], and JobDigest [2] provide a good understanding of the parallel program quality, and students need to be taught how to use them. Moreover, using such quality maintenance tools for parallel applications must become an integral element of the software development cycle.

While discussing the properties of a parallel algorithm, it is also important to evaluate its potential for the specific architecture of various classes of computing systems, and the specific parallel programming technologies used. The computational kernel is compact and has high computational intensity—a good condition for using accelerators. The interaction between parallel processes is intensive and results in major overhead on delivering messages—this means that developers should look at computers with shared memory or use one-sided data transfer functions like Put/Get. An algorithm has SIMD parallelism, so graphics processors can efficiently implement it. By adding these touches to the description of algorithm's properties, we pursue the main goal of showing students a strong connection between the parallel algorithm properties and the computer architecture features, which form the grounds for creating high-quality parallel applications to efficiently solve real-life problems.

Our experience in analyzing algorithm properties and implementations not only determined the key issues that are addressed in this work, but also became the foundation for the AlgoWiki project [3, 4]. The project's main goal is to provide a description of the fundamental algorithm properties that enable full understanding of their theoretical potential and their implementation features regarding various classes of parallel computing systems. The project is expected to result in the development of an open encyclopedia based on wiki technologies and available to the entire academic and educational community. The first version of the encyclopedia is available at [13], where users can describe both their own pedagogical experience and their knowledge of specific parallel algorithms. The encyclopedia already contains many useful and live examples that can be used in lectures and seminars to explain the particular features of parallel algorithms.

How Does One Make a Training Curriculum Parallel?

Parallel algorithms, parallel computing systems, parallelism—all of these are fundamental concepts that must be at the foundation of any curriculum on computational mathematics, applied mathematics, and generally of any curriculum on Computer Science. We are implementing this approach at the Faculty of Computational Mathematics and Cybernetics at the Lomonosov Moscow State University (CMC MSU), starting with Bachelor's degree coursework and continuing with many Master's degree and post-graduate education programs. In this section we would like to show the great diversity of methods and techniques that can be used in the educational process to support parallel computing topics in various training programs. The choice of particular materials needed in each specific case is up to teachers. Our goal is to help selecting the most suitable methods for including this topic in a curriculum, to inspire the teacher to explore various directions, and to suggest potential pedagogical techniques for implementing this in practice.

Parallelism Concepts: In Every Lecture Course

We analyzed the structure of lecture courses in the Bachelor's degree programs at CMC MSU, and it turned out that parallelism concepts can be added to each one rather easily and gracefully. Some courses only require parallelism to be mentioned; for some, examples would need to be replaced with those having parallel specifics, but without any impact on the logical flow of presentation; sometimes, a course would need to add new lectures. However, this modification, turning the classical "serial" curriculum into a parallel one based on the ideas of parallel computing, does not present any specific challenges. Let's give some examples of how individual training courses can be modified; this will make it clearer regarding how parallelism can be introduced into any training course.

Algorithms and algorithmic languages (1st year). The concept of parallel execution for an algorithm needs to be introduced right in the very first semester of study (at an intuitive level).

Linear algebra and analytical geometry (1st year). For simple examples (summing vector elements, dot product), introduce the idea of computational complexity, discuss the possibility of parallel execution (without any theoretical justification), and perform assessments of parallel execution timing. The information structure of a classical matrix multiplication algorithm should be shown, pointing out the possible options for parallel execution.

Computer architecture (1st year). Introduce the principles of pipeline and parallel data processing, explain the architecture of multi-core processors, introduce the notions of independent functional units, vectorization, peak and sustained performance. Explain organization of memory subsystems in modern processors, and memory hierarchies.

Physical basics of designing computers (2nd year). Computer representations of numbers, accuracy, rounding, parallel bitwise execution of arithmetic operations. Rounding, the laws of exact and machine arithmetic.

Operating systems (2nd year). Parallel processes, the fork/join model, synchronization methods, process interaction methods, deadlocks, determinacy, correctness, viability, fault tolerance.

Discrete mathematics, graph theory (2nd year). Two-dimensional grid, n-dimensional torus, n-dimensional hypercube, etc. as examples of multi-processor system topologies. The shortest path between nodes, critical path length, the outcome degree of graph vertices, routing and fault-tolerance. The need for and complexities of processing ultra-large graphs.

Algorithm complexity (3rd year). The serial and parallel complexity of an algorithm, computational intensity of an algorithm.

Introduction to numerical methods (3rd year). Information structure of implicit and explicit numerical methods, the trade-offs between computational complexity and the parallel structure of algorithms (explicit methods have good parallel structure but can have slow convergence, while implicit methods are serial in nature but can have better convergence).

Mathematical physics equations (3rd year). Parallel problem-solving methods, key steps in problem-solving on computers: task—algorithm—programming technology—program—compilation—execution, the need to preserve parallelism at every problem-solving step.

Databases (4th year). Parallel DBMS, ultra-large DBMS, load balancing in parallel query execution.

Optimization methods (4th year). Examples of parallel optimization methods.

Distributed systems (4th year). Metacomputing concepts, grid, cloud technologies, distributed processing technologies.

A detailed description can be provided for each training course, but this is not in the scope of this work. The idea is to show how naturally the ideas of parallelism can be included into almost every lecture course in a bachelor's degree program.

Emphasis on Parallelism in Exam and Test Questions

Questions on exams and tests for various disciplines must be formulated in such a way as to explicitly mention the ideas of parallelism, rather than conceal them. Accents in the questions can vary and they can change a flavour of the questions—

several possible question formulations are shown below, which could be used, with minor modifications, for exams within many disciplines.

1. Types of parallel data processing, their features.
2. Evaluating the computational complexity of large tasks.
3. Memory hierarchy, computational locality, data locality.
4. Amdahl's law, its corollaries, and superlinear speedup.
5. Quality indicators of parallel programs: speedup, efficiency, scalability.
6. Strong scalability, wide scaling, weak scalability. Isoefficiency.
7. Parallel implementation of problems characterized by high computational complexity with matrix multiplication as an example.
8. Graph models of programs, their relationship.
9. The concepts of information dependency and information independence. An algorithm's resource of parallelism.
10. Information structure of algorithms. The critical path of an information graph.
11. Equivalent transformations of programs. Elementary loop transformations.
12. Types of parallelism: finite parallelism, massive parallelism, coordinate parallelism, skewed parallelism.
13. Parallel form of an information graph, its width and height. Canonical parallel form. Parallel complexity of algorithms.
14. Application efficiency dependence on the choice of data structures.

Online Testing: Knowledge Check and Continuing Education

A properly formulated question does more than just test the level of knowledge: it allows objects and ideas to be observed and studied from another perspective, showing alternative sides of the material just learned. An extensive bank of questions was built with the authors' active involvement in the Sigma student knowledge online testing system for parallel computing [14]. Importantly, a system like this not only allows one's knowledge level to be checked for taking a test or exam. It can be successfully used for self-testing, providing students an opportunity to test how well they understand the material learned in the classroom.

All questions in the system can be divided into several categories, as we will illustrate below with a few examples ("+" marks the correct answers).

1. *Questions for testing the correct understanding of the definitions.*
   Mark the correct statements:

   • Superlinear speedup can be achieved by a large number of functional units.
   • Program's scalability means that the program is suitable for parallel execution.
   • A parallel program that does not possess strong scalability can possess weak scalability. +
   • Efficient parallelization can be measured in terms of the number of processes running at a given moment in time.

2. *Simple calculation questions.*

What is the minimum time it will take to add up 512 numbers on 200 processors using pairwise summation, if two numbers are added in 1 second and the time for a transfer data between the processors is negligible:

- 1 second
- 8 seconds
- 9 seconds
- 10 seconds +
- 11 seconds
- 200 seconds
- 384 seconds
- No correct answer.

3. *Questions for analyzing the structure of algorithms or program fragments.*

The multiplication of two matrices is programmed using the following fragment:

```
for(i = 0; i < n; ++i)
   for(j = 0; j < n; ++j)
      for(k = 0; k < n; ++k)
         A[i][j] = A[i][j]+B[i][k]*C[k][j];
```

What statements regarding the structure of this fragment are correct:

- The height of the canonical parallel form of the information graph for this fragment equals n. +
- The critical path length of the informational graph for this fragment equals n.
- The information graph for this fragment consists of n^2 independent computational branches. +
- This fragment cannot be executed on a computer with shared memory.
- Using any loop order in this fragment will yield the same result up to the rounding error. +
- Reordering the loops will not change the program's execution time.

4. *Questions for understanding serial and parallel algorithm complexity.*

A computer executed a program that multiplies two square dense matrices using a standard algorithm (without using fast multiplication methods) in 4 seconds at a performance of 32 GFlops. What were the sizes of the arrays?

- 500×500
- 1000×1000
- 2500×2500
- 4000×4000 +
- 5000×5000
- No correct answer.

From Theory to Practice

Parallel algorithms, like parallel computing in general, are an area where practice is the key. All of the theoretical ideas introduced can be easily illustrated with examples, tasks, and exercises from parallel programming practice, and by all means this should be done. Parallelism came from practice and should be explained in practical examples all the time. The experience of teaching such disciplines shows this is completely feasible even for "purely theoretical" concepts.

Various graph models of programs are considered as part of the "Structure of algorithms and programs" topic. Let's consider two types of vertices—operators (V1) or separate executions of operators (V2). For example, if a statement is executed three times in a loop, we'll get 1 operator or 3 executions of operators.

Also there are two types of edges—operational (E1) or information (E2) relationship. Vertex A is connected to vertex B with operational relationship if and only if vertex B can be executed right after vertex A. Vertex A is connected to vertex B with information relationship if and only if vertex B uses as an argument some value obtained in vertex A.

By using different methods of choosing vertices and types of relationships between them, we can derive four basic graph models: program control graph (V1 + E1), program information graph (V1 + E2), operational history (V2 + E1) and information history (sometimes referred to as information or dependency or dataflow graph (V2 + E2)). Despite the abstract nature of these concepts, they can easily be illustrated with simple examples. In particular, for one and the same example:

```
for(i = 0; i < n; ++i) {
    A[i] = A[i-1] + x;          (1)
    B[i] = B[i] + A[i];         (2)
}
```

all the four basic graph models are presented in Fig. 15.

To understand potential and properties of the models and to relate these ideas to program features, presentation of the material can be accompanied by a number of questions or tasks.

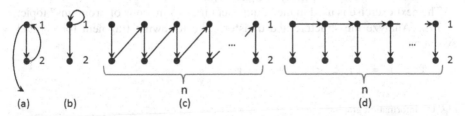

Fig. 15 Four basic graph models: program control graph (**a**); program information graph (**b**); operational history (**c**); information graph (**d**)

Task. Can the information history of a certain fragment be represented by an empty graph?

The answer is "yes". A fragment possessing this property is shown below:

```
for(i = 0; i < n; ++i)
   A[i] = A[i] + B[i]*c;
```

Interestingly, an empty graph is far from "exotic"; on the contrary, it reflects an exceptionally important property of the program—its high degree of parallelism.

It is also important to show potential limits for the introduced concepts.

Task. Can the information history of a certain program fragment have 20 vertices and 200 edges?

Indeed, information history can be structured in different ways. But to answer this question correctly, one needs to remember its two main properties: the absence of multiple edges and its acyclic nature. This means that an information graph with the maximum number of edges will look as follows (Fig. 16).

It follows from here that the maximum number of edges for a graph of n vertices can be calculated as $(n-1)+(n-2)+(n-3)+\cdots+2+1 = n(n-1)/2$, which equals 190 for $n = 20$, so the correct answer to the question above is "no".

Another type of task can help to assess the understanding of basic concepts such as the serial and parallel complexity of an algorithm.

Task. Determine the serial and parallel complexity of an algorithm implemented using the following fragment:

```
for(i = 2 ; i <= n ; ++i)
   for(j = 2 ; j <= m ; ++j)
      A[i][j] = A[i][j] * A[i][j-2];
```

The loop body will be executed $(n-1)(m-1)$ times, and each execution of the operator placed in the loop body corresponds to one multiplication operation; therefore, the total number of operations and the serial complexity equal to $(n-1)(m-1)$.

Parallel complexity is the critical path length of the fragment's information graph plus 1, where each vertex corresponds to one execution of the operator in the loop body. To determine it, we need to build an information graph, which looks as follows for this fragment (Fig. 17).

Obviously, parallel complexity is equal to $\lfloor m/2 \rfloor$.

The next exercise is used in the "Equivalent transformations of programs" topic.

Task. Analyze the structure and transform the following fragment for parallel execution:

```
   for(i = 1; i < n; ++i) {
```

Fig. 16 Information graph with the maximum number of edges

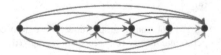

Fig. 17 Information graph of
the fragment

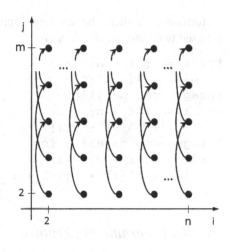

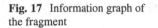

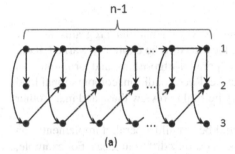

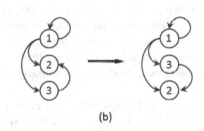

(a)   (b)

Fig. 18 Information graph of the fragment (**a**) and transformation of the loop body (**b**)

```
1        A[i] = A[i-1]*p + q;
2        C[i] = (A[i] + B[i-1])*s;
3        B[i] = (A[i] - B[i])*t;
    }
```

The entire fragment's information structure is shown in the Fig. 18a: the fragment definitely has a substantial resource of parallelism. First let's break down the loop body to identify the serial and parallel parts within the loop body operators. The respective training course contains a statement: the necessary and sufficient condition for loop distribution is that the parts being distributed must be located in different, strongly-connected components of the information graph that represents the loop body. All three operations represent individual, strongly-connected components of the information graph, so loop distribution is possible for all three operators. The execution order for the three new loops is determined by the information structure of the information graph for the loop body: first 1, then 3, then 2, as shown on the right side of Fig. 18b.

Executions of operator 1 are connected with an informational dependency, so vertex 1 has a self-loop, and the corresponding loop requires serial execution. Operations 2 and 3 are not self-connected, so all iterations of these loops can be

performed in parallel. The resulting fragment is shown below (OpenMP directives
are used to declare parallel loops):

```
for(i = 1;  i < n;  ++i)
   A[i] = A[i-1]*p + q;
#pragma omp parallel for
for(i = 1;  i < n;  ++i)
   B[i] = (A[i] - B[i])*t;
#pragma omp parallel for
for(i = 1;  i < n;  ++i)
   C[i] = (A[i] + B[i-1])*s;
```

Parallel Programming Features

Moving students from serial programming to writing parallel programs usually
does not cause major issues. Nevertheless, they need to be explicitly warned about
the prospective issues that are typical for parallel algorithms and programs, to
prevent them from occurring again in the future. Race condition, computational load
disbalance, Amdahl's law impact—it is very helpful to review these and many other
concepts.

A possible assignment for this topic would be to build a parallel implementation
of a simple algorithm, while the task can focus on very different ideas. For example,
build a parallel implementation of an algorithm optimized for a certain parameter,
such as the utilized resource of parallelism, execution time on a specific computer,
amount of memory used, scalability, implementation efficiency, etc.

Task. A program fragment is given:

```
for(i = 1 ;  i <= n ;  ++i)
   for(j = 1 ;  j <= n ;  ++j)
      A[i][j] = A[i][j] * A[i][j] * A[i-1][j-1] ;
```

Create an implementation that uses the maximum resource of parallelism for this
algorithm at any given moment.

The first thing one needs to do before building a parallel implementation is
to determine the information structure of a code and identify its resource of
parallelism. The information graph for this fragment will look as follows (Fig. 19a).

To use the maximum possible resource of parallelism, one needs to build the
canonical parallel form for this graph (it is shown in the Fig. 19b). Individual levels
are shown using dashed lines. The required parallel implementation must use the
entire resource of parallelism in a fragment by going through the levels in the
parallel form: the total number of levels is n (this number represents the parallel
complexity), and the number of vertices on each level, which can be executed in
parallel, varies from $2n - 1$ to 1. This approach can be implemented as follows:

```
for(k = 1 ;  k <= n ;  ++k) {
```

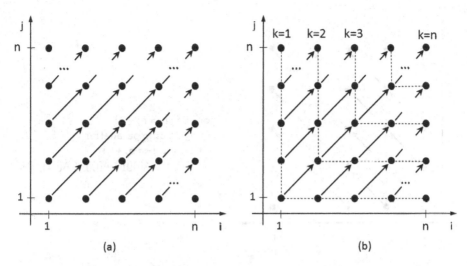

Fig. 19 Information graph of the fragment (**a**) and its canonical parallel form (**b**)

```
#pragma omp parallel sections
    {
#pragma omp parallel for
      for(i = k; i <= n ; ++i)
          A[i][k] = A[i][k] * A[i][k] * A[i-1][k-1];
#pragma omp section
#pragma omp parallel for
      for(i = k+1 ; i <= n ; ++i)
          A[k][i] = A[k][i] * A[k][i] * A[k-1][i-1];
    }
}
```

This implementation also has its own nuances: the potential for using the entire resource of parallelism with this program depends on whether the OpenMP programming system supports nested parallelism.

One should recognize that this implementation is not necessarily optimal for other criteria such as execution time or code simplicity. For example, the parallel execution of the fragment can be organized using skewed parallel branches, even though this type of implementation increases parallel complexity from n to $2n - 1$. At the same time, the structure of the dependencies in this example allows a very simple form of parallelism to be used for any coordinate, i or j, as in the example shown in the Fig. 20.

The parallel complexity of the implementation equals n, just like the first case, but a more convenient regular parallelism type is used with the same number of operations at every step. This version will likely be used in practice.

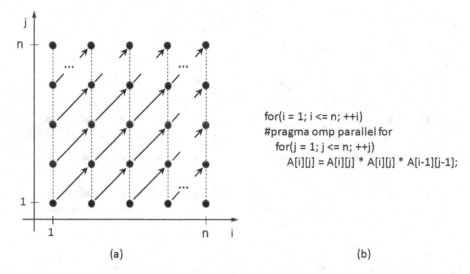

```
for(i = 1; i <= n; ++i)
#pragma omp parallel for
   for(j = 1; j <= n; ++j)
      A[i][j] = A[i][j] * A[i][j] * A[i-1][j-1];
```

(a)   (b)

Fig. 20 An implementation of the fragment using coordinate parallelism

The Efficiency of Parallel Applications: A Matter of Special Attention

Creating efficient parallel applications is one of the objectives for training students in this field. It requires knowledge of parallel problem-solving methods, experience in using parallel programming technologies, and an understanding of parallel computing system architecture. A number of techniques and methodologies can be used to reinforce the material with focus on various stages of supercomputer co-design, which is the central element ensuring the efficiency of parallel applications.

Let's look at one possible version of this task which we used at the Summer Supercomputing Academy [1] held at the Lomonosov Moscow State University.

Task. Implement a parallel program on a supercomputer that multiples dense square matrices with double precision, and examine its scalability. A description of the algorithm can be found at [7]. The implementation needs to be written in a high-level programming language (C or Fortran) using MPI technology (for extra points—write a hybrid version using OpenMP inside a computing node and MPI for communication between nodes). The task requires that none of the matrices a, b or c can be stored as a whole at any given node. Moreover, auxiliary arrays at each node can only be used to store portions of the original matrices, but not the whole matrices. The total RAM at all nodes is sufficient for storing all relevant data. The size of matrices to be multiplied in this experiment is $n = 4096$. Matrix elements of the type double (DOUBLE PRECISION) can be initialized as follows:

$$a_{ii} = b_{ii} = (n - 2)/n$$

$$a_{ij} = b_{ij} = -2/n \text{ if } i \neq j$$

With these inputs, the output shall be an identity matrix, which is easy to verify. The task is to determine the correlation between program execution time (excluding initialization) and the number of processors. The number of processors p shall equal to powers of four.

First let's describe the approach to parallelization. A review of the information dependency graph for this algorithm provided in section "What Knowledge of Algorithm Properties Is Needed in Practice?" shows that all elements of the matrix c can be computed independently from each other. Therefore, the parallel program can use a procedure to determine a single element in the resulting matrix c as the basic building block. In this case, various methods of distributing elements within the matrix c between various processes will determine the different versions of the resulting program.

Generally, an absence of information dependencies makes any distribution of the elements in the matrix c possible; however, the most natural ones are row, column and block distributions. To improve data locality, it is best to use a block distribution, where the basic block is a set parts of adjacent rows or columns (see Fig. 21).

In the case of MPI implementation, the distribution cannot just apply to the resulting matrix c (and related operations), but must also be determined for the original matrices a and b. This determines how much data transfer the program will need. Different versions of parallel implementation for this algorithm can be found here:

- Version 1 [8] (row distribution of matrices a and c, column distribution of matrix b);
- Version 2 [9] (block distribution of all three matrices).

In version 1, the whole row of matrix a needed to compute a certain matrix element is stored in the memory allocated to that process, while a column in matrix b may not be present in the memory for the same process, but rather is stored entirely in the memory for another process. In fact, each column in matrix b is involved in

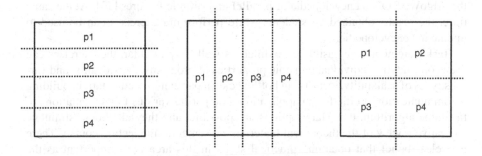

Fig. 21 Row, column and block distributions of the matrices between four processes

Table 1 Comparison of two versions of matrix multiplication implementations

	Number of arithmetic operations	Amount of data transferred	Volume of data stored for each process
Version 1	$2n^3$	$n^2(p-1)$	$\frac{4n^2}{p}$
Version 2	$2n^3$	$2n^2(\sqrt{p}-1)$	$\frac{3n^2+2n^2\sqrt{p}}{p}$

Table 2 Comparison of execution time (in seconds) for two versions of the matrix multiplication implementations on the MSU "Lomonosov-2" supercomputer

Number of processes	1	4	16	64	256	1024
Version 1	44.19	12.86	4.49	1.43	0.50	49.05
Version 2	44.66	12.76	4.44	0.84	1.05	3.05

computing the entire column of matrix c, so it must be sent to all processes of the application.

In case of a block distribution into a two-dimensional process grid (version 2), each matrix dimension requires $\sqrt{p}$ processes. Matrices a, b and c are distributed between the processes in uniform blocks. For the process to be able to execute its part of operations associated with the elements of matrix c, it needs to receive data from the processes containing the rows of matrix a and columns of matrix b for the respective block.

A comparison of the two versions by the number of arithmetic operations, total volume of data transferred and the volume of data stored by each process is shown in the Table 1. Even though both versions have the same number of arithmetic operations, version 2 requires transferring less data, but more data needs to be stored for this implementation.

A comparison of each implementation's scalability on the MSU "Lomonosov-2" supercomputer (1280 nodes using Intel Xeon E5-2697v3 processors connected with an InfiniBand FDR communication network) is shown in the Table 2.

The examples above illustrate just a small portion of the wide variety of techniques and methods that can be used in the educational process for this topic. A lot of materials for exercises can be found in the numerous algorithms described in the AlgoWiki Open encyclopedia of parallel algorithmic features [13]. At the same time, the results obtained by students while performing exercises can be used to update the encyclopedia.

Task statements can easily be modified as well. In particular, the exercises can focus on studying individual dynamic properties of programs, data locality, and various types of scalability, identifying bottlenecks in parallel implementation, building a communication profile for an application, comparing various implementations of the same algorithm, etc. Many options are possible, and they all should stimulate a creative review of the theoretical materials presented in the lecture course. There is a clear belief that obtaining practical skills in this area is as important as the theoretical study.

References

1. Summer Supercomputing Academy. http://academy.hpc-russia.ru/en. Cited 26 Jan 2018
2. Adinets, A.V., Bryzgalov, P.A., Voevodin, V.V., Zhumatii, S.A., Nikitenko, D.A., Stefanov, K.S.: Job Digest: an approach to dynamic analysis of job characteristics on supercomputers. Computational Methods and Software Development: New Computational Technologies, vol. 13, pp. 160–166 (2012)
3. Antonov, A., Voevodin, Vad., Voevodin, Vl., Teplov, A.: A study of the dynamic characteristics of software implementation as an essential part for a universal description of algorithm properties. In 24th Euromicro International Conference on Parallel, Distributed, and Network-Based Proceedings, pp. 359–363 (2016)
4. Antonov, A., Voevodin, V., Dongarra, J.: Algowiki: an Open encyclopedia of parallel algorithmic features. Supercomputing Frontiers and Innovations, vol. 2, no. 1, pp. 4–18 (2015)
5. Big Data and Extreme-scale Computing (BDEC). http://www.exascale.org/bdec. Cited 26 Jan 2018
6. Computer Science Curricula 2013 (CS2013). http://ai.stanford.edu/users/sahami/CS2013. Cited 26 Jan 2018
7. Dense matrix multiplication. http://algowiki-project.org/en/Densematrixmultiplication. Cited 26 Jan 2018
8. Dense matrix multiplication example, version 1. https://github.com/srcc-msu/CDER-2016/blob/master/dgemm/mpi_1d_grid.c. Cited 26 Jan 2018
9. Dense matrix multiplication example, version 2. https://github.com/srcc-msu/CDER-2016/blob/master/dgemm/mpi_2d_grid.c. Cited 26 Jan 2018
10. Dongarra, J., Beckman, P., Moore, T., Aerts, P., Aloisio, G., Andre, J., Barkai, D., Berthou, J., Boku, T., Braunschweig, B., et al.: The international exascale software project roadmap. International Journal of High Performance Computing Applications, vol. 25, no. 1, pp. 3–60 (2011)
11. Supercomputing Education in Russia, Supercomputing Consortium of the Russian Universities, Tech. Rep. (2012) http://hpc.msu.ru/files/HPC-Education-in-Russia.pdf. Cited 26 Jan 2018
12. Future Directions in CSE Education and Research. Workshop Sponsored by the Society for Industrial and Applied Mathematics (SIAM) and the European Exascale Software Initiative (EESI-2), Tech. Rep. (2015) http://wiki.siam.org/siag-cse/images/siag-cse/f/ff/CSE-report-draft-Mar2015.pdf. Cited 26 Jan 2018
13. Open Encyclopedia of Parallel Algorithmic Features. http://algowiki-project.org/en. Cited 26 Jan 2018
14. Parallel computing collective test bank "SIGMA". https://sigma.parallel.ru/BankTest/Start/index.php?lang=en. Cited 26 Jan 2018
15. Prasad, S.K., Chtchelkanova, A., Dehne, F., Gouda, M., Gupta, A., Jaja, J., Kant, K., La Salle, A., LeBlanc, R., Lumsdaine, A., Padua, D., Parashar, M., Prasanna, V., Robert, Y., Rosenberg, A., Sahni, S., Shirazi, B., Sussman, A., Weems, C., and Wu, J.: NSF/IEEE-TCPP Curriculum Initiative on Parallel and Distributed Computing — Core Topics for Undergraduates, Version I. 55 pages (2012) http://www.cs.gsu.edu/~tcpp/curriculum. Cited 26 Jan 2018
16. Sadovnichy, V., Tikhonravov, A., Voevodin, V., Opanasenko, V.: Lomonosov: Supercomputing at Moscow State University. In: Contemporary High Performance Computing: From Petascale toward Exascale, ser. Chapman & Hall/CRC Computational Science. Boca Raton, United States: Boca Raton, United States, pp. 283–307 (2013)
17. Scalasca. http://www.scalasca.org. Cited 26 Jan 2018
18. Tau Performance System. http://www.paratools.com/tau. Cited 26 Jan 2018
19. Vampir — Performance Optimization. https://www.vampir.eu. Cited 26 Jan 2018
20. Voevodin, V.: Mathematical Foundations of Parallel Computing. World Scientific Publishing Co., Series in computer science, vol. 33 (1992)

21. Voevodin, V., Gergel, V.: Supercomputing education: the third pillar of HPC. Computational Methods and Software Development: New Computational Technologies, vol. 11, no. 2, pp. 117–122 (2010)
22. Voevodin, V., Voevodin, Vl.: Parallel Computing. BHV-Petersburg, St. Petersburg (2002)
23. Wirth, N.: Algorithms + Data Structures = Programs. Prentice Hall PTR (1978)

Modules for Teaching Parallel Performance Concepts

Apan Qasem

Abstract This chapter introduces three teaching modules centered on parallel performance concepts. Performance related topics embody many fundamental ideas in parallel computing. In the ACM/IEEE curricular guidelines (ACM2013), an entire knowledge unit has been devoted to parallel performance. In addition, performance topics pervade every knowledge area within PDC and can be found across other knowledge areas including Algorithms, Architecture and Systems Fundamentals. The three modules presented in this chapter cover a range of parallel performance topics. Since power savings have become an important consideration from hand-held devices to supercomputers, energy efficiency is also emphasized in each module. The modules focus more on architectural and algorithmic issues rather than the programming aspects. The modules are constructed to illustrate parallel performance issues primarily through code examples and experimental studies. This approach makes the modules accessible to students who do not *yet* have a strong background in parallel programming. Thus, the target audience for this chapter are instructors who are teaching CS1, with or without parallel programming, and also instructors who are teaching upper-level electives where their students may already have taken a semester of parallel programming.

Relevant core courses: CS1, Operating Systems, Computer Architecture

Relevant PDC topics: speedup (C), efficiency (C), Amdahls Law (A), space vs. time (C), power vs. time (C), synchronization and communication (C), task granularity (A), scheduling and mapping on multicore (A), load balancing (A), trade-offs in performance and power (C), Analysis and Evaluation: linear and super linear speedup (C), latency and bandwidth trade-offs, data locality, SMP (C), NUMA (C), strong and weak scaling (C), (Bloom classification in parentheses)

A. Qasem (✉)
Texas State University, San Marcos, TX, USA
e-mail: apan@txstate.edu

© Springer International Publishing AG, part of Springer Nature 2018
S. K. Prasad et al. (eds.), *Topics in Parallel and Distributed Computing*,
https://doi.org/10.1007/978-3-319-93109-8_3

Context for use: CS1 fundamentals, operating system thread scheduling, parallel architecture performance evaluation

Learning outcomes:

- list and define parallel performance metrics: speedup, efficiency, linear speedup, super linear speedup, latency and bandwidth
- describe the implications of Amdahl's law on parallel performance
- recognize the use of parallelism to achieve strong scaling and weak scaling
- analyze the effects of load imbalances on performance and power
- apply techniques to balance load across threads or processes
- explain the need for inter-thread synchronization and communication
- apply techniques to pin and schedule threads on multicore systems for improved performance
- describe how cores share memory resources, such as DRAM and cache
- recognize the importance of exploiting data locality in parallel applications

Introduction

This chapter introduces three teaching modules centered on parallel performance concepts. Performance related topics embody many fundamental ideas in parallel computing. In the ACM/IEEE 2013 curricular guidelines (ACM2013), an entire knowledge unit has been devoted to parallel performance [1, 2]. In addition, performance topics pervade every knowledge area within PDC and can be found across other knowledge areas including Algorithms, Architecture and Systems Fundamentals.

The three modules presented in this chapter cover a range of parallel performance topics. Since power savings have become an important consideration from hand-held devices to supercomputers, energy efficiency is also emphasized in each module. The topics provide at least 3.5 h of Core-Tier 1, Tier 2 and Elective hours from ACM2013. The modules are designed to be introduced in CS1 and two upper-level electives, namely, Operating Systems and Computer Architecture. They are, however, designed with enough flexibility to enable adoption in a number of undergraduate courses at various levels.

The modules focus more on architectural and algorithmic issues rather than the programming aspects. The modules are constructed to illustrate parallel performance issues primarily through code examples and experimental studies. This approach makes the modules accessible to students who do not *yet* have a strong background in parallel programming. Thus, the target audience for this chapter are instructors who are teaching CS1, with or without parallel programming, and also instructors who are teaching upper-level electives where their students may already have taken a semester of parallel programming.

Elementary Concepts

This module is designed to introduce fundamental concepts in parallel computing in a CS1 course. The concepts are illustrated with no particular binding to any programming language and therefore can be introduced in different flavors of CS1 courses.

Recommended Length 1 lecture (1:15 min)
Recommended Course CS1, CS2

Organization and Content

The major topics in this module include (i) overview of parallel computation on a multicore processor, (ii) data dependence and need for synchronization in parallel programs, (iii) parallel performance and Amdahl's law and (iv) energy efficient computing. The topics are introduced through lectures slides, an in-class activity, code examples and a program demo. The following subsections describe how these topics are explained and the order in which they are introduced.

Parallelism in Real Life

The module begins with an in-class activity that engages the students and demonstrates the benefits of parallelism. An activity that works quite well with CS freshman is a live simulation of the word search problem where students act as processing threads. In this activity, the class is split into k groups. Each group is assigned the task of finding a collection of words in a book and reporting the page numbers where the words occurred. Each group gets a copy of the book. But the copies are sectioned into different-sized segments. Thus, one group might get the entire book in one chunk while another may be assigned one page per group member. The students are then asked to try to find an efficient method of solving the problem with resources they are given. Naturally, the teams with fewer pages per student (thread) are likely to get to the results first. However, care must be taken in selecting the words and their positions and in segmenting the text.

Parallel Computing and Its Importance Today

Following the in-class example, a set of lecture slides defines parallel computing and discusses its importance in today's world. A high-level definition of a parallel computer is presented. Student familiarity with basic Von-Neumann architecture is assumed (not an unrealistic expectation for CS1 students). The discussion of the definition of a parallel computer is followed by some history of parallel computing.

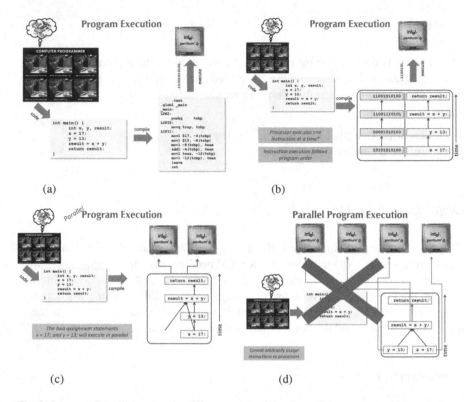

Fig. 1 Lecture slides illustrating the differences in serial and parallel program execution. Animation is used for the different blocks in the slideshow

The point is made that parallel computing has been around for a long time, ever since the beginning of computing. Notwithstanding, it has only become mainstream in the last decade. Brief descriptions of mainframe, vector computers and clusters are presented. This is followed by a discussion of multicore computers of today. The importance of energy efficiency and the role it has played in the evolution of computer chips and given rise to multicore systems is discussed. The lecture slides emphasize the need for achieving higher performance at lower power consumption or at specified power budgets. The ubiquity of parallel computers is also discussed. Students are asked to guess/comment on the number of processing cores on their smartphones and tablets. Their guesses are then validated against actual numbers. A discussion follows on the need for more parallel processing cores.

Sequential vs. Parallel Program Execution

A major portion of the module is spent introducing the student to the fundamental difference in sequential and parallel program execution. A walk-through example

Fig. 2 A simple parallel
code written in SimPar

```
int add() {
  int x, y, result;
  #PARALLEL {
    x = 17;
    y = 13;
  }
  result = x + y;
  return result;
}
```

Fig. 3 Incorrectly
parallelized code

```
int add() {
  int x, y, result;
  #PARALLEL {
    x = 17;
    y = 13;
    result = x + y;
  }
  return result;
}
```

is used for this purpose. Figure 1 shows a subset of the slides that are used to explain this topic. The slides are accompanied by a set of examples written in SimPar [3]. Two such examples are shown in Figs. 2 and 3. SimPar is a simple macro language that uses an intuitive pragma based syntax. Since students are generally not expected to be familiar with any parallel programming language in CS1, SimPar is an effective tool to discuss parallelism with real examples without getting bogged down in syntax minutiae. SimPar contains only one kind of parallel statement, a directive in the form of #PARALLEL { ... }. This implies that all high-level statements enclosed in the subsequent block will be executed concurrently. SimPar processes such directives by taking each statement in the block and converting it into a Pthread function. Supplementary materials for this chapter includes a SimPar parser that can be used to create other simple examples. The instructor should be aware that SimPar is not a realistic parallel language and is very limited in ability. Thus it should not be used for creating extended examples beyond CS1. During the walk-through of the example, students are asked to list the order in which the statements will execute on the processor. A parallel directive is then inserted for the two assignment statements and the meaning is explained to the students. The program is then extended to include array assignments instead of just simple assignments. This program is compiled and executed and the result examined in class. Students are then asked to comment on what other statements could be parallelized. The instructor leads them to an example where the result statement is put in the PARALLEL block along with the two assignment statements. This program is run, potentially several times, and the error demonstrated to the students. The students are then asked to describe the problem in the code. This is followed by a discussion of data dependence and the challenges with parallel programming.

Parallel Programming Tools

Students are told that SimPar is not a real language. The syntax for real languages are more complex and so is the programming model. Some of the currently available parallel languages and tools, including OpenMP, Pthreads, MPI are presented. The suitability of each is *briefly discussed*. The slides include example codes for each of these parallel languages. However, students are told they are not expected to learn the syntax at this stage.

Performance Metrics

In this segment of the module, performance issues in parallel computing are reiterated. This is followed by definitions and examples of sequential and parallel performance metrics. A simple parallel search code written in SimPar is used to do an in-class demo to show the differences in the performance metrics. Sequential and parallel (OpenMP) versions of the code are also shown in class. The code is compiled and executed with different data sets. Execution time and energy are measured for each run. A convenient tool for measuring power consumption on Intel processors is Likwid [4], freely available for download. The specific performance metrics and definitions that are discussed include

- Execution time
- Energy
- Speedup and Greenup
- Amdahl's Law
- Linear speedup
- Scalability

Pedagogical Notes

The author has used this module in CS1 courses in three semesters at Texas State University. In all three cases, it was helpful to introduce this module towards the end of the semester when students are somewhat more confident with the syntax of the sequential language that is being used in the class.

For the in-class activity, we found that a group size of four and a section size of two pages per member for the most *parallel* group is ideal. Making groups larger, makes the *sequential* group not as engaged. More than two pages of dense text makes the example run too long. We also found that, it is helpful to assign some form of reward to the team finishing first. This motivates the teams to be more engaged in the activity. Our experience also showed that it is better to place the stronger and more vocal students in the sequential group. Since the activity is framed as a competition and the sequential group is almost certain to not win, putting under-performing students in that group is not advisable.

It is advisable that instructors practice the live coding examples ahead of lecture time. Students often raise questions and suggest alternate approaches. The instructor should be fairly comfortable with the examples in order to incorporate these suggestions on-the-fly. The instructor should also take care to use the same system for the demo as the one used for practice. Variations in system configuration can make some examples not work as expected.

Sample Exercises

1. Computer A has 4 processors and Computer B has 8 processors. A parallel program P, takes 16 s to run on A and 12 s to run on B. Is this the type of performance you would expect out of P? Give one explanation as to why P does not achieve more/less performance.
2. Execute simple programs written in SimPar. Compare their performance with performance of sequential versions.
3. Download the C++ implementations of (i) knapsack and (ii) quicksort from http://tues.cs.txstate.edu. Consider the opportunities for parallelism in these two codes. Insert SimPar directives to parallelize the two applications. Execute the parallel applications and compare their performance with the sequential version of the code.

Task Orchestration

This module focuses on performance issues related to communication and synchronization of parallel applications. It is intended to be introduced in the Operating Systems course, as it provides the most context for the material covered.

Recommended Length 1.5 lectures (2 h)
Recommended Course Operating Systems

```
#define PI 3.141

int main() {
    double radius, area;
    radius = get_radius_from_circle();
    area = PI * radius * radius;
    printf("Circle area = %f\n", area);
}
```

```
    #define PI 3.141

    int main() {                                         S2 needs
        double radius, area;                              radius
S1      radius = get_radius_from_circle();               from S1
S2      area = PI * radius * radius;                         S3
S3      printf("Circle area = %f\n", area);               needs
    }                                                      area
                                                         from S2
```

Fig. 4 Code example illustrating data dependence

Organization and Content

This module begins by introducing students to some fundamental concepts in parallel programming. Notions of data dependence, synchronization, race condition, load balance and task granularity are explained. Architecture-specific performance issues such as those that occur on shared and distributed-memory parallel computers are also covered. A producer-consumer application is used as a running example to illustrate various performance issues. Power-performance trade-offs are highlighted in each context.

Data Dependence

After a quick review of parallel computing (two slides, as used in CS1 module), the module introduces the students to the notion of data dependence in parallel programs. Sequential and parallel versions of a simple function is presented. The example in Fig. 4 uses the computation of an area of a circle. But many other examples are possible. The parallel version of the example code is written in SimPar [3].

Students are asked to predict the outcome of the code when executed with certain input. The code is run several times in sequential and parallel mode. Results are discussed and students are asked to comment on the discrepancy. Following this discussion, the annotated code is presented as a slide, highlighting the dependencies in the code. The formal definition of data dependence is then presented. Various forms of dependence are also discussed briefly. The point is

Fig. 5 Sequential version of
producer-consumer code

```
int main() {
    while (!done) {
        fill_buffer(buf);          // produce
        if (buf_is_full(buf))
            empty_buffer(buf);     // consume
    }
}
```

```
#pragma omp parallel {
    #pragma omp section {
        fill_buffer(buf);
    }
    #pragma omp section {
        empty_buffer(buf);
    }
}
```

Fig. 6 Incorrectly parallelized producer-consumer code

made that both sequential and parallel programs must preserve all dependencies
in the code for semantically correct execution. For sequential programs this is
trivial since instructions are executed in program order. If students have already
taken the Architecture course, then the notion of instruction-level parallelism (ILP)
can be brought into this discussion. An example can be used to convey that the
degree to which ILP can be performed is determined by the dependencies between
the statements in question. Re-ordering transformations performed by compilers
can also be discussed to further illustrate the importance of data dependence in
semantically correct program execution.

Following this, the running example, a produce-consumer application is presented. The one shown in Fig. 5 uses the bounded-buffer problem as an example.
But many other examples for parallel producer-consumer can be created with slight
modifications. The supplementary material for this module includes an example
with the knapsack problem. The sequential code is then explained to the class.
(Figure 5 omits the actual producer-consumer functions). The parallel version of
the code is then presented. Figure 6 shows the example in OpenMP. The instructor
may continue the parallel example in SimPar but then it cannot be used later in the
module for performance experiments, as the results would prove non-intuitive. If
an OpenMP example is used, a brief review of OpenMP syntax may be required
at this point. Alternatively, this can be handled off-line with the aid of tutorials or
handouts, as discussed in section "Pedagogical Notes". The parallel version of the
code is executed several times to produce incorrect results. Again, students are asked
to identify the cause of the problem. Class discussion ensues, until the dependencies
in the code have been identified and clearly articulated.

Fig. 7 Another incorrectly
parallelized
producer-consumer code

```
full = 0;
#pragma omp parallel {
    #pragma omp section {
        fill_buffer(buf);
        full = 1;
    }
    #pragma omp section {
        while (!full) {
            /* wait */
        }
        empty_buffer(buf);
    }
}
```

```
flag = 0;
#pragma omp parallel {
    #pragma omp section {
        fill_buffer(buf);
        #pragma omp flush
        flag = 1;
        #pragma omp flush(flag)
    }
    #pragma omp section {
        #pragma omp flush
        while (!flag)
        #pragma omp flush(flag)
            empty_buffer(buf);
    }
}
```

Fig. 8 Correctly parallelized producer-consumer code

Synchronization

After it has been established that the code in Fig. 6 is producing incorrect results due
to data dependence violation, students are then asked if it is possible to correctly
parallelize the code and if so what conditions must hold. This discussion leads to
the notion of synchronization in parallel programs. The example in Fig. 7 is then
constructed in-class by editing the example from Fig. 6. This code is compiled and
executed several times to show the code still has not been correctly parallelized.
The students are then asked to identify the dependence that caused this problem.
This brings up the need for atomic operations, the idea of a critical section and race

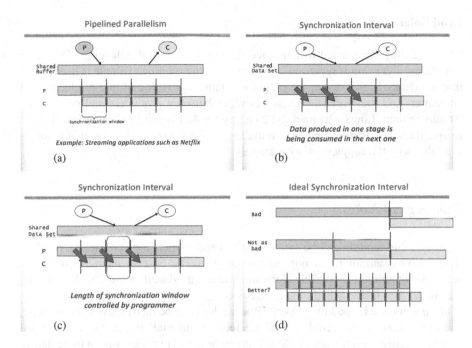

Fig. 9 Lecture slides illustrating pipelined parallelism and the role of synchronization interval on performance

condition. The code is then fixed in-class by placing *guards* around the operations on the flag. This version of the code is shown in Fig. 8. Finally, the code is executed a few times to show that it indeed now produces correct results.

The pragmas are then modified to parallelize the example code in a pipelined fashion. Figure 9 shows a subset of the animated slides that explains pipelined-parallelism, the synchronization interval and its effect on performance.

Task Granularity

Task granularity and how it is controlled by the synchronization interval is introduced using a set of lecture slides. The impact of task granularity on performance is also explained. Following this the pipelined-parallel producer-consumer example is revisited. Students are asked to identify the amount of work performed per thread (i.e., task granularity). The amount of work is expressed in number of items read/written to the buffer. The code is then executed with different task granularity by using the BLOCK parameter in the OpenMP pragma. The results of these executions demonstrate to the student the significance of task granularity and cost of synchronization to parallel performance.

Load Balancing

OS scheduling is revisited to introduce the concept of load balancing. The basic scheduling algorithm is reviewed and once again the running example is used for an in-class demo. In this demo, the program is launched with multiple producers and consumers and the work is broken un-evenly between producers and consumers. At launch time, Linux `thread_affinity()` API is used to pin certain threads to specific cores to illustrate load imbalance. The script to perform this demo is available with the supplementary materials.

Pedagogical Notes

Although this module can be introduced in other upper-level courses (e.g., Unix Systems Programming), in our experience it works best in the OS course. A seamless integration is possible if the module is introduced in the OS class during the week when thread scheduling is discussed.

To provide background for OpenMP, a handout can be distributed ahead of time. A sample handout is included with the lecture material. Furthermore, there are several excellent online tutorials. Students can be asked to review one of these before the lecture. The supplementary material contains urls for online tutorials.

To increase student engagement, lecture slides related to load balancing for energy efficiency can be presented interactively as problem sets. The problems can be drawn out on the board or the slides can be animated and students can be asked to come up with a thread mapping solution as a group.

It is advisable that instructors practice the live coding examples ahead of lecture time. Students often raise questions and suggest alternate approaches. The instructor should be fairly comfortable with the examples in order to incorporate their suggestions into the demo.

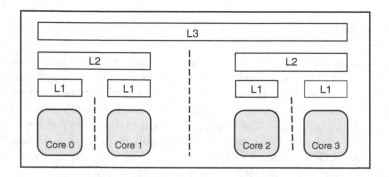

Sample Exercises

1. Consider the high-level block diagram of a multicore system as shown in the figure above. A multi-threaded producer-consumer application is executing on this system. The application has 4 threads with 2 producers (p0 and p1) and 2 consumers (c2 and c1). Data produced by p0 is consumed by c1 and data produced by p1 is consumed by c0.

   * Describe a suitable schedule to improve the overall performance of the application. Explain why your schedule is likely to deliver improved performance.
   * Would your schedule change if the primary objective is to reduce power? Why or why not?

2. Implement a feedback queue scheduler using the OS framework used in the class. The scheduler should aim to minimize power consumption on a multicore system.

3. Parallelize the provided n-body simulation code using OpenMP and then derive an optimal affinity-based schedule. The scheduler can be implemented using affinity support in either Pthreads or GNU OpenMP.

Analysis and Evaluation

This module concentrates on performance estimation and measurement of parallel systems, including efficiency, linear and super-linear speedup, throughput, data locality, weak and strong scaling, and load balance. Performance estimation of sequential architectures and the implications of Amdahl's law are typically part of current computer architecture courses. This module extends these concepts and investigates parallel performance in light of Amdahl's law. It explores modern parallel benchmark suites such as PARSEC (task, data, and pipelined parallelism) and Lonestar (amorphous parallelism) and demonstrates how to write benchmark programs to measure the performance of parallel hardware. It discusses how to identify potential for speedup as well as upper speedup bounds and performance obstacles.

Recommended Length 1 lecture (1:15 min)
Recommended Course Compilers, Computer Architecture, Upper-level CS elective

Organization and Content

This module starts with a review of elementary performance concepts and OpenMP syntax. This is followed by discussion of several advanced performance concepts.

The lecture slides for this module are complemented with a series of micro-benchmarks written in OpenMP. Alternate implementations in Pthreads are also provided in the supplementary materials. Each benchmark highlights a particular performance issue. Each benchmark will also include several *student* versions. The student versions expose parameters in the code that students can alter in various ways to impact the performance of the code. The student versions of the code also includes omitted code blocks that the students are expected to fill in as an exercise. A set of scripts measure various performance metrics including execution time, cache misses and processor power consumption.

Review of Elementary Performance Concepts

This section is similar to the module section described in section "Performance Metrics". The main difference is that the examples used are more involved and written in OpenMP.

Review of OpenMP Syntax

This segment of the module provides a quick review of basic OpenMP syntax and semantics. It is assumed the students are familiar with parallel program execution but not necessarily with any programming language. Therefore, this introduction is very basic. Only the `parallel` regions and `parallel for` constructs are covered. The goal is to give students enough knowledge for them to modify existing code but not necessarily for them to be able to write efficient parallel programs on their own. If a student comes in with OpenMP programming experience, this module is still very useful as it will train her to tune the OpenMP pragmas to extract better performance from her code. As was done with the task orchestration module, the OpenMP tutorial can also be done offline to save some lecture time.

Strong and Weak Scaling

The notion of scalability of parallel programs is introduced in this segment. The distinction between strong scaling and weak scaling is discussed. The code shown in Fig. 10 is used as a running example. The code is explained and then executed with 1, 2, 4, and 8 threads on an 8 core machine. Other configurations are feasible based on computer availability. Before each run of the code, students are asked to guess the execution time. As the code is written the program will achieve strong scaling on up to 16 cores on current-generation processors. To observe scaling effects beyond 16 cores, the data set needs to be >4 GB. This introduces NUMA effects and page faults that prevent the application from achieving linear speedup.

```
pixel *src_images = (pixel *) malloc(sizeof(pixel) * PIXELS_PER_IMG * IMGS);
pixel *dst_images = (pixel *) malloc(sizeof(pixel) * PIXELS_PER_IMG * IMGS);

initialize(src_images);

DATA_ITEM_TYPE gs;
omp_set_num_threads(THREADS);                    // fix number of threads
start = omp_get_wtime();
int i;
#pragma omp  parallel for private(i)
for (i = 0; i < IMGS; i++) {                 // process images in parallel
  int img_index = i * PIXELS_PER_IMG;
  for (int k = 0; k < ITERS; k++) {
    for (unsigned j = img_index; j < img_index + PIXELS_PER_IMG; j++)
      dst_images[j] = (0.3 * src_images[j].r + 0.59 *
                       src_images[j].g + 0.11 * src_images[j].b;
  }
}
```

Fig. 10 Example parallel code to demonstrate scaling

The code in Fig. 10 is then used to conduct a weak scaling experiment. The data set size is increased progressively until performance stops to scale. How much the data set needs to be increased depends on the particular platform where the code is being run. On some machines, runs for larger data sets can take up several minutes. So this needs to be weighed in when doing the demo. However, the code is designed in a way such that on most machines, memory bound behavior will show up for runs that take no more than 30 s. Similar to the strong scaling demo, before each run students are polled for the execution time. Following these demos the notions of strong scaling and weak scaling are formalized. A set of lectures slides and charts illustrating scaling trends are used for this purpose.

Linear and Super Linear Speedup

The code from Fig. 10 is re-used to explain the concepts of linear and super-linear speedups. The single-threaded version is labeled as the baseline and then speedup is calculated for 2, 4, and 8 thread versions. The obtained speedup is correlated with the number of threads/cores and shown to match the definition of linear speedup. The image processing example code is then transformed using tiling to improve data locality, as shown in Fig. 11. If time permits, this can be done live in class, as the technique is explained. Otherwise the example can be created ahead of time. The tiled version of the code is re-run with 2, 4, and 8 threads to demonstrate super-linear speedup. The working set size is orchestrated to exceed most L2 caches on current generation processors. A tiling size of 16–24 would keep the working set in cache. Some trial and error may be necessary prior to the demo to determine the exact size.

```
pixel *src_images = (pixel *) malloc(sizeof(pixel) * PIXELS_PER_IMG * IMGS);
pixel *dst_images = (pixel *) malloc(sizeof(pixel) * PIXELS_PER_IMG * IMGS);

initialize(src_images);

#define TILESIZE 64

DATA_ITEM_TYPE gs;
omp_set_num_threads(THREADS);                   // fix number of threads
start = omp_get_wtime();
int i;
#pragma omp  parallel for private(i)
for (i = 0; i < IMGS; i++) {          // process images in parallel
  int img_index = i * PIXELS_PER_IMG;
  for (usigned j = img_index; j < img_index + PIXELS_PER_IMG; j = j + TILESIZE)
    for (int k = 0; k < ITERS; k++) {
      for (unsigned jj = j; jj < j + TILESIZE; jj++)
        for (unsigned j = img_index; j < img_index + PIXELS_PER_IMG; j++)
          dst_images[j] = (0.3 * src_images[j].r + 0.59 *
                        src_images[j].g + 0.11 * src_images[j].b;
    }
}
```

Fig. 11 Tiled version of image processing parallel code used to demonstrate data locality effects

Latency vs. Bandwidth

The concepts of memory bandwidth and latency and their effects on parallel performance is discussed next. Sequential versions of the code in Fig. 12 are first used to demonstrate the importance of locality in performance. The code on the left exploits spatial locality while the code on the right does not. The parallelization of the two codes is then explained and the parallel versions of the codes are executed. A second example with a tiled computation is also introduced briefly to illustrate the notion of temporal locality and its impact on performance. This demo establishes the fact that parallelism alone cannot overcome limitations with memory locality. The code in Fig. 10 is then run with a larger data set where the data set is large enough to exceed the available memory bandwidth per socket. After the execution of the program, the point is reiterated that scalable performance can be limited by memory factors.

SMP vs. NUMA

The discussion on latency and bandwidth leads to a discussion in parallel architectures and the main considerations for programming such systems. This discussion is left at a very high-level and uses slides to illustrate the differences between the architectures. Programming models and tools for the different systems is also discussed. GPUs and heterogeneous systems architectures with CPUs and GPUs are also touched on.

Fig. 12 Parallel code with
and without spatial exploited
spatial locality

```
int main() {
    int **a;
    omp_set_num_threads(12);
    a = (int **) malloc(sizeof(int *) *
                                    DIMSIZE);
    int i,j;
    for (i = 0; i < DIMSIZE; i++)
        a[i] = (int *) malloc(sizeof(int) *
                                    DIMSIZE);
#pragma omp parallel for private(i,j)
    for (i = 0; i < DIMSIZE; i++)
        for (j = 0; j < DIMSIZE; j++)
            a[i][j] = 17;
    return 0;
}
```

```
int main() {
    int **a;
    omp_set_num_threads(12);
    a = (int **) malloc(sizeof(int *) *
                                    DIMSIZE);
    int i,j;
    for (i = 0; i < DIMSIZE; i++)
        a[i] = (int *) malloc(sizeof(int) *
                                    DIMSIZE);
#pragma omp parallel for private(i,j)
    for (j = 0; j < DIMSIZE; j++)
        for (i = 0; i < DIMSIZE; i++)
            a[i][j] = 17;
    return 0;
}
```

Power vs. Performance

This module ends with a discussion on energy efficiency of parallel applications.
The importance of saving power and attaining high-performance at specified power
budgets is explained.

Pedagogical Notes

It is advisable to run the experiments a few times before the actual in-class demo.
This will allow the codes to *adapt* to the execution environment and the instructor
will be able to make any necessary changes. Details on how to tune the parameters
of the code so that they exhibit the expected behavior are provided with sample
codes and scripts.

In the default configuration, the slowest code in the examples runs for a few seconds. This is done to not take up too much class time. Nonetheless, if time permits, the longer versions of the codes should be used as the performance differences make more of an impression on the students. During these long runs the instructor may further elaborate on the topics.

Sample Exercises

```
#include<stdlib.h>
#include<stdio.h>
#include <omp.h>

#define DIMSIZE 80

int main(int argc, char *argv[]) {
    int **a;

    omp_set_num_threads(THREADS);
    int dim = atoi(argv[1]);
    a = (int **) malloc(sizeof(int *) * dim);
    int i,j,k;
    for (i = 0; i < dim; i++)
        a[i] = (int *) malloc(sizeof(int) * dim);

    int BLOCK = 220;
    int jj;
    for (j = 1; j < dim; j = j + BLOCK)
#pragma omp parallel for private(i,j)
        for (k = 0; k < 100; k++)
            for (jj = j; jj < (j + BLOCK); j++)
                for (i = 1; i < dim; i++)
                    a[i][j] = 17;
        return 0;
}
```

1. Set the DIMSIZE, THREADS and BLOCK variables in the above code to different values (select values based on class discussion) and execute the code on a server X with 8 cores and server Y with 16 cores. Record performance statistics using perf. Prepare a report and explain the performance variations you observe on the two machines.
2. Download the PARSEC benchmark suite (http://parsec.cs.princeton.edu). Select one application from the group: *canneal, dedup* and *streamcluster* and another

application from the group: *swaptions*, *bodytrack*, *facesim*. Conduct a performance study of the two selected applications on a compute server with at least 16 cores. Use the `parsecmgmt` package to execute the applications with input data sets: small, medium, large and native, and with different thread counts: 2, 4, 8, 16, 32 and 64. Record performance statistics using `perf`.

What are the main performance trends you observe? What does that say about the characteristics of the two selected programs? Relate the performance trends to scalability concepts discussed in this module and prepare a report.

References

1. The Joint Task Force on Computing Curricula Association for Computing Machinery (ACM)/IEEE Computer Society, "Curriculum Guidelines for Undergraduate Degree Programs in Computer Science," 2013.
2. S. Prasad, A. Chtchelkanova, F. Dehne, M. Gouda, A. Gupta, J. Jaja, K. Kant, A. La Salle, R. LeBlanc, A. Lumsdaine, D. Padua, M. Parashar, V. Prasanna, Y. Robert, A. Rosenberg, S. Sahni, B. Shirazi, A. Sussman, C. Weems, and J. Wu, "2012 NSF/IEEE-TCPP Curriculum Initiative on Parallel and Distributed Computing - Core Topics for Undergraduates, Version I," http://www.cs.gsu.edu/~tcpp/curriculum/, accessed: 2018-02-11.
3. A. Qasem, "SimPar : A macro language for introducing parallel concepts to CS 1 students," https://github.com/apanqasem/simpar.git, accessed: 2018-02-11.
4. J. Treibig, G. Hager, and G. Wellein, "LIKWID: A Lightweight Performance-Oriented Tool Suite for x86 Multicore Environments," in *39th International Conference on Parallel Processing Workshops*, 2010.

Scalability in Parallel Processing

Yanik Ngoko and Denis Trystram

Abstract The objective of this chapter is to discuss the notion of *scalability*. We start by explaining the notion with an emphasis on modern (and future) large scale parallel platforms. We also review the classical metrics used for estimating the scalability of a parallel platform, namely, speed-up, efficiency and asymptotic analysis. We continue with the presentation of two fundamental laws of scalability: Amdahl's and Gustafson's laws. Our presentation considers the original arguments of the authors and reexamines their applicability in today's machines and computational problems. Then, the chapter discusses more advanced topics that cover the evolution of computing fields (in term of problems), modern resource sharing techniques and the more specific issue of reducing energy consumption. The chapter ends with a presentation of a statistical approach to the design of scalable algorithms. The approach describes how scalable algorithms can be designed by using a "cooperation" of several parallel algorithms solving the same problem. The construction of such cooperations is particularly interesting while solving hard combinatorial problems. We provide an illustration of this last point on the classical satisfiability problem SAT.

Relevant core courses: This material applies to ParAlgo courses.

Relevant PDC topics: Scalability in algorithms and architectures, speedup, Costs of computation, Data parallelism, Performance modeling

Learning Outcome: Students at the end of this lesson will be able to:

- Perform a classical speed-up analysis,
- Perform an efficiency and isoefficiency analysis,
- Understand the complementarity between Amdahl's and Gustafson's laws,

Y. Ngoko
Qarnot Computing, Montrouge, France
e-mail: yanik.ngoko@qarnot-computing.com

D. Trystram (✉)
Université Grenoble-Alpes, Grenoble, France
e-mail: Denis.Trystram@univ-grenoble-alpes.fr

© Springer International Publishing AG, part of Springer Nature 2018   79
S. K. Prasad et al. (eds.), *Topics in Parallel and Distributed Computing*,
https://doi.org/10.1007/978-3-319-93109-8_4

- Analyze the benefits of parallel processing,
- Determine the best approaches for designing a parallel program,
- Identify limitations in the parallelism of a program,
- Perform a trade-off analysis between time and energy consumption, and
- Envision alternative approaches for the design of scalable algorithms.

Context for use: This chapter is intended to be used in intermediate-advanced courses on the design and analysis of parallel algorithms. The material covers data parallelism, performance metrics, performance modeling, speedup, efficiency, Amdahl's law, Gustafson's law, and isoefficiency. It also presents an analysis of Amdahl's and Gustafson's laws when considering resource sharing techniques, energy-efficiency and problem types. The analysis could be too advanced for a CS2 student because it requires a background in modern parallel systems and computer architectures.

Introduction

Parallel machines are always highly powerful and complex (See the history of computing in [20]). This is obvious when we consider the evolution in the number of cores of top supercomputers in recent years.[1] This progression is driven by the conviction that with more powerful machines, we could reduce the running times in the resolution of challenging, compute-intensive problems such as real-time simulations (climate, brain, health, universe, etc.). On this latter point, let us emphasize that the experts are convinced that such simulations could be undertaken only with future exascale platforms.[2]

At first glance, it might seem obvious that given a parallel algorithm and a machine, the running time of the algorithm while using x CPUs will be greater than the one we could expect with more than x CPUs. However, this is not necessarily true; indeed, the computation of a parallel algorithm is split between a computational part required for creating parallelism (a set of workers corresponding to threads or processes), computations required for running the concurrent workers, and those necessary for communication and synchronization. Given a more powerful machine (in term of cores or CPUs), the main option for reducing the running time of a parallel algorithm would consist in increasing the number of independent computations. However, this will probably induce more communication, synchronization, and a more important overhead for the creation of parallelism. Hence, it cannot be totally guaranteed that the gain induced by the *increase in parallelism* will be balanced by these additional operations.

---

[1]Details are available at http://top500.org
[2]http://www.exascale-projects.eu/

This observation shows that we need a conceptual support to justify why super-computers with more computational units could serve to tackle more efficiently compute intensive problems. Historically, the notion of scalability was introduced for this purpose. Roughly speaking, it describes the capacity of an algorithm to efficiently solve larger problems when it is executed on a machine with *more parallelism*.

The purpose of this chapter is three-fold. First, we intend to provide an understanding of scalability, deeper than the intuitive one. We define the concept, discuss its interest and introduce key metrics used for its quantification. The concept of scalability is also associated with two fundamental laws: Amdahl's and Gustafson's laws. Our second objective is to put these laws in perspective with the computability of problems, modern resource sharing techniques and the concept of energy-efficiency. Finally, we introduce a new statistical approach for improving the scalability of parallel algorithms.

Background on the Scalability

We conclude that an algorithm is scalable from an analysis of its *behavior* when it is used in the processing of larger problems with more parallelism in the machine. For this purpose, we need metrics to characterize the behavior of a parallel algorithm. In this section, we will first introduce some classical metrics. Then, we will show how they can be used to analyze scalability.

Speedup and Efficiency

Definition 1 (Speedup) Let us consider a parallel machine made of p computing units and a computational problem P. Let us assume an instance of P for which the sequential algorithm has a running time equal to T_1. Finally, let us assume a parallel algorithm $\mathscr{A}$ whose running time in the resolution of the instance on p computing units is T_p. Then, we define the speedup achieved by $\mathscr{A}$ when solving the problem instance as

$$S_p = \frac{T_1}{T_p}$$

The notion of computing units will depend both on the parallelism of the underlying machine and on the implementation of $\mathscr{A}$. Thus, these units might consist of cores, processors and even containers. For the sake of simplicity, in the rest of this chapter, unless otherwise stated, we will consider that computing units correspond to processors. This choice is debatable as a parallel algorithm might support different types of parallelism (cores, processors, etc.) However, the resulting conclusion might still hold in choosing a lower level of parallelism. Another question in this definition is how to define T_1. There are at least two choices: the execution time of the parallel algorithm on one processor, or the execution time of the best sequential algorithm for solving P. It is this latter metric that we will consider.

Given an instance of P, let us consider that the number of sequential operations to perform in its resolution is W. We will also refer to W as *the work*. In general, we could expect to have $1 \leq S_p \leq p$. The argument derives from the common sense since with p processors, we could divide the number of sequential operations to perform into no more than p pieces of work, which leads to an *acceleration* in the resolution time of at most p. However, for several reasons, it might be possible to have $S_p > p$. One reason is that we might have more cache faults in the sequential algorithm when it processes an instance $\mathscr{I}$ whose total work is W than in the case where it processes sub-instances of $\mathscr{I}$ of work $\frac{W}{p}$.

Definition 2 (Efficiency) The efficiency of a parallel algorithm on a problem instance is as follows:

$$E = \frac{S_p}{p}$$

In general $E \in [0, 1]$. But, as S_p could exceed p, E could be greater than 1.

In order to compute the speedup or the efficiency, we need to consider a specific instance of problem P. However, for the sake of clarity, we will consider that if two instances have the same size, they also have the same work and execution time. For instance, let us consider that P consists of multiplying two (dense) square matrices. Let us also assume two problem instances $A \times B$ and $C \times D$ where $A, B, C, D \in \mathbb{R}^{n \times n}$. The size of the first instance is the number of elements of A added to the number of elements of B, which is $2n^2$. The size of the second instance is also $2n^2$. Thus, $A \times B$ and $C \times D$ hold the same amount of work. This conclusion is confirmed in practice since the processing of both instances will require the same number of floating-point operations.

Asymptotic Analysis of Speedup and Efficiency

The asymptotic analysis is a central concept in the study of parallel algorithms. Given a metric that depends on a set of parameters, its objective is to state how the metric behaves when the parameter values become infinite. The first model of asymptotic analysis that we consider focuses on speedup. Its objective is to capture the speedup behavior when the number of processors and problem size increase. This model can be used for a theoretical or experimental analysis of the parallel algorithm. The theoretical analysis is discussed in the next section by means of Amdahl's and Gustafson's laws.

In order to capture the speedup behavior through experiments, a classical tool consists in generating a 2D chart, which states the speedup reached for the various instances depending on the number of processors. An example is depicted in Fig. 1. The curves have been obtained using following the function:

$$T(n, p) = \left(\frac{n}{p} + p \right) .4 \times 10^{-8}$$

This function is representative of the running time we could observe on the problem of finding the maximum of a vector of real numbers. Indeed, with p processors, the problem can be solved as follows. First the vector is partitioned into p pieces. Local maxima are then computed in parallel for each sub-vector. Finally, the maximum of the local maxima is returned. If we proceed this way, then the number of comparisons is $\frac{n}{p}$ for each sub-vector and p for finding the maximum among local maxima. If a comparison takes 4×10^{-8} s, then we have the above function.

In Fig. 1, one can notice that the greater the size of the problem, the higher the speedup we can reach with multiple processors. This is because, when n increases, the speedup curve becomes close to the identity line ($y = x$). We obtain a speedup close to p and hence, an efficiency close to 1. In such a situation, we conclude that the parallel algorithm is *scalable*. More generally, *we say that an algorithm is scalable if its efficiency can be kept constant when increasing the size and the number of processors*. In this example, we make a projection of the speedup in establishing that for large values of n and p, it remains close to the identity function.

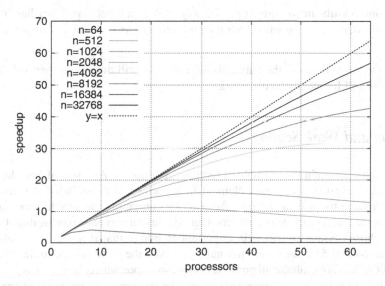

Fig. 1 Practical example of speedup distribution in the search of the maximum element of a vector of size n

An interesting question is then to know whether or not it is only such a function that can be achieved in an asymptotic analysis. This is discussed in the next section.

Types of Speedups

Definition 3 (Linear speedup) We say that a parallel algorithm has a linear speedup if the speedup S_p converges towards p when both the problem size and the number of processors increase.

The problem size is not included in the definition. This means that, whatever the problem instance, the parallel algorithm efficiently shares the amount of work among the processors. Linear speedups will typically be observed in parallel algorithms composed of workers that do not need to communicate. This is the case for instance of Monte-Carlo simulations.

Definition 4 (Super-linear speedup) We say that a parallel algorithm has a super-linear speedup if $S_p > p$ when both the problem size and the number of processors increase.

Super-linear speedups will typically be observed in parallel search algorithms based on backtracking. Indeed, assuming that the sequential depth first search space is represented as a tree, we could avoid a deep exploration of the paths that do not lead to optimal results in splitting the tree in the case where the solution is at the beginning of another path. The order of cache accesses may also play a role in super-linear speedups phenomena.

Definition 5 (sub-linear speedup) We say that a parallel algorithm has a sub-linear speedup if $S_p < p$ when both the problem size and the number of processors increase.

Due to some limits in the parallelization that we will discuss further, sub-linear speedups will frequently be observed.

Strong and Weak Scaling

Let us consider again the example of section "Asymptotic Analysis of Speedup and Efficiency" (finding the largest element in a vector). We concluded that there is a convergence towards the identity function by computing the speedup for various problem sizes and processors numbers. Our conclusion was based on the distribution of the chosen points (n, p). It is important to notice that if n was only selected between 64 and 512, we would not have observed the convergence to the identity line. As we cannot evaluate all possible points, an important challenge in asymptotic analysis is to make an appropriate selection. For this purpose, two types of speedup analysis are considered in practice: weak scaling and strong scaling analysis.

In weak scaling analysis, we evaluate the speedup, efficiency or the running time of a parallel algorithm in points (n, p) where we ensure that the problem size per processor remains constant. A common practice in weak scaling analysis consists in doubling both the size of the problem and the number of processors. If the running time or the efficiency remains constant, then the algorithm is scalable. In strong scaling analysis, we are interested in determining how far we can remain efficient given a fixed problem size. Therefore, for a fixed problem size, we increase the number or processors until we observe a change in the efficiency.

Isoefficiency

Given the running time function of section "Asymptotic Analysis of Speedup and Efficiency", we computed in Fig. 2, different values of the efficiency assuming that the problem size per processor (denoted by n_p) is 64 and 1024. As one can notice, if a linear speedup is clearly visible for $n_p = 1024$, it is not the case for $n_p = 64$. An important question in scalability analysis is then to know how to increase the problem size per number of processors. The *isoefficiency* [10] concept was introduced for this purpose.

More generally, for a given efficiency, the isoefficiency function of a parallel algorithm shows how to increase the problem size with respect to the number of processors in order to keep a given value of the efficiency. Given a fixed efficiency value, all parallel algorithms will not have the same isoefficiency function. In general we will distinguish between algorithms for which the size must be increased

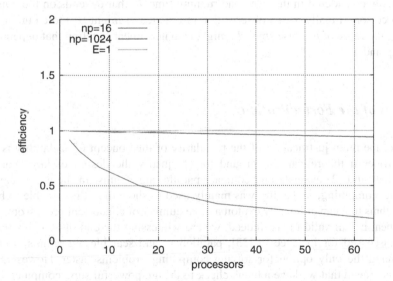

Fig. 2 Efficiency depending on the work per processor

as an exponential function of the number of processors (poorly scalable algorithms) and those for which the size must be increased linearly (highly scalable algorithm).

There is no generic method for the computation of the isoefficiency function. However, in the case where an analytic formulation of the sequential and parallel algorithm execution time is available, such a function could be computed. For instance, let us assume a basic sequential algorithm whose running time is expressed as $T_1 = n \cdot t_c$ where n is the problem size and t_c a computing time. This corresponds to a simplified case where the sequential algorithm consists in performing n times a given operation. Let us also assume that the parallelization of this algorithm has a running time $T_p = \frac{T_1 + T_o}{p}$. The idea in this latter formula is that given p processors, we can divide the time of the sequential algorithm by a factor of p. However, we must consider the overhead induced by communication, synchronization, and other artifacts such as the creation of parallelism. Assuming the previous expression, the efficiency becomes:

$$E = \frac{1}{1 + \frac{T_o}{n \cdot t_c}}$$

Therefore, we can relate the problem size to the number of processors and the efficiency with the following formula:

$$n = \frac{E \cdot T_o}{t_c (1 - E)}$$

Thus, if we want to keep efficiency constant, we must use the equation $n = K \cdot T_o$ where $K = \frac{E}{t_c(1-E)}$. In this formula, p does not explicitly appear; however, it is implicitly considered in the overhead running time T_o that depends on the number of processors. Finally, let us observe that in practice, it might be more complex to derive the value of n to use since T_o might be a non-linear function that depends on both p and n.

Limits of the Formalization

One of the main justification of the popularity of the concept of scalability is that it provides a theoretical background to: (1) justify the design of large parallel machines and (2) evaluate and compare parallel algorithms. In the early stage of parallel computing, scalability was mainly used to show that it is possible to build algorithms that can efficiently exploit a huge number of concurrent processors. This argument is still valid today. Indeed, we are witnessing the end of Moore's law as it is discussed for instance in [23]; parallelism and scalable algorithms are then becoming the only option for solving computing problems faster. However, this does not mean that we have a blank check to design powerful supercomputers. It is important to notice that we are also in another era of computation where the quality of algorithms is no longer strictly based on the running time. Other dimensions such as energy consumption have gained in importance, which has led to the definition of new metrics like *energy-efficiency*. Roughly speaking, the energy efficiency of a parallel program captures the ratio between the amount of energy it uses and the time it takes. Somehow, the term *power efficiency* is more precise because this ratio corresponds to the amount of Watts used by the algorithm. We do believe that the concept of scalability must be directly extended to deal with this new notion. For instance, a scalable energy-efficient algorithm could be an algorithm that can maintain its energy-efficiency when both the problem size and the number of processors increase. In other words, the first limit of the formalization described previously is not to take into account the other qualitative dimensions of the behavior of an algorithm.

The second limit we observe is that the proposed formulation does not handle the specificity of several computing problems and algorithms. Indeed, we implicitly assumed here that the size of a problem instance determines (or at least is related to) its hardness. In addition, we also assumed that all instances with the same size are similarly hard. These assumptions are not true for many computing problems, in particular those which are NP-hard. For instance, on the satisfiability problem SAT, the execution time of a backtracking algorithm will depend on the distribution of exact solutions in the search space. The difference between the running times of two instances of the same size could be huge. A direct consequence of this inability to relate work and problem size is that the concept of asymptotic analysis as described previously could no longer be applied.

Finally, we considered in our presentation that computing units correspond to processors. We also implicitly assumed that these processors have the same performance. Today however, the architecture of parallel machines has greatly evolved. Computing units could correspond to containers, virtual machines, cores or any combination of hybrid components (heterogeneity). In addition, with the complexity of machines nowadays, many other elements are related to the machine configuration could play a role in the performance of an algorithm (and its implementation). The question is then to determine whether or not the scalability results observed on a specific machine are valid on another one. In the past, similar interrogations led to the introduction of theoretical models of parallel machines like the well-known PRAM model. We encourage the reader who wish to learn more about it to read the seminal paper [7] or the dedicated chapter in the book of Cosnard and Trystram [4].

Scalability Laws

In the previous section, we showed how to use speedup and efficiency in an experimental evaluation of the scalability of a parallel algorithm. In this section, we will present how to theoretically estimate these metrics.

Amdahl's Law

Somehow, it might be counterintuitive to consider Amdahl's law as a scalability law. In its original paper [1], Amdahl introduced the law to explain that most actual problems do not have enough parallelism that could use the full potential power of supercomputers. Amdahl's argumentation was originally based on a statistical analysis. He showed that there is little benefit in parallelizing some computing problems and particularly, those for which we only have irregular algorithms.[3] Amdahl's analysis was right and even today, there are several computing problems on which the best parallel algorithms only achieve poor speedups. The idea to make a statistical analysis of the parallelism in term of computational problems was also ingenious. We will come back to this point and briefly introduce the P-completeness theory whose aim is to capture the problems of the P class that are hard to parallelize [11].

Although Amdahl showed that the usefulness of supercomputers might be overestimated, he proposed a simple but powerful model for the analysis of parallel algorithms. This model shows how to characterize the speedup and efficiency of a parallel algorithm as mathematical equations.

Mathematical Formulation of Amdahl's Law

Let us assume a sequential algorithm that solves a problem instance in W operations. The first assumption in Amdahl's law is to partition W into two fractions: namely, a sequential fraction f_{seq} and a parallelizable part f_{par} such that $f_{seq} + f_{par} = 1$. The sequential part is composed of operations that must be done one after the other and the parallelizable part corresponds to operations that can be performed simultaneously. Let us denote by t_c the execution time of a basic operation (all instructions are assumed to be identical). Then, the sequential running time of the algorithm is

$$T_1 = (f_{seq} + f_{par})W \cdot t_c$$

Given p processors, the second assumption in Amdahl's law is that we will have to distinguish between two types of computations: computations of the sequential part that will be executed on a single processor and the ones from the parallel fraction that will be shared (ideally) among all processors. This leads to the following expression:

---

[3]Irregular algorithms are characterized by non-uniform memory pattern accesses. For such algorithms, we will *frequently* be in the situation where the data we want to access are not in the caches. Some such well-known irregular algorithms include: Cholesky factorization, finite differences algorithms, agglomerative clustering, Prim's algorithm, Kruskal's algorithm, belief propagation.

$$T_p = f_{seq} \cdot W \cdot t_c + \frac{f_{par} \cdot W \cdot t_c}{p}$$

Consequently, the speedup is

$$S_p = \frac{1}{f_{seq} + \frac{1-f_{seq}}{p}}$$

It is important to observe here that the assumptions underlying Amdahl's law are debatable. In particular the speedup in this model is at most linear whereas super-linear speedups can be observed in practice. This situation happens because Amdahl's law assumes a parallel algorithm issued from the parallelization of the instructions of a sequential one. But in practice, the parallel algorithm could be issued from a completely new design of the problem.

Limits to Scalability

Amdahl's work pioneered several researches on the limit to scalability. In 1973, Stephen Cook introduced the P-completeness theory. This branch of the complexity theory aims at identifying problems for which there is no parallel algorithm that takes a poly-logarithmic time in the problem size, while using a polynomial number of processors. One of the objective of the P-completeness theory is to identify problems that are inherently sequential. This means that there is no efficient parallel algorithm for their resolution. In their book, Greenlaw, Hoover and Ruzzo give a compendium of P-complete problems [11] which includes several classical problems including scheduling, minimum set cover, and linear programming.

Another important limit to scalability is the memory wall. The memory wall is due to an imbalance between the memory bandwidth, latency and the processor speed [26]. On several machines, the running time to perform a Load/Store operation in DRAM exceeds the time of a multiplication. There are several techniques that were introduced in computer machines to avoid such a wall. A possible solution is to recover data loading with computations: the processor can start another instruction if the data of a prior one are not available. With this approach, given a same parallel program, the execution order of its instructions could change from one machine to another (out-of-order execution [16]). However, even with such a solution, we can still remain constrained by the DRAM access time.

The third limit is the energy consumption. Indeed, the power consumption of a supercomputer grows with processor utilization. This consumed energy is transformed into heat that must be dissipated. Several studies showed that the cooling can account for up to 40% of the energy consumed in a datacenter[6]. To reduce this cost, the Power Usage Efficiency metric (PUE) was introduced to estimate the efficiency of datacenters. Roughly speaking, the PUE is the ratio between the total energy consumed by a datacenter and the one devoted to computations. The closer

PUE to 1, the better the datacenter. In such a context, it is important to keep the parallel efficiency of an algorithm under a threshold where it does not consume too much energy in the perspective of PUE minimization.

Gustafson's Law

The concept of scalability as it is known today owes much to the work of Gustafson [12]. Indeed, the original Amdahl's paper showed that given a fixed problem size, we will always reach a limit in its parallelization. This view is what we refer today as the strong scaling perspective. Without contradicting Amdahl's observation, Gustafson showed that this does not mean that huge parallel machines are useless. Indeed, the greater the numbers of resources, the faster the solution of large problems. Thus, he introduced the weak scaling analysis and the idea of evaluating the efficiency of the algorithm in both increasing problem sizes and number of processors. The work of Gustafson also revisited the analysis proposed by Amdahl to show how large speedups can be obtained in parallel algorithms. The general analysis he proposed is reviewed below.

Mathematical Formulation of Gustafson's Law

Just like with Amdahl's law, Gustafson's law is based on the concept of serial and parallelizable fraction of work (the global work is denoted by W as before). However, instead of considering these proportions in the sequential algorithm, Gustafson's analysis assumes that we know them in the parallel algorithm. Let us assume that for a parallel algorithm that runs with p processors, the serial and parallel fractions are f'_{seq} and f'_{par} respectively. The algorithm running time is

$$T_p = (f'_{seq} + f'_{par}) \cdot W \cdot t_c$$

For the equivalent sequential algorithm, the running time is

$$T_1 = (f'_{seq} + f'_{par} \cdot p) \cdot W \cdot t_c$$

This leads to a *scaled* speedup equal to

$$S_p = p + (1 - p) f'_{seq}$$

In order to determine the difference between the scaled speedup and the speedup as formulated by Amdahl, let us assume that $p = 1024$ and half of the work is parallel ($f_{seq} = f'_{seq} = 0.5$). Then, while Amdahl's speedup is equal to 1.998, the scaled speedup is equal to 512.5. The difference is huge but it is easy to explain. Indeed, Amdahl's and Gustafson's analyzes are implicitly based on two different approaches in the design of parallel algorithms. In Amdahl's case, the parallel algorithm will execute (in parallel) instructions of a sequential algorithm. This envisions automatic parallelization and instruction level parallelism. In Gustafson's case, the parallelism is created depending on the number of processors. This envisions data parallelism. Further, the scaled speedup is biased by the fact that the sequential algorithm was considered as a *degenerated version* of the parallel algorithm.

Discussion About Generic Laws

In this section, we will extend the discussion of the Amdahl's and Gustafson's laws. The objective is to put in perspective these laws with respect to the problem types, modern resource sharing techniques and energy-efficiency.

Problem Types in Amdahl's and Gustafson's Law

As already mentioned, the type of addressed problems is a central notion in both Amdahl's and Gustafson's analysis. Indeed, the original Amdahl's paper targeted a set of problems that are hard to efficiently solve with a parallel algorithm because of irregular boundaries or non-homogeneous data distributions. In the same spirit, Gustafson introduced the notion of scaled speedup, emphasizing problems on which he obtained near-linear speedups. We do believe that the notion of problem type has received too little attention in parallel programming studies.

One of the most important theory developed for classifying the problem types in parallelization is the P-complete theory [11]. Let us recall that a problem is P-complete if: (1) it can be solved by a parallel algorithm in polynomial time, but (2) it cannot be solved in poly-logarithmic time with a polynomial number of processors, although $P = NC$. Here, NC is the class of problems that can be solved in poly-logarithmic time using a polynomial number of processors [19]. The fundamental question of the P-completeness theory reflects the pessimistic Amdahl's view on the parallelization (if $P \neq NC$) and the optimistic Gustafson's view (if $P = NC$). Indeed, if $P = NC$, then we can develop a highly parallelizable algorithm for all polynomial-time problems. Notice that a way to improve the algorithms is to consider the randomized version RNC (which aims at determining an efficient parallel solution with high probability). For instance, the problem of finding a maximal matching is in RNC and not in NC. Despite its great interest, the P-complete theory does not completely cover the class of all computational problems. In particular, there are hard problems that we can only practically address with heuristics. This includes problems of the NP and $PSPACE$ class [8].

Another interesting view of problems type in parallelization was introduced in [2]. In their paper, the authors considered 13 key techniques or kernels to implement parallel algorithms. The techniques/kernels cover several computing domains like dense and sparse linear algebra, databases, machine learning, etc. Contrary to the P-completeness theory, NP-hard problems can be considered here since backtracking and branch-and-bound are part of these techniques. Despite the interest of this work, it does not however discuss one of the main aspect of Amdahl's and Gustafson's analyzes where the notion of problem types was considered within the perspective of investigating the limits we can expect from parallelization.

To conclude this "philosophical" section, we could say that when dealing with parallelization, the problem type under consideration is crucial. For instance, it is more likely to have an efficient parallel algorithm on a numerical problem than on NP-hard combinatorial problems. Hence, we feel that, to fully complete the vision of problem types in both Amdahl's and Gustafson's works, a statistical evaluation of the most frequent parallel computing kernels implemented in parallel systems is necessary.

Amdahl's and Gustafson's Law Revisited for Modern Resource Sharing

Amdahl's and Gustafson's discussions were about the *usefulness* of a massive parallel machine. In his original paper, Amdahl wrote: *"Demonstration is made of the continued validity of the single processor approach and of the weaknesses of the multiple processor approach in terms of application to real problems and their attendant irregularities"*. As an answer, Gustafson concluded with *"Our work to date shows that it is not an insurmountable task to extract very high efficiency from a massively-parallel ensemble"*.

The original papers of Amdahl and Gustafson share at least two common assumptions regarding the usefulness of parallel machines. The first feature is the interest in speedup optimization. Somehow, they considered that a parallel machine is useful if it can help to solve problems faster. As already mentioned, such a vision is debatable nowadays since computing has an energetic price. We will return to this point in section "Amdahl's and Gustafson's Law and Energy-Efficiency". The second feature shared by both works is to consider the usefulness of parallel machines in an algorithmic/application centered viewpoint that does not account on the margin, we could have at the operating system and middleware levels.

Today, most parallel machines are associated with a resource managing system, most often based on a client/server model. Here, each user (client side) can concurrently submit, deploy and run several parallel algorithms on a subset of processors of a parallel machine. To manage this concurrency, new concepts have emerged like the notion of job and job scheduler. A job refers to an instance of a parallel algorithm composed of features (the source algorithm to run, the input data files, the output data files, the number of processors, etc.) Each job is routed towards a job scheduler that will determine the compute nodes on which it will be executed.

Depending on the parallel algorithm under consideration, a job could be parallel, moldable or malleable [5]. In the former, the requirements in term of processors for the job is fixed (typically like in MPI applications). In the moldable case, the job could run with different number of processors. In the malleable case, we additionally consider that the processor assignment of a job can change during its execution.

With modern resource sharing techniques, it does not matter whether a job does not fully use the total number of available processors or not. In such cases, we could deploy another job that will be concurrently executed with it. It is also important to notice that important progress was done to ensure that a concurrent run will not negatively impact another one. Finally, resource sharing even goes further with virtualization [18]. With virtual machines and containers, we can artificially duplicate the physical resources of a machine that will be shared between several parallel algorithms. In addition, we can adaptively remove or add physical resources to any parallel algorithm [13].

To conclude this discussion, we argue that with modern resource sharing techniques, parallel machines became systems that are exploited to varying degrees depending mainly on the activity of the users and the interactions with the system. This does not mean that the question of the scalability of a parallel algorithm is no longer important, but that there is a complementary answer to the question of parallel machine utility. Modern resource sharing techniques have also introduced a new consideration regarding scalability. As alluded to earlier, the greater the number of requests for job processing, the more useful the parallel machine because it can be maintained *full*. However, this reasoning holds only if we are able to quickly take appropriate scheduling decisions for the submitted jobs. This means that job scheduling algorithms should also be scalable.

Amdahl's and Gustafson's Law and Energy-Efficiency

When considering the question of energy-efficiency, we tend to focus excessively on the huge consumption of supercomputers while neglecting the progress made in the design of processors. At this point, it is important to recall that for several years, the design of processors followed Koomey's law which states that the number of computations per joules of energy dissipated has been doubling every year [15]. This means that for the past several years, efforts have been made to bear on improving computer hardware regarding the energy-efficiency. Unfortunately, the same is not true for computer software and algorithms. This is because energy-efficiency was not taken into account in Amdahl's and Gustafson's laws. These laws should be revisited since energy consumption could not be neglected any longer.

A naive belief is that, in order to minimize energy consumption, it is sufficient to minimize the running time. This is because the energy consumed by a parallel algorithm on a given machine can be estimated as the sum of the instantaneous power consumed throughout the execution of the algorithm [17]. Unfortunately, this reasoning does not hold for modern large scale platforms. The instantaneous power consumption is not always a fixed quantity. It includes a variable part that depends on the algorithm run. This variable consumption will depend on the load, the frequency and voltage at which the machine is run. Therefore, we could have a faster algorithm that finally consumes more energy. However, the story does not end here. Indeed, let us observe that faster algorithms will in general be also the ones which are more compute-intensive. However, at the processor level, compute-intensive algorithms generally produce more heat. Consequently, we could even need a more sophisticated cooling mechanism to lower the temperature of a supercomputer on which we run a compute-intensive algorithm.

One of the most interesting metrics for energy-efficiency in the vision of Amdahl's law is the speedup per Watt or performance per watt ratio used in the top green 500 list (See the green500.org site for details) and well conceptualized by Woo and Lee [25]. An algorithm that scales on the speedup per Watt is able to maintain the same speedup and average watt consumption when both problem size and number of processors increase. Woo and Lee also proposed a theoretical estimation of the speedup per Watt on several types of multi-core architectures. In the proposed expressions, the speedup is defined as in Amdahl's law. The concept of speedup per Watt has some weakness, for instance if the Watts refer to a unit, it is not the case for the speedup. In addition, we could criticize the fact that it is hard to isolate the consumption of a parallel program from the one induced by the run of an operating system or middleware. Despite these weaknesses, it is certainly one of the most promising option to extend Amdahl's and Gustafson's speedups to the minimization of energy consumption.

Finally, let us observe that the isoefficiency could be used as a powerful tool to reduce energy consumption. Indeed, a parallel algorithm will not lead to the same energy consumption depending on the number of processors it uses. An interesting question is then to determine the right number. Thanks to isoefficiency, we could answer as follows: depending on the efficiency we want to maintain, we can compute for each problem size the number of processors we want to use. This observation also suggests new ideas. For instance, it might be interesting to formulate the isoefficiency while considering an average Watt consumption we wish to maintain. Such models could in particular use the important progress made these last years on the theoretical modeling of the power consumption [17].

Designing Scalable Algorithms in Modern Large Scale Platforms

We propose in this section a general method for building scalable parallel algorithms. The proposed method is in particular motivated by the desire to automate the parallel resolution of NP-hard problems. However, it can be applied to a larger range of problems. It is based on three main pillars that are presented as follows.

Background

Pillar I: The Need of New Strategies for Strong Scaling

Fifty years ago, parallelism mainly focused on supercomputers dedicated to scientific computing, it is now available on any general purpose computer and applications. At the same time, supercomputers are always increasingly more powerful and alternative parallel systems like computational grids or clouds have emerged. This constant increase of parallel processing capabilities is challenging for the design of strong scaling algorithms. We illustrate this point on the following example.

Let us assume a machine with a huge number of processors (p_{max}). Let us also assume that we want to solve three instances I_1, I_2, I_3 of the problem P. For their resolution, we have two parallel algorithms A and A'. A has the best average execution time on the three instances while A' has the best execution time on I_1. In such a context, a rough asymptotic projection would consist in recommending A for the resolution of P.[4]

Let us now assume that in the run of A on I_1, a number of processors (denoted by p_{ssl}) offers no gain in term of parallelism. We refer to this point as the *strong scaling limit* and formally define it as the smallest number of processors p_{ssl} such that:

$$\forall p > p_{ssl}, T_p \leq T_{p_{ssl}}$$

As one can notice, the strong scaling limit is not the same depending on the instance we are solving. This is clearly visible in Fig. 1 where the limit is reached more quickly for $n = 64$ than for $n = 512$. Due to the sequential part of any parallel algorithm, we could expect such a limit to be determined with $p_{max} \longrightarrow +\infty$. Returning to the asymptotic projection we made, the existence of a strong scaling limit suggests that for optimizing the efficiency in the resolution of I_1, there are $p_{max} - p_{ssl}$ processors that we should not use. As machines are increasingly powerful, we can expect in the future to have another machine whose processors are similar to the ones of the first one but with a greater number of processors p'_{max}.[5] In this latter machine, the previous asymptotic projection will still hold. However, in the resolution of I_1, we will now have $p'_{max} - p_{ssl} > p_{max} - p_{ssl}$ that are not useful. In conclusion, if the evolution of architectures leads to generations of machines with more processors, it could be inefficient to use these *additional processors* in the resolution of simpler computational instances.

---

[4]The idea to compare algorithms based on their average running time on a set of representative computational instances is used in international competitions between algorithms. One of the most famous is the SAT competition where one goal is to solve the maximal number of SAT instances given a maximal time limit. SAT refers to the boolean satisfiability problem.

[5]This was observed on multicore machines where generations of machines integrate more cores.

To face this situation, let us assume that the strong scaling limit of A' on I_1 is $p'_l > p_{ssl}$. We could have been able to scale on $p'_l - p_{ssl}$ additional processors if it was A' instead of A that was run for solving I_1. The fact of having two algorithms with different strong scaling limits was observed in the resolution of several hard combinatorial problems, including the boolean satisfiability problem that we will present in section "Case Study".

As the parallelism of machines increases, we should invest in the design of cooperative executions of algorithms solving the same problem. In 1976, John Rice paved the way for a general theory of cooperative algorithms in theorizing the *algorithm selection problem* [21]. The Rice conceptualization latter inspired several studies on the automatic composition of algorithms and automatic tuning. Rice also introduced a methodology for the algorithm selection problem that we will not present here. The method we will present is inspired by the work of Huberman, Lukose and Hogg [14] on the formulation of a general theory for cooperative parallelism based on algorithm portfolios. In particular, given k parallel algorithms $A_1, \ldots A_k$, we propose to define a cooperative execution of the various algorithms as a concurrent run of each algorithm A_i on p_i processors that is stopped as soon as an algorithm finds a solution. Here, $p_i \in \{0, \ldots, p_{max}\}$ and

$$\sum_{i=1}^{k} p_i \leq p_{max}$$

We will refer to such cooperative executions as *resource sharing schedules*. Since 2010, resource sharing schedules have been successfully applied to the parallelization of the boolean satisfiability problem. In particular, several resource sharing-based solvers won the competition.

It should be noted that resource sharing schedules are not the only model of cooperative parallelism based on algorithm portfolios. Alternatives like time and malleable sharing schedules were proposed [9]. Nonetheless, they will not be discussed in this chapter. In this part, we will show how with resource sharing schedules, we can envision a new method for the design of scalable algorithms.

Pillar II: Benchmark Instead of Problem Size

In the previous weak scaling analysis, our conclusions on the general behavior of a parallel algorithm were based on observations made on a subset of instances characterized by their problem size. As already mentioned, however, the notion of problem size is not meaningful with all types of problems. On an NP-hard problem, we could have the following situations:

- the running time of a small instance exceeds the time of a larger instance
- the running times of two instances of the same size completely differ

This suggests that the idea of projecting the general behavior of a parallel algorithm based on a subset of instances, chosen mainly on their sizes, could be wrong. In addition, such a selection might not have any sense if we consider the problem resolution in a business perspective. Indeed, in this context, each problem will be associated with a context or domain that will constraint the types of practical instances. For instance, a delivery company that frequently solves the traveling salesman problem in France will not necessarily be interested in problem instances coming from Africa or the USA. When designing an algorithm for such a company, the question is not to be able to scale on any problem instance but on those representative of its business activity.

For these reasons, we do believe that in the evaluation of a parallel algorithm, we should constitute a reference benchmark of instances that might be representative of the context in which the problem will be solved. Fortunately, such benchmarks exist for several classical computational problems like the resolution of sparse linear systems or the satisfiability problem.

A main drawback while considering benchmarks is that we need another definition of the scalability. Indeed, the previous definition was based on maintaining a value of the efficiency when both the problem size and the number of processors increase. With this definition, the design of a parallel algorithm has a clear objective, that is to target linear or super linear speedup in an asymptotic analysis. What should then be the objective if we restrict ourselves to a (limited) benchmark viewpoint?

To address this question, we propose to proceed as follows. Let us assume that U is the universe of problem instances on which we want to be efficient. Here, U could be infinite. Let us also assume that there exists a finite set B of *representative problem instances* (i.e., our benchmark). Let $T(I, p)$ be the running time for solving instance I with p processors. We define the average efficiency in the resolution of B as:

$$\phi(B) = \frac{1}{|B|} \sum_{I \in B} \phi(I), \text{ where } \phi(I) = \left(\frac{T(I,1)}{p_{max} \cdot T(I, p_{max})} \right)$$

We then say that we correctly scale if

$$\zeta(B) = \frac{1}{|B|} \sum_{I \in B} |\phi(I) - \phi(B)| \longrightarrow 0$$

This definition of scalability shares a core idea with the prior one we considered in the previous sections of the chapter. It is to maintain an average efficiency over the benchmark. Indeed, we scale when the efficiency on any benchmark instance get close to the average efficiency. However, there is a major difference since the proposed definition is *machine-aware* in the sense where the speedup is always computed on the total number of available processors. Thus, this new definition somehow combines features of strong scaling (behavior of a single instance on large number of processors) with those of weak scaling (general behavior of several

instances). However, there is a weak point in the proposed definition: given U, what
we really want is to have

$$\frac{1}{|B|} \sum_{I \in U} |\phi(I) - \phi(B)| \longrightarrow 0$$

Thus, the choice of B is critical because it must be representative of the instances
we have in U. An open question at this stage is to know how we apply this new
definition of efficiency to cooperative executions. We will return to this point in
section "Computation of Cooperative Executions".

Pillar III: The Need of Auto-Tuning-Based Approaches

The increasing complexity of current machines makes auto-tuning unavoidable [22].
Any auto-tuning approach aims at solving a fundamental problem whose abstract
view is the following: we assume an algorithm that can be configured on a set
of parameters θ. We also consider a performance criteria (running time, energy
consumption, etc.) on which we want to optimize the run of the algorithm. Each
parameter θ_i is associated with a definition domain $dom(\theta_i)$ that defines the values
it can take. The goal is then to decide on the values to set for each θ_i (in the run of the
algorithm) in the perspective of optimizing the performance criteria we considered.
Nowadays, auto-tuning is unavoidable because on modern parallel architectures,
there are several architectural parameters that can be configured to optimize the
implementation of an algorithm. For a short view on such parameters, we refer
the interested reader to works related to the optimization of dense linear algebra
kernels [24].

We are convinced that the design of a parallel algorithm could no longer be
restricted to the formulation of a computational process that states how to generate
a correct output from a given input. The algorithm designed must be associated
with a search optimization process that will state how to automatically tune the
algorithm in a particular machine. The method we will propose formulate such
a search process in the case of a parallel algorithm thought as the cooperative
execution of several other ones.

Computation of Cooperative Executions

Given a computational problem P, we propose the following method to design an
efficient parallel algorithm for its resolution.

- **Phase 1:** Collect a set of parallel algorithms $A_1, \ldots, A_k$ that solves P.
- **Phase 2:** Create a set B of reference instances in the resolution of P.

- **Phase 3:** Compute the running times $T_{A_j}(I, p)$ for solving any instance I by algorithm A_i when using p processors ($1 \leq p \leq p_{max}$).
- **Phase 4:** Determine the resource sharing schedule for which $\phi(B)$ is maximized and

$$\zeta(B) = \frac{1}{|B|} \sum_{I \in B} |\phi(I) - \phi(B)| \longrightarrow 0.$$

- **Phase 5:** Encode the resource sharing schedule as a new parallel algorithm.

In this method, we assume that the running time of the resource sharing schedule is defined according to the equations:

$$T(I, p_{max}) = \min_{1 \leq j \leq k} T_{A_j}(I, p_i)$$

$$T(I, 1) = \min_{1 \leq j \leq k} T_{A_j}(I, 1)$$

As mentioned earlier, this is because, in resource sharing schedules, the execution is halted as soon as an algorithm finds a solution. Summarizing our method states how from a set of parallel algorithms and a benchmark, we can tune and build a cooperative executions of algorithms. We state how to optimize the cooperation of algorithms on the running time. But, the method could be extended to other performance criteria like the minimization of energy consumption. In this case, one challenge is to define aggregation rules that state how to deduce the energy consumed by a cooperative execution on p processors, from the one measured on $p' < p$ processors.

In the proposed method, we could have several choices in Phase 4. For instance, let us assume that we have 2 processors and 5 parallel algorithms. Then, there are 20 valid resource sharing schedules we could consider. This result is obtained as follows: we have 5 potential schedules where only one processor is used, 5 potential schedules where one algorithm strictly uses the two processors and 10 schedules in which two algorithms are run concurrently, each with one processor. A challenging question is to choose between all these schedules. For this choice, our conviction is that in the case where p_{max} is not too large, a brute force algorithm might be used. However, in the general case, we do believe that the optimal solution can be found from a search that uses the strong scaling limits as the frontiers of the search. We can also use the different heuristics proposed in [3].

To illustrate the different options in the choice of a resource sharing schedule, let us consider that we have the running time distribution of Fig. 3 on a basis B of 3 representative instances. Such running times are assumed to be collected in Phase 3 of the proposed method. It is important to notice that this phase can be extremely time consuming. Indeed, given k algorithms and p_{max} processors, we have $k \times p_{max}$ running time values to compute. Let us assume that any estimation takes in average t_a seconds. Then, the expected duration for data collection is $k.p_{max}.t_a$ seconds. In

Fig. 3 Example of running time on 1 and 2 processors for 5 algorithms and 3 instances

#processors	1					2				
	A_1	A_2	A_3	A_4	A_5	A_1	A_2	A_3	A_4	A_5
I_1	40	60	60	80	41	37	56	51	68	38
I_2	100	130	70	150	125	89	100	65	125	100
I_3	10.2	10.6	10.5	10.3	8.4	10	10.5	10.4	10	9

addition, we must repeat the execution of the algorithms in order to have an accurate estimation. If we repeat s times, then the expected duration is $k.p_{max}.t_a.s$ seconds. On NP-hard problems, t_a could be high. This means that it is interesting to study how we could reduce the duration of the data collection processes. We will not discuss on these aspects on this chapter.

Let us come back to the running times of Fig. 3. On 2 processors, our method clearly shows that the resource sharing we consider will lead to different behavior. For instance, if we deploy only A_1 on two processors, then we obtain

$$\phi(B) = \frac{40}{(37 \times 2)} + \frac{70}{(89 \times 2)} + \frac{8.4}{(10 \times 2)}$$

$$\simeq 0.45$$

$$\zeta(B) = \frac{1}{3} \left(\left| \frac{40}{(37 \times 2)} - \phi(B) \right| + \left| \frac{70}{(89 \times 2)} - \phi(B) \right| + \left| \frac{8.4}{(10 \times 2)} - \phi(B) \right| \right)$$

$$\simeq 0.059$$

If now we consider the schedule that runs A_3 on one processor and A_5 on another processor, then we have: $\phi(B) = 0.49$ and $\zeta(B) = 0.0054$. As one can remark, this latter schedule is preferable to the prior one on both objectives.

For the method to work, we need to already have several algorithms solving the same problem. Fortunately, this is the case for most computational problems. We also need to be able to estimate the running times of an algorithm on problem instances. Unfortunately, such estimations are not easy to obtain on some algorithms like those based on random choices. In the next section, we will describe in detail a practical case study of this method.

Case Study

The Boolean Satisfiability Problem

The objective in the boolean satisfiability problem (known as SAT) is to determine whether or not, a propositional formula written in Conjunctive Normal Form (CNF) is true (satisfiable) or not (unsatisfiable) [8]. Let us recall briefly the context of this classical problem: a CNF formula is defined as a conjunction of clauses over a finite set of boolean variables. More precisely, let us consider n boolean variables $x_1, \ldots, x_n$, a literal has either a variable x_i or its negation $\neg x_i$. A clause is a disjunction of literals. For instance, $C_1 = x_2 \vee \neg x_4$ is a clause and $(x_2 \vee \neg x_4) \wedge (\neg x_1 \vee x_2 \vee x_3)$ is a CNF formula.

SAT is a good candidate for illustrating the previous method for the following reasons:

- this problem is NP-complete and the hardness to solve an instance is not always correlated to its size; For instance, if x_u in a clause and $\neg x_u$ in another clause, it is easy to remark that the formula is unsatisfiable (whatever its size);
- it is rather easy to find several algorithms for solving SAT. Indeed, new solvers are proposed each year in the SAT competition (see www.satcompetition.org);
- there exist several benchmarks on the problem. The SAT competition regularly proposes a set of benchmark instances. The benchmarks are grouped in different classes reflecting practical scenarios and/or hardness to solve some instances;
- resource sharing schedules have already been applied successfully on SAT. Several winners of the SAT competition implemented a portfolio of solvers based on the resource sharing schedule model. However, to the best of our knowledge, such schedules were not tuned according to the method we proposed in the previous section.

Building Resource Sharing Schedules for SAT

To illustrate our method, let us consider the data of the SAT competition available at.[6] It is composed of a benchmark of 300 instances (corresponding to set B) and 4 parallel SAT solvers ($A_1, \ldots, A_4$). The running times of the solvers are known for all the instances for both 8 and 32 cores. Using these data, we were interested in studying if the proposed method could be used to build a better solver on a larger number of cores.

Since we already have a set of solvers and SAT instances, the requirements of phases 1 and 2 of our method are met. For phase 3, we should have performed a benchmark evaluation. However, we choose to only use the running time estimation we already have. These data are available from the website of the Penelope solvers. The drawback of this choice is that first we do not have the estimation for all numbers of cores, and second, we cannot compute the efficiency because the estimation of the sequential run is not known. On the first point, we assumed that any of the available solver could only be run with 8 or 32 cores. Regarding the second point, we propose to minimize the cumulative running time $\sum_{I \in B} T(I, p_{max})$.

Let us remark that *this goes in the direction of the maximization of* $\phi(B)$. Finally, we did not encode the parallel algorithm corresponding to the cooperative execution of our resource sharing schedule. Figure 4, shows the cumulative time of the different

---

[6]http://www.cril.univ-artois.fr/~hoessen/penelope.html

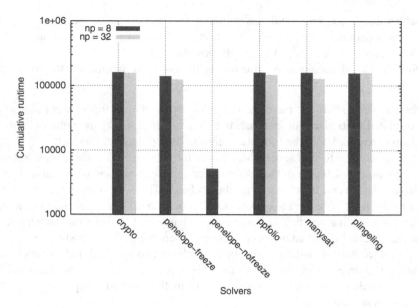

Fig. 4 Cumulative runtime

solvers. As we can notice, there is a gain in the running time when increasing the number of cores. However, this gain is far from what could be expected in a linear speedup.

On 8 cores, the best solver suggested by this figure is the *penelope-nofreeze* solver. As it was run only with 8 cores (the results for 32 cores are not available), an interesting question is then to know if we could obtain a better solver on 32 cores in combining the prior ones. The answer is yes, as shown in Fig. 5a, we were able in combining 4 different solvers to compute a better resource sharing schedule on 32 cores. The gain obtained here is 946.64 s. The gain here is the difference between the cumulative runtime of the best solver and the best resource sharing schedule.

The second question is to know if in combining the solvers, we could obtain a better solution on more than 32 cores. The answer again is yes. In Fig. 5b, we depicted the absolute running time difference between the best solver we found on $p > 32$ cores and the best solver on 32 cores. The increase in running time could be observed until reaching $p = 208$ cores.

These results showed that the proposed method can be used to build an efficient parallel algorithm by composing several algorithms solving the same problem SAT. However, it is important to notice that we only provided here a theoretical validation. An effective implementation of the resource sharing schedule can in practice add a runtime overhead. However, the gain states that there is still an important margin.

(a)   (b)

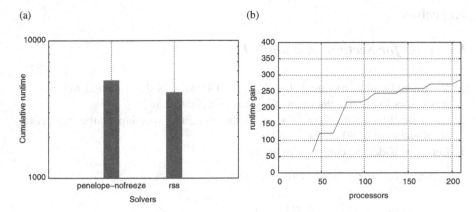

Fig. 5 Running times of resource sharing schedules. (**a**) Resource sharing schedules on 32 cores. (**b**) Gain using more than 32 cores

It is also important to notice that we did not handle the maximal time limit set in the run of the solvers and the correctness of the results. This latter point is important since some of the solvers were based on a heuristic search.

Conclusion

Amdahl's and Gustafson's law are still valid for modeling the performance of parallel algorithms. However, as discussed in this chapter, we need to extend them to other qualitative dimensions (such as energy-efficiency) and to the specificity of modern parallel platforms (virtual machines or containers, etc.) We also need a general formulation of scalability that could handle a larger class of problems; in particular problems for which it is not reasonable to assume that the larger the problem size, the more compute intensive the instance.

In this text, we introduced a candidate solution for the modern design of scalable algorithms. It is based on cooperative parallelism; it shows how to define a parallel run based on a computable optimization model. The model can be adjusted to optimize the run on several criteria like the efficiency, the runtime or even the energy consumption. It is also noteworthy that the proposed method focuses on strong scaling that will become a major issue, as the parallelism available in modern machines will continue increasing. Finally, our method is based on automatic tuning that is inescapable as the complexity of machines continues increasing. Improvements to our proposed method include the choice of the benchmark of instances or the reduction of the runtime required to measure the running time of the algorithms on instances.

Exercises

Exercises for Section "Amdahl's Law"

1. In Amdahl's law, assuming that $f_{seq} = 0.4$, what is the maximal number of processors to use to achieve an efficiency of at least 0.38?
2. On a machine with 8 cores, what is the maximal speedup in the multicore parallelization of 90% of a program?
3. Let us consider the computer program

```
for (i = 0; i < 100; i++) {
    a[i] = b[i] + c[i];
    d[i] = a[i] + d[i-1]/2
}
```

   where a,b,c and d are arrays of integers of size 100.

   (a) In considering only the inner loop instructions, what is the fraction of additions of the program (b[0] + c[0] is an addition)?
   (b) What is the maximal speedup we can expect from the parallelization of the algorithm?

4. With the MapReduce paradigm, we can count the number of words in a document by the means of a process that includes four steps: splitting, map, shuffle, reduce.

   At the beginning, given a document of n lines, the master node splits it into n sub-documents, each corresponding to a line. It then assigns these sub-documents to different workers. The map step follows where each worker runs a map function that consists of sending the pairs ("key", 1) where "key" is a word found in the sub-document it processes. After the completion of the map step, the master node groups the emitted pairs by keys and sends all the data of a given key to a distinct worker. The process ends with the reduce step where each worker adds up the number of keys it has and returns the cumulative value.

   In Fig. 6, a graph illustration of this process is provided with an input document of 3 lines. The objective is to count the number of occurrences of each of the words. This number is obtained after the reduce step.

   In a MapReduce process, a step is only started if the prior one is completed. Let us assume that in the map step, we have p workers, each deployed on a distinct processor. Let us also assume that the map function given a line l_i of length $|l_i|$ will run in $\Theta(|l_i|)$. We consider for the sake of simplicity that all computations are done using a shared memory.

   (a) In the worst case, what is the completion time of the map step assuming that each worker will get $\frac{n}{p}$ lines?
   (b) How many workers should we use in this phase to achieve a speedup of c (in the step)?

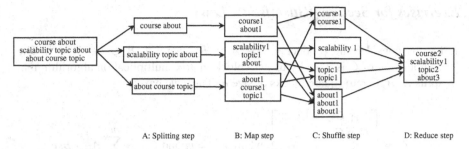

| A: Splitting step | B: Map step | C: Shuffle step | D: Reduce step |

Fig. 6 MapReduce example

(c) Assuming that in each line l_i, we have the same probability to have 1, 2 or $|l_i|$ distinct words, how many workers should we use in the reduce step to maximize the efficiency of this step?

(d) Assuming that the time of the splitting and shuffle steps are known, given p workers in the map phase and q workers in the reduce step, propose a theoretical estimation of the execution time.

5. The well-known Fibonacci numbers are defined by the recurrence

$$F(0) = 0 \,,\, F(1) = 1 \,,\, F(n) = F(n-1) + F(n-2) \text{ for } n > 1$$

Let us consider a multi-threaded program that proceeds as follows: given a value $n > 1$, it creates two threads that respectively computes $F(n-1)$ and $F(n-2)$. It then adds the value produced by the thread and returns it.

(a) How many threads are created in the computation of $F(n)$?

(b) Propose an asymptotic analysis of the scalability of this program.

(c) What are the limits to the scalability of this program?

(d) Propose a better parallelization for the computation of the Fibonacci numbers.

6. Given a square matrix of size n and a vector of n elements, let consider an algorithm for matrix-vector multiplication whose cost on p processors is given by the following equations:

$$T_1 = n^2 t_c$$

$$T_p = t_c \left(\frac{n^2}{p} \right) + t_s \log p + t_w n$$

where, t_c, t_s and t_w are constants.
Determine the isoefficiency function of this algorithm.

Exercises for Section "Gustafson's Law"

1. Let us consider the product $C = A \times B$ where $A, B, C \in \mathbf{R}^{m \times m}$. The computation of C is done by using a block matrix multiplication algorithm that splits A, B, C into square blocks of size q. Thus, we have:

$$C = \begin{bmatrix} C_{11} & C_{12} & \cdots & C_{1n} \\ C_{21} & C_{22} & \cdots & C_{2n} \\ \vdots & & \ddots & \vdots \\ C_{n1} & C_{n2} & \cdots & C_{nn} \end{bmatrix} \text{ where } nq = m \text{ and } C_{ij} = \sum_{k=1}^{n} A_{ik} \cdot B_{kj}$$

   Let us consider a multi-threaded implementation on a p cores machine. At the beginning of the execution, a master thread creates n^2 tasks and put them in a stack S. Here, each task corresponds to the computation of a block C_{ij}. p other threads are next created; their processing consists of iteratively removing a task from S that they then process. All threads are stopped as soon as S becomes empty.

   (a) Propose a parallel implementation of this algorithm.
   (b) Develop an asymptotic analysis of the scaled speedup of the proposed implementation in the case where $q = n$
   (c) Assuming 4 threads, deduce the value of q that optimizes the scaled speedup (Use experimental results of the proposed implementation for this).

2. In the prior algorithm, let us now consider that the sum $\sum_{k=1}^{n} A_{ik} \cdot B_{kj}$ is parallelized. This means that when a thread steals a task corresponding to the computation of C_{ij}, it next creates d other sub-threads such that the sub-thread l will compute

$$C_{ij} = C_{ij} + \sum_{k=(l-1)(\frac{n}{d})+1}^{l(\frac{n}{d})} A_{ik} \cdot B_{kj}$$

   Propose an asymptotic analysis for $(q = n, d = 2)$ and $(q = n, d = \sqrt{n})$.

3. Let us consider a vector of real numbers $\bar{x} = (x_1, \ldots x_n)$ on which we want to compute the standard deviation and mean. Here,

$$\sigma(\bar{x}) = \sqrt{\frac{1}{n} \sum_{i=1}^{n} (x_i - \mu)^2}, \text{ and } \mu(\bar{x}) = \frac{1}{n} \sum_{i=1}^{n} x_i$$

   For the parallelization of this computation we consider a two-phase algorithm. The first phase computes the mean in parallel. Assuming that we have p cores, one subdivides the list into p near-equal partitions. Each thread then adds up the number in its partition and a final thread adds up the sub-sums of the other threads. In the second phase, one proceeds in the same way as for the standard deviation. The threads compute the squared differences of the numbers in their partitions and a last thread computes the standard deviation.

(a) What is the computational complexity of this algorithm?
(b) What is its theoretical speedup?
(c) Could we do better while parallelizing the sums?

4. Let us consider a web server associated with a queue of incoming requests. The server processes each request in Δ seconds.

(a) Which minimal number of requests per second ensures that the size of the queue will be always greater than 1?
(b) In order to reduce the processing time, one decides to create p instances of the web server. All the instances are associated with the same queue. What is the number of requests that will be processed per second?
(c) At which date a request that enters in the queue at date t_0, with d predecessors in the queue, will be processed? What is the efficiency of the processing?

Exercises for Section "Designing Scalable Algorithms in Modern Large Scale Platforms"

1. Let us assume a benchmark of instances B and the running time $T(I, p)$ for $I \in B$ and $1 \leq p \leq p_{max}$

(a) Write a brute force algorithm that computes the resource sharing for which $\phi(B)$ is maximized and

$$\frac{1}{|B|} \sum_{I \in B} |\phi(I) - \phi(B)| \longrightarrow 0$$

(b) What is the computational complexity of your algorithm?
(c) Discuss the desirability of computing such resource sharing schedules on matrix multiplication algorithms (consider the case of dense and sparse matrices).

2. Let us reconsider the case study of the satisfiability problem. Assuming that $T(I, 1) = T(I, 8)/8$, apply the algorithm of the previous exercise to determine the best resource sharing schedule.
3. In this case study, compute the best resource sharing schedule on 16 processors and propose a parallel implementation of this algorithm.

References

1. Gene M. Amdahl. Validity of the single processor approach to achieving large scale computing capabilities. In *Proceedings of the April 18–20, 1967, Spring Joint Computer Conference*, AFIPS '67 (Spring), pages 483–485, New York, NY, USA, 1967. ACM.
2. Krste Asanovic, Rastislav Bodik, James Demmel, Tony Keaveny, Kurt Keutzer, John Kubiatowicz, Nelson Morgan, David Patterson, Koushik Sen, John Wawrzynek, David Wessel, and Katherine Yelick. A view of the parallel computing landscape. *Commun. ACM*, 52(10):56–67, October 2009.
3. Marin Bougeret, Pierre-François Dutot, Alfredo Goldman, Yanik Ngoko, and Denis Trystram. Approximating the discrete resource sharing scheduling problem. *Int. J. Found. Comput. Sci.*, 22(3):639–656, 2011.
4. Michel Cosnard and Denis Trystram. *Algorithmes et Architectures parallèles (english version by Intenat. Thomson publishing 1995)*. InterEditions, France, 1993.
5. Pierre-Francois Dutot, Grégory Mounié, and Denis Trystram. Scheduling Parallel Tasks: Approximation Algorithms. In Joseph T. Leung, editor, *Handbook of Scheduling: Algorithms, Models, and Performance Analysis*, chapter 26, pages 26–1–26–24. CRC Press, 2004.
6. Richard Brown et al. Report to Congress on Server and Data Center Energy Efficiency: Public Law 109–431. Technical report, Lawrence Berkeley National Laboratory, 2008.
7. Steven Fortune and James Wyllie. Parallelism in random access machines. In *Proceedings of the Tenth Annual ACM Symposium on Theory of Computing*, STOC '78, pages 114–118, New York, NY, USA, 1978. ACM.
8. Michael R. Garey and David S. Johnson. *Computers and Intractability: A Guide to the Theory of NP-Completeness*. W. H. Freeman & Co., New York, NY, USA, 1979.
9. Alfredo Goldman, Yanik Ngoko, and Denis Trystram. Malleable resource sharing algorithms for cooperative resolution of problems. In *IEEE Congress on Evolutionary Computation*, pages 1–8. IEEE, 2012.
10. Ananth Y. Grama, Anshul Gupta, and Vipin Kumar. Isoefficiency: Measuring the scalability of parallel algorithms and architectures. *IEEE Parallel Distrib. Technol.*, 1(3):12–21, August 1993.
11. Raymond Greenlaw, H. James Hoover, and Walter L. Ruzzo. *Limits to Parallel Computation: P-completeness Theory*. Oxford University Press, Inc., New York, NY, USA, 1995.
12. John L. Gustafson. Reevaluating amdahl's law. *Commun. ACM*, 31(5):532–533, May 1988.
13. Benjamin Hindman, Andy Konwinski, Matei Zaharia, Ali Ghodsi, Anthony D. Joseph, Randy Katz, Scott Shenker, and Ion Stoica. Mesos: A platform for fine-grained resource sharing in the data center. In *Proceedings of the 8th USENIX Conference on Networked Systems Design and Implementation*, NSDI'11, pages 295–308, Berkeley, CA, USA, 2011. USENIX Association.
14. Bernardo. A. Huberman, Rajan. M. Lukose, and Tad. Hogg. An economic approach to hard computational problems. *Science*, 27:51–53, 1997.
15. Jonathan Koomey, Stephen Berard, Marla Sanchez, and Henry Wong. Implications of historical trends in the electrical efficiency of computing. *IEEE Ann. Hist. Comput.*, 33(3):46–54, July 2011.
16. Bich C. Le. An out-of-order execution technique for runtime binary translators. *SIGPLAN Not.*, 33(11):151–158, October 1998.
17. Tao Li and Lizy Kurian John. Run-time modeling and estimation of operating system power consumption. *SIGMETRICS Perform. Eval. Rev.*, 31(1):160–171, June 2003.
18. Susanta Nanda and Tzi-cker Chiueh. A survey of virtualization technologies. Technical report, SUNY at Stony Brook, 2005.
19. Nicholas Pippenger. On simultaneous resource bounds. In *Proceedings of the 20th Annual Symposium on Foundations of Computer Science*, SFCS '79, pages 307–311, Washington, DC, USA, 1979. IEEE Computer Society.

20. S. K. Prasad, A. Chtchelkanova, F. Dehne, M. Gouda, A. Gupta, J. Jaja, K. Kant, A. La Salle, R. LeBlanc, A. Lumsdaine, D. Padua, M. Parashar, V. Prasanna, Y. Robert, A. Rosenberg, S. Sahni, B. Shirazi, A. Sussman, C. Weems, and J. Wu. *NSF/IEEE-TCPP Curriculum Initiative on Parallel and Distributed Computing - Core Topics for Undergraduates, Version I.* Online: http://www.cs.gsu.edu/~tcpp/curriculum/,55pages,USA,2012.
21. John R. Rice. The algorithm selection problem. *Advances in Computers*, 15:65–118, 1976.
22. Walter Tichy. Auto-tuning parallel software: An interview with thomas fahringer: the multicore transformation (ubiquity symposium). *Ubiquity*, 2014(June):5:1–5:9, June 2014.
23. Moshe Y. Vardi. Moore's law and the sand-heap paradox. *Commun. ACM*, 57(5):5–5, May 2014.
24. R. Clint Whaley, Antoine Petitet, and Jack Dongarra. Automated empirical optimization of software and the ATLAS project. *Parallel Computing*, 27(1–2):3–35, 2001.
25. Dong Hyuk Woo and Hsien-Hsin S. Lee. Extending amdahl's law for energy-efficient computing in the many-core era. *Computer*, 41(12):24–31, December 2008.
26. Wm. A. Wulf and Sally A. McKee. Hitting the memory wall: Implications of the obvious. *SIGARCH Comput. Archit. News*, 23(1):20–24, March 1995.

Energy Efficiency Issues in Computing Systems

Krishna Kant

Abstract The purpose of this chapter is to introduce energy efficiency issues in computer systems and its importance to the PDC curriculum. This is done mostly at a basic level, i.e., definitions of terms and basic concepts (K and C Bloom levels), so that the students get a broad overview of the entire field as it applies from very low hardware level up to software and service level issues. Energy management in parallel and distributed systems are also covered. The chapter attempts to convey the idea that the energy is ultimately consumed by transistors and wires, and a thorough understanding of the hardware issues is essential to effectively deal with the energy efficiency and adaptation issues. Some of the material can be considered at the A Bloom level as well.

Relevant core courses: Systems, Arch 2, ParAlgo

Relevant PDC topics: Cross-cutting topics: Power Consumption

Learning outcomes: Know basics of energy, power and thermal issues in computing, importance of and technology trends in power consumption, power-performance tradeoffs, power states and their use at HW and SW level, power adaptation, energy efficiency of parallel programs

Context for use: Traditionally, computing has focused only on performance at all levels including circuits, architecture, algorithms, and systems. With power consumption and power density playing a central role at all these levels, it is crucial to teach students about power and power-performance tradeoffs at all these levels.

K. Kant (✉)
Temple University, Philadelphia, PA, USA
e-mail: kkant@temple.edu

© Springer International Publishing AG, part of Springer Nature 2018   111
S. K. Prasad et al. (eds.), *Topics in Parallel and Distributed Computing*,
https://doi.org/10.1007/978-3-319-93109-8_5

Why Does Energy Efficiency Matter?

As the world gets increasingly fused with information technology, the energy consumption associated with the information technology continues to rise. For the mobile and embedded devices such as mobile phones and sensors, energy efficiency is crucial because of the increasing demands on the batteries, which have not scaled well in capacity as compared to the energy demands placed on them. In many embedded applications, particularly the emerging Internet of Things (IoT), battery replacement or charging is infeasible or very costly, and there is increasing trend towards energy harvesting from the environment in a cost effective way. This places severe limitations on energy use and thus makes energy efficiency the foremost issue in the design of both hardware and software. On the other end of the scale, data centers continue to grow in size in order to handle increasing client demands for running complex queries requiring substantial computation and processing of large amounts of data, often in real time. This has a direct impact on increasing energy consumption of data centers, and in turn requires greater emphasis on energy management and energy efficient processing algorithms.

Energy Related Challenges

The high energy consumption in data centers poses numerous challenges which make the energy efficiency increasingly important. First, there is the cost of electricity and electric infrastructure. Large data centers consume 10 MW or more power, and the corresponding energy cost may amount to 40% or more of its operating cost. (We will shortly discuss relationship between power and energy.) Large power consumption implies costly and costly-to-maintain electric infrastructure that includes large step-down transformers (along with their cooling costs), uninterruptible power supplies (UPS), relays/switch-gear, backup generators, multiple stages of AC-DC converters, etc. All of this can be scaled down by making the data center more energy efficient. Second, large data centers require large cooling infrastructures (including water chiller plants, air conditioners, fans, etc.), which too can be scaled down with more efficient operations. It is estimated that 1 W saved by using energy efficient operation of the servers can result in up to 3 W of total saved power.

Other less obvious side effects of high energy consumption come into play at the basic architectural levels. Ever since its inception in early 1970s, the semiconductor technology has thrived on the steady reduction of "feature size", which refers to a basic measure used for designing and patterning transistors and on-chip connections between them (the "wires"). For example, the current Intel processors are based on the 14 nanometers (nm) technology, the latest one being the so called 8th generation Coffeelake. Traditionally, the feature size has decreased by a factor of $\sqrt{2}$ every 2–3 years, however, the huge difficulties in going down to very small feature sizes has slowed this trend. One critical issue arising due to continued reduction in feature size is that in addition to the power consumption, we also need to pay attention to *power density*, or the power consumed per square cm of chip area.

Power density was already threatening to become unsustainable in early 2000s, and a variety of techniques were essential to keep it in check, and these continue to be crucial. The techniques range from the lowest level of semiconductor device physics to circuit, logic, architecture, and beyond. The current state of the technology can easily put more than a billion transistors on a cm^2. For example, Intel's core i7-6950 extreme edition built using the 14 nm technology has 10 cores and consumes 140 W, and packs 3.4B transistors in 246 mm^2 die. The power density is (140/2.46) or 72 W/cm^2. Since all power consumed ultimately appears as heat, this means that we need to have enough cooling capacity to remove 72 W power from each cm^2 area. With transistor layers already being stacked vertically in the emerging 3-D architectures, this becomes an extremely challenging task.

The net result is that it is possible to put lot more processing cores on a die than can be simultaneously powered on due limitations in removing the heat. This leads to the so called problem of "dark silicon" where it becomes necessary to keep some of the transistors unpowered in order to meet the heat dissipation requirement. In other words, energy efficient operation of cores has direct implications in terms of usable cores per package.

Making Computing Energy Efficient

Achieving energy efficiency involves a multi-level effort that includes an interplay of energy efficient semiconductor materials, transistor and circuit designs, hardware architecture, and software design along with suitable *power management techniques* that are engaged at various levels in order to provide suitable trade-offs between performance and energy consumption. Unfortunately, the recent trends in the manufacturing and circuit design introduce the third element – reliability. One reason for reliability issues arises from atomic level feature sizes. In particular, the latest technology already operates at the feature size of 10 nm, which amounts to only 30 Silicon atoms! Such small sizes allow for quantum mechanical tunneling and make it impossible to shape the boundaries of transistors accurately, thereby leading to unreliable operation. Another reason for unreliability is the continuing decrease in the operating voltage. As we shall see later, decreasing operating voltage can reduce the power consumption substantially, and there are emerging trends of operating the chips close to voltage levels where the transistors may not even switch reliably. This unreliability is handled by error detection and repetition of errored operation at a very low level; however, there may be some danger of residual undetected errors. These residual errors can be handled at higher levels by additional redundancy; however, this results in both additional power consumption and loss of performance. In other words, increasingly we need to consider a 3-way tradeoff:

performance, power, and reliability. In this chapter, we do not address this tradeoff; however, this is an important emerging issue in the context of energy efficiency.

Basic Concepts

Power vs. Energy vs. Heat

Power and energy are often used interchangeably in informal discussions; however, they are different and it is crucial not to mix up the two. *Energy*, measured in Joules, is defined for the entire computation of interest whereas *Power*, measured in Watts, is the rate of energy consumption. That is Watts = Joules/s. For charging purposes, electric energy is often measured in units of "Kilo-watt hours" (KWH). Obviously, 1 KWH = 3.6 Mega Joules.

For a given program running from start to finish, the important parameters are its total energy consumption and its running time. The ratio of the two gives the average power consumption. If it is important to complete the program quickly, it may be possible to run it at a higher power level and thereby finish it quickly. Conversely, if we can wait to get the results, it may be possible to run the program at a lower power. Since the electrical infrastructure at all levels (from data center to individual servers) has limits on how much power can be drawn, the maximum allowable power consumption is also often limited. Higher power also generates more heat (as discussed below) which may further require limits on power consumption. Smart power management techniques, discussed later, could reduce the power consumption, so that the power circuit limits are not violated and the electronics does not overheat. In terms of electricity cost, however, it is the total energy consumed that is of primary interest. Many power management techniques can also reduce the total energy consumption, as discussed later.

Power consumption increases the average kinetic energy of the atoms of the material (e.g., silicon) which heats the material. Since energy cannot be destroyed, nearly all of the power consumed is eventually converted into heat, which manifests itself as higher temperature. When two materials (e.g., silicon and surrounding air) are at different temperatures, there is a heat flow from higher temperature material to the lower temperature material until the average temperature of the two is equalized. Thus, in the long run, the temperature of a substance is determined by the balance between power consumption (which raises temperature) and cooling (forced or natural) due to the surrounding substance (e.g., air) at lower temperature. The rate at which the temperature approaches the final steady state value depends on the thermal properties of the materials which determine heat transfer via conduction, convection, and radiation. The important point to note is that even if the steady state temperature is moderate, over short periods the temperature within the material (e.g., transistor junction) may become high enough to cause damage or errors. Thus effective cooling at the points of highest heat generation is crucial but becomes harder and harder to achieve as the transistor density grows.

Idle vs. Active Power Consumption

In the traditional semiconductor technology, there are two types of power consumptions that are of interest: (a) static (also known as *idle* or passive) power, and (b) dynamic (also known as *active*) power. A conventional transistor consists of a silicon "channel" from *Source* to the *Drain* terminal. This channel is controlled by a "gate" that can apply a positive or negative voltage to the channel. In a so called npn transistor, the positive voltage causes a flow of electrons through the channel, which turns the transistor on, whereas a negative voltage cuts off the electron flow and turns the transistor off. Unfortunately, even with negative voltage, at the gate there is a small flow of current in the channel, known as "leakage current", which tends to increase steadily as the transistor feature size shrinks. A higher current flow means higher power consumption at the same voltage, since Power = Current × Voltage. The leakage current is the primary source of static or idle power, since this consumption happens even if there is no activity.

In particular, let V denote the Source-Drain voltage (often denoted in the literature as V_{cc} or V_{dd}). Let I_L denote the leakage current. Then the idle power consumption $P_{idle} = V \times I_L$. The idle power may form 20–40% of the total power consumption of a CPU or a memory module, and is expected to increase as feature sizes shrink. Thus, it is important to devise mechanisms to reduce it via a set of *Idle power management* techniques.

Explaining the dynamic power consumption requires a bit more understanding of the CMOS (complementary metal oxide semiconductor) technology which is used almost universally in current digital designs. Suffice to say that a CMOS device actually uses two basic transistors such that while one of them is on, the other is off. Thus the functioning of the CMOS device amounts to switching back and forth between these two complementary configurations, which can be identified as representing the logical 0 and 1. Each switch from 1 to 0 or 0 to 1 consumes power that is over and above the static (idle) power. This is the dynamic power and it clearly is proportional to the rate of switching – or the frequency at which the electronic component operates, henceforth denoted as f and measured in Hertz. Every transistor has an inherent capacitance, denoted C, and each switching amounts to charging or discharging this capacitor. The charge held by a capacitor is given by the product of capacitance and operating voltage, and the current, by definition, is the rate at which charging (or discharging) happens. That is, dynamic current = Capacitance × Voltage /switching_time = $C \times V \times f$. Since, dynamic power = Voltage × dynamic current, the dynamic power consumption of a CMOS transistor, $P_{dynamic} = 1/2 \times C \times V^2 \times f$. Here, the factor 1/2 results from the fact that half the energy is cycled in 0–1 transition and the other half in 1–0 transition.

It is important to note that the $P_{dynamic}$ computed here will be consumed only when the transistor switches in every cycle. In general, the transistor will switch only some fraction of time, denoted U. Then the total power consumption of a transistor is given by:

$$P_{total} = P_{idle} + U \times P_{dynamic} = V.I_L + 1/2\, UC.V^2.f \qquad (1)$$

Notice that reducing V reduces both idle and dynamic power, and hence is an attractive way to reduce the total power consumption.

The above equation can be used to compute power consumption of the entire core or CPU by considering all of the transistors that comprise the core or CPU. For example, I_L can be thought of as the total leakage current from all the transistors comprising the core/CPU. Furthermore, we can think of U as the *utilization*, defined as the fraction of cycles for which the core/CPU is busy. The capacitance C then is an effective value that corresponds to capacitance of all those transistors that switch, on the average, in one cycle.

Let us illustrate this with an example of a core with operating voltage (V) = 1.2 V, Leakage current (I_L) = 7.5 Amp, Effective Capacitance (C) = 10 nanoFarad, and operating frequency (f) = 2 GHz. In this case, the idle power is $1.2 \times 7.5 = 9.0$ W, and maximum dynamic power = $0.5 \times 10 \times 1.44 \times 2 = 14.4$ W. Thus, the core will consume 23.4 W when 100% busy, or 16.2 W when 50% busy.

The above equation applies to not only the CPUs but also to other components of the system (e.g., cache, DRAM, links connecting various components, etc.), although the details vary. For example, many links within the chip are "synchronous" in that they continuously transmit either real frames or small "fill-in" frames. For such links, the power consumption remains the same irrespective of their utilization. The actual power consumption also depends on circuit level power management via "power gating", or not supplying power to any entities that do not need to be powered up. Aggressive power gating can significantly reduce the transistors that need to be powered up and thereby reduce the overall power consumption. Of course, power gating is not free, since the power gating itself requires transistors that consume extra power. In this chapter, we will largely talk about power management on the coarser scale than targeted by power gating, and this may be suitably accomplished in hardware or software.

Power States and Their Management

Almost all major components in a modern server offer control knobs in the form of power states – a collection of operational modes that trade off power consumption for performance in different ways. Power control techniques can be defined at multiple levels within the hardware/software hierarchy with intricate relationships between knobs across layers. We call a power state for a component *active* if the component remains operational while in that state; otherwise we call the state *inactive* or *idle*. These active and inactive states offer *temporal* power control for the associated components. Another form of power control is *spatial* in nature, wherein only a subset of a set of identical components are operational.

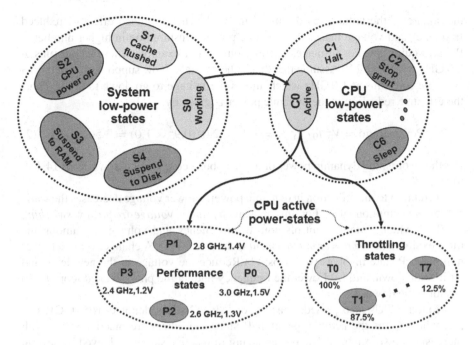

Fig. 1 System and device power states

At the highest level, the OS-directed Power Management (OSPM) defines a set of power states for the entire machine that most users are already familiar with. There are six such "system" states denoted S0...S5, with the following being the most relevant: S0 (working), S3 (standby – or inactive with state saved into the DRAM), S4 (hibernating – or inactive with state saved into the secondary storage), and S5 (essentially turned off and requiring reboot, but rebooting possible remotely). For example, when you press the sleep button on a laptop, it enters the S3 state (unless hibernate or hybrid sleep-hibernate states are chosen in the options). In the system state S0, individual devices (e.g., CPU, memory, links, etc.) have further power states defined. In particular, the CPU offers three types of states, C states (inactive) and P and T states (active) as shown in Fig. 1.

Processor Power States

The most commonly known processor states are the P (performance), where P0 refers to the highest frequency state and P1, P2, etc. refer to progressively lower frequency states. Lower frequency allows operation at a lower voltage as well and thus each P state corresponds to a supported (voltage, frequency) pair as illustrated in Fig. 1. The active power consumption is proportional to frequency but goes as the square of the voltage, as discussed in Eq. 1. The combined effect of reduced frequency and voltage makes the dynamic power consumption in higher numbered P states quite low. For instance, consider our earlier example where a core runs at 2.0 GHz with V=1.2 V. Assume that this is the P_0 state. Now suppose, that in the P_2 state the core runs at 1.0 GHz which allows the voltage to go down to 0.9 V. Thus, the effective reduction in the dynamic power is given by:

$$\text{Power Ratio} = V_0^2 f_0 / V_1^2 f_1 = 1.2^2 \times 2.0/(0.9^2 \times 1.0) = 3.56 \qquad (2)$$

In other words, the dynamic power of a 100% busy core will go down from 14.4 W to 4.05 W!

In addition to the reduction in dynamic power, a lower voltage decreases the static power consumption also. For this reason, *dynamic voltage-frequency switching* (DVFS), which suitably controls both the frequency and voltage, are among the most explored *active power management* technologies [1]. A changeover to a higher numbered P state involves two steps: (1) Reduce the voltage to the new level and let it settle down, and (2) Lower the frequency and let the phase-locked loop (PLL) circuit settle down.

Modern CPUs also provide inactive or "sleep" states denoted as C0, C1, C2, ..., where C0 is the active (operational) state and others are inactive states with increasing power savings. The power saving in inactive state is achieved by several techniques including turning off clocks, lowering the voltage (to lower the leakage power), and taking some additional actions such as flushing the cache and powering it off. The precise action taken is architecture dependent. For example, C1 and C2 states only turn off clock, C3 flushes the lowest level cache as well, and C6 flushes the next level cache and powers down some links. It is expected that future processors will have even deeper inactive states. The deeper C state provides higher power savings but at the cost of longer transition times into and out of the state. With actions such as cache flushing, the impact of the sleep persists even beyond the point when the CPU is active again, since much of the flushed data may need to be fetched again from lower level cache or memory.

The *idle power management* techniques must intelligently decide when to put the CPU in a C state, which one, and when to exit it. With multiple C states, there is also the question of transitioning between different C states. Unfortunately, the CPU cannot directly go from one inactive C state to another – instead it must first become active (i.e., transition to C0 state) and then choose another inactive state to go into.

In case of a multi-core CPU, many of the above states apply to individual cores as well. However, depending on the architecture, certain features may not be available independently to each core. For example, if all (or a group of) cores lie on the same voltage bus, it is not possible to change their voltages independently. For sensible power management, it is necessary to relate the core states to the CPU (i.e., entire package) states. The general rule is that the state of the entire package corresponds to the state of the core that is in the shallowest C state. For example, if even one core in a package is active (in C0 state), the entire package must be considered to be active, so that it can function normally in terms of data transfer to and from the package level cache or the memory.

Processors often also implement T (throttling) states, but these are entirely intended to handle thermal emergencies by introducing gaps in the clock (to allow the processor to cool down). We will not discuss them further here.

Memory Power States

The ever increasing appetite for more memory is already making memory power consumption rival CPU consumption. Thus aggressive management of memory power is essential. A memory stick or DIMM (Dual inline memory module) of DRAM (dynamic random access memory) consists of several memory devices (or chips), each of which typically provides 8 bits of data in parallel. There is considerable internal structure to the modern DDR (dual data rate) DIMMs. In particular, each DIMM is divided into "ranks", with 1, 2 or 4 ranks per DIMM. A "rank" is usually a set of 9 memory devices (8 for data and 1 for parity). The 8 devices of a rank collectively and in parallel provide the entire 64 bit (8 bits from each of 8 devices) "chunk" over the memory channel. Memory controllers often support multiple channels, each allowing one or more DIMMs and capable of independent data transfer. Since the data from all ranks of all DIMMs on a channel must flow over that channel, the ranks can be lightly utilized even if the channel is quite busy.

As its name implies, a DDR DRAM transfers 8 bytes (64 bits) of data on both edges of the clock. It thus takes 4 cycles to transfer a typical 64 byte cacheline. The total power consumption of a 2-rank DIMM is in 3–4 W range with idle power of about 1 W. Several sleep states are available for the *idle power management* of the DRAM. The two shallowest sleep states are called "fast" and "slow" CKE (clock enable) states; these allow an inactive rank to deactivate parts of the circuitry such that it is possible to reactivate them in a few ten's of nanoseconds. As the name implies, the slow CKE is somewhat slower, but achieves higher power savings. For a much deeper sleep, it is possible to put a DIMM in "self-refresh" mode. The essential characteristic of the DRAM technology is that the stored data must be constantly "refreshed", i.e., read and then written back every 64 ms or less; otherwise, it will be lost. Normally, the refreshing is done by the *memory controller*, which is the intelligent entity that interfaces CPU and the DRAM. In self-refresh mode, the DRAM refreshes itself, so that the memory controller and the link between the memory controller and DRAM can become inactive. Self-refresh is typically used when the CPU is placed in C6 state. The memory can also use *active power management* provided that it can run at several different frequencies. As with CPU, these active states are most useful if the lowering of frequency is also accompanied by a suitable lowering of the voltage.

Link Power States

Modern computer systems use a variety of networking media both "inside-the-box" and outside. The best known outside-the-box networking technology is Ethernet, but there are others as well, such as Fiber-Channel (used for storage networking), and InfiniBand (for low latency interconnection in high performance computing). Ethernet and other technologies can consume a substantial percentage of the IT power in a large data center, and their power management is becoming essential. These technologies continue to increase in speed – for example, 10 Gb/s Ethernet is becoming quite popular in data centers, and the higher end data centers are moving to 40 and 100 Gb/s Ethernet. Although the power consumption of a 10 Gb/s Ethernet is only about 3 times that of 1 Gb/s Ethernet, the increased speed often results in power inefficiency. The main reason for this is that most network links carry very little traffic most of the time, and high bandwidth is required only sporadically. Thus an upgrade from 1 to 10 Gb/s Ethernet increases the power consumption by a factor of 3, with very little additional traffic carried on the average.

Moden computer systems have several internal interconnects, that run at much higher speeds than outside-the-box interconnects, and can collectively consume a significant percentage of platform power. With many cores per CPU, a significant number of links and/or link interfaces are required for interconnecting the cores, and this interconnect must support extremely low latency and very high bandwidth. The IO interconnects, including PCI-Express, SATA, SAS form other prominent inside-the-box interconnects. There are still others such as interface between memory controller and DIMM or interconnect between CPU and IO complexes. An intelligent power management of such interconnects also becomes crucial for platform power reduction. Such links are invariably "synchronous" which means that the power consumption does not depend on the data rate.

Most current links support at least two PHY (physical) layer low power states, called L0s and L1, respectively which can be used for *idle power management*. The L0s power state is unidirectional, in that the transmitter for each direction of the link can independently decide to go into low power mode when it has nothing to transmit, whereas the receiver side remains active. The L1 power state involves a handshake between transmitter and receiver, and thus allows both of them to go into low power when there is nothing to transmit. The L1 state can reduce the idle power quite substantially but this comes at the cost of substantial latency; therefore, L1 is typically used with the C6 CPU state.

As with other devices, links can also be operated at lower speeds in order to reduce their active power, and thus allow for their *active power management*. Depending on the type of link, the speed change may be either a matter of simply changing the clock rate or a switch-over to a different PHY. An example of the latter is the 40 Gb/s Ethernet operating at 10 or 1 Gb/s. Such a PHY switch can be extremely slow, and the power reduction may not be significant. Furthermore, the lower speed means longer data transmission time and may not provide any gains in energy consumption.

The energy efficient Ethernet (EEE), also known as Green Ethernet, provides a software controlled low-power idle (LPI) mode initiated by the transmitter, and thus can be used independently for each direction of the link. An LPI enabled transmitter sends a LP_Sleep signal to the receiver, so that the receiver can place its side also in low power mode. The transmitter can tell the receiver how long it will be in LPI mode. On wake up, the transmitter sends a LP_Wakeup signal to the receiver. When the transmitter is in the LPI mode, it continues to send periodic refresh or heart beat signal to maintain the synchronization while consuming only about 10% of the normal power. Wakeup from LPI involves a significant exit latency since in addition to the exit delay, the transmitter needs to wake up the receiver before transmitting anything. The two relevant parameters in this regard are Sleep time (T_s) and Wake-Up time (T_w). For a 10 Gb/s link, with 1500 bytes packet size T_s and T_w will be 2.88 and 4.48 μs respectively. This amounts to transmission time of several packets and thus the mechanism is useful when the traffic shows significant gaps between packet bursts. If the workload does not have such characteristics and there is no traffic shaping to make it behave so, LPI can be useful only at very low utilization levels. Consequently, we will also study the usefulness of basic L0s state based control assuming that is provided by the Ethernet interface. Such a control keeps the receiver side always awake and only the transmitter can sleep; however, the low transition latencies and lack of handshake between transmit and receive sides makes the mechanism suitable at higher utilization levels as well.

Collective Power Management

In the above, we considered the power states of a single entity (e.g., a CPU core, memory rank, or a link) or a composite entity considered as a single unit (e.g., entire CPU with all its cores, entire DIMM, or a set of parallel links between two devices). However, whenever we have a composite entity or a set of entities (e.g., a set of servers), it is possible to power manage them together without significantly hurting the performance. The basic idea is to consolidate the load on a certain subset of devices, so that the others can go into low power mode. What is important here is that the vacated devices can generally go into a deep sleep state and stay there for long periods of time. In contrast, if each individual device is power managed separately, it may be able to sleep only briefly, and thus it cannot go into a deep sleep mode. The reason for the latter is that a deeper sleep mode invariably comes with long latencies to transition in and out of the sleep mode.

A special case of collective power management occurs in modern interconnection links, which are made up of several multiple "lanes" of "serial" links. A lane of a serial link carries only 1-bit of data at a time and uses mechanisms (e.g., differential signaling) to make it very robust and noise free. Current systems have generations 1, 2, or 3 of the technology, which supports respectively, 2.0, 4.0, and 8 Gb/s bandwidth per lane. Thus, an 8-lane link, referred to as x8 link, can support 2, 4, and 8 GB/s bandwidth depending on the generation. For example, graphics cards typically use x8 or x16 PCI-E links.

Serial links allow *dynamic width management* wherein certain lanes can be put in low power mode to reduce power consumption when the traffic (i.e., the bandwidth requirement) is low. A highly desirable feature of width control is that so long as some lanes are active and traffic is low enough for the available bandwidth, there is very little delay impact of the power management on the traffic. A dynamic width control algorithm has to operate within the constraints of supported widths associated with the underlying link hardware.

Dynamic width management can also be used for DRAM by keeping only some of the ranks active at a time, as discussed in [2]. For example, if a server has two DIMMs (presumably on two different memory channels) and each is a 2-rank DIMM, we have a total of 4 ranks and we could rotate among them so that, say, only 2 ranks are active on the average. Such a mechanism trades off memory access latency (and hence performance) against the power consumption since the inactive ranks can be put in one of the sleep modes.

For more general collective power management, let us consider the set of cores in a CPU. In general, only some of the cores may be needed to handle the current workload. In this case, the other cores can go into a deep sleep state such as C6 or could even be powered off depending on how quickly we want to be able to turn on those cores. Even if all cores in the CPU are identical, it does matter which cores are turned on or off by the power management algorithm. One reason for this is that the heat produced by one core affects the adjacent cores, which means that it may be undesirable to simultaneously operate two adjacent cores. A well-known scheme in this regard is called "core hopping", where the cores are used in a cyclic fashion to ensure that all cores generate approximately the same amount of heat. Another reason why the choice of core to turn on/off matters is that each active core typically accumulates its "working set" (i.e., the data most essential to the operation of the program) in its cache. Turning off this core and restarting the computation on another core would force that core to fetch its working set from the shared cache or memory, and thus slow it down. This is a particularly important issue for core hopping – while systematic cycling through different cores may balance out the heat generation, it may also hurt performance due to disturbances to the working sets. Yet another form of power management for a multicore CPU involves the so called "turbo mode" operation. If only some of the cores are active, they can run at a higher frequency and still maintain the desired thermal envelope.

Collective power management also applies at higher levels. For example, if a server rack has 10 servers, each running at a utilization of at most 20% (a fairly typical situation), a significant amount of power can be saved by consolidating all of the workloads and distributing it to only 3 servers. In this case, each server will run at 67% utilization, which may be reasonable. (Except in case of long-running, CPU bound tasks, it is generally not possible or desirable to run a server close to 100% utilization on a sustained basis.) The remaining 7 servers could then be put in one of the system sleep states such as S3, S4, or S5 depending on how much restart delay we are willing to tolerate. For example, suppose that each server consumes

100W of idle power, 5 W in S4 state (note that in S4 only a small wakeup circuitry is powered on), and the active power at 100% utilization is 150 W. Then, in the original configuration the total power consumption is $10x(100 + 0.2 * 150) = 1300$ W. With consolidation, we instead have $3 * (100 + 0.667 * 150) + 7 * 5 = 635$ W, a more than 50% savings.

Energy Management Algorithms

A smart energy management may involve use of several mechanisms used at different time scales. In the above, we discussed the idle and active power states, which can be controlled either individually for each device or as a set. The time scale aspect is crucial since the idle durations of any resource can vary over an extremely large range. For example, the CPU may experience stalls at the level of individual instructions while it is waiting for data from memory or last level cache – these stalls would be in the range of 10's to 100's of ns. Larger idle periods – in the range of microseconds to milliseconds could occur due to wait for network or storage devices. Even longer idle periods may be governed by user demands (e.g., queries that require processing) which itself involves variations over 1–100's seconds (e.g., gaps between successive queries) and over hours or longer (e.g., hourly and daily variations).

The duration of idle period is crucial when using *idle power management*, since a longer idle period allows deeper sleep, less latency impact of state transition, and more sophisticated control. However, with *active power management*, the duration of the idle period does not matter; instead, what matters is how quickly the activity level (e.g., device utilization level) changes. We discuss these in the following for both short time scales (fine grain power management) and longer time scales (medium and coarse grain power management).

Fine Grain Power Management

Fine-grain power management refers to power management actions that can capture traffic intensity changes and idle periods ranging from 10's of ns to 10's of us. In this space, active power management may not be useful, and will not be discussed here. As for the idle power management, the very short sleep durations demand very simple, hardware based solutions. Most of the hardware sleep states discussed above (e.g., C0, C3, C6 for CPU, L0s/L1 for links, fast/slow CKE for DRAM, etc.) can be exploited here by a hardware algorithm. Note that collective control using sleep states, such as link width control, are also useful here.

The top part of Fig. 2 shows a device (CPU core, link, DRAM rank, etc.) without any power management. In this case, the device becomes idle (IDL) once its transaction queue empties out, and then busy (BSY) again when a request (or

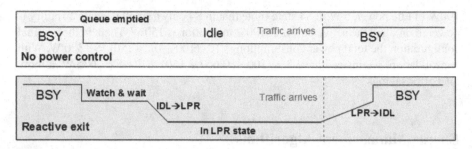

Fig. 2 State illustration without and with power control

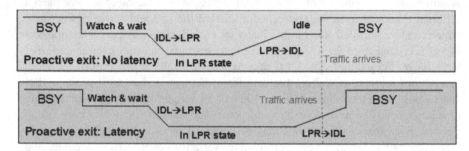

Fig. 3 State illustration with proactive power control

"traffic") arrives. The bottom part of Fig. 2 shows the simplest possible approach to using the low-power (LPR) mode. Since we do not know when the next request will arrive, we wait for a while (called "runway"), and then transition to LPR mode. The runway is a parameter of the algorithm and can be set to a fixed value or adjusted dynamically based on the device utilization. The transition takes some time as shown. Eventually, when the traffic arrives, we exit the LPR mode and become busy again. Notice that this is a *reactive* control with respect to exit from LPR mode. It has the property to let the device stay in LPR mode as long as possible, but the incoming traffic is always delayed by the exit delay. Clearly, this algorithm is trivial to implement in HW and will have negligible overhead.

Figure 3 shows a *proactive* variation of the algorithm where the device stays in LPR state for a certain amount of time, say T_L, and then exits proactively. Of course, if the duration T_L has not expired and traffic arrives, the device will still exit reactively from LPR mode. The figure shows two cases. In case (a) the T_L estimate falls too short, in which case the device exits LPR state prematurely, which is not good for power savings. In case (b), T_L estimate is too long, which means that the traffic experiences some delay (although the delay will always be bounded by exit latency). In order to make the proactive algorithm work properly, we need a good estimate of T_L, and this estimate needs to be continuously updated. A simple way to do so is to update T_L based on the idle periods observed in the recent past, but more detailed information about the traffic can provide better predictions.

As discussed above, a device can have multiple sleep states with increasing power savings and exit delays. The normal way to use multiple states is to first enter the shallowest state (with smallest exit latency), and then progressively move to deeper sleep states if no traffic arrives. Unfortunately, it is not possible to directly switch from one sleep state to another; instead, it is necessary to wake up the device to full idle mode and then transition it to the desired deeper sleep mode. This limits the usefulness of successive "promotion" to deeper sleep states; instead, it may be better to simply choose the most suitable state upfront based on current utilization level, and promote it further only if the idle period turns out to be extremely long.

In addition to sleep state control, the width control can also be exercised easily by the HW. An algorithm for this is described in [3]. The basic idea is to monitor the device utilization, and if the utilization crosses a predefined threshold, change the width. The algorithm needs to avoid ping-ponging (i.e., a rapid switching between high and low widths), but this is easily accomplished by introducing some hysteresis in the algorithm.

Medium and Coarse Grain Power Management

At medium/coarse grain, the algorithms can be implemented in firmware or software and can be more and more sophisticated as the time granularity increases. The algorithms can make use of both active and idle state controls, as discussed below.

The active state control such as DVFS reduces power by matching the throughput capability of the device to the current needs. The best known active controls are DVFS controls for CPUs – they are an integral part of power management in current systems. For example, Intel's *SpeedStep technology* and AMD's *PowerNow technology* make use of P states to dynamically switch the processor to higher numbered P state when the CPU utilization is low, and to a lower numbered P state when the utilization increases. The net effect is to keep the effective CPU utilization after change over to suitable P state at a fairly high level (say, around 70–80%) irrespective of the actual load. This allows a significant reduction in the average CPU power consumption. Both the *SpeedStep* and *PowerNow* can be characterized as reactive in nature in that they change state based on the utilization in recent past. It is possible to make use of predictive algorithms here, but may not provide any significant advantages.

It is important to note that a change in P state is not instantaneous and could require a few microseconds or more depending on how it is done. This is because a voltage change needs time to settle down, and a frequency change requires to phase locked loop (PLL) circuitry to lock the new frequency. Intel's improved implementation, called *Enhanced SpeedStep*, tries to reduce this latency with a small increase in power. If a software based algorithm is used to decide and switch among P states, the delays could be in milliseconds or more. An important point about P states is, however, that multiple cores may be supplied voltage from the same "rail" and thus all of those cores must use the same voltage. This considerably limits the flexibility of DVFS. Furthermore, with voltage levels shrinking and getting closer to thresholds for reliable switching, there may not be much scope for reducing voltages at lower frequencies. Note that simply lowering the frequency without lowering the voltage may be detrimental from the energy perspective. To see this, consider an entirely CPU bound task which runs for 10 s on a 2.0 GHz processor. Then the task would take 20 s to complete if the processor frequency is reduced to 1.0 GHz. Obviously, the active power in this case will be 1/2 of the original, but the active energy will be the same (twice as long at half the rate). Moreover, the processor will now be consuming idle power for 20 s, instead of only 10 s (and perhaps placed in a low power mode rest of the time). Thus, the total energy consumption is actually larger!

The idle state control can be much like the fine grain HW algorithms described above, except that there is a scope for more sophisticated decision making here. For example, if the workload is rather stable, one could learn its characteristics and use them to determine the "runway" and the low power state to be used. In some cases, it may be possible to even eliminate the runway since a more detailed understanding of the application behavior may tell us when the application is unlikely to use a particular resource.

Although medium and coarse grain power management are often integrated together, it is worth making a clear distinction between the use of lower speeds and sleep modes (usually done at the time granularity of minutes or lower) and simply powering off the resources (which becomes attractive at the granularity of 10's of minutes or longer). Two prominent examples of the latter are: (a) consolidating servers by "packing" the workload on as few a servers as possible and shutting down the rest, and (b) copying actively used data to certain disks and spinning down the rest [4, 5]. Such consolidation can save a substantial amount of energy when the usage pattern is easily predictable or known. For example, once we know the low usage periods during a day (typically late night and early morning), we can decide how many servers and disks to keep active. This number may be somewhat overestimated to deal with uncertainties, and then medium grain controls can extract further savings by making use of active and passive controls.

Software Energy Efficiency

In the above, our focus has been almost exclusively on hardware energy efficiency and management techniques. Energy is ultimately consumed by transistors and wires; in fact, energy is consumed even when the transistors are not switching (i.e., not computing). This makes software level energy consumption discussion rather difficult. For example, it is not possible to associate a fixed amount of energy with basic operations such as add, multiply, data copy, etc. The problem is that the energy consumption associated with any operation depends on many hardware level details, including the number of operation units, how they are used, the location of the instruction and data (core level cache, shared cache, or memory), and how any given operation relates to others around it because of pipelining, prefetching, speculative execution, etc. Furthermore, even if one could estimate per operation energy consumption, it is not necessarily meaningful because a hardware unit that does not perform any operation may still consume some idle power, and this power consumption depends on how the unit is power managed. Because of these difficulties, it is usually not possible to consider energy consumption in the same simple way that we use for estimating the complexity of the algorithm. Nevertheless, it is of great interest to consider how to reduce the energy consumption of the system during the time the program of interest is running. In this section we discuss this issue in the context sequential programs. Parallel program related issues are discussed in the following section.

Algorithmic vs. Energy Efficiency

In general, if we can restructure the program or change the underlying algorithm so that it finishes more quickly, it will likely also consume less energy. In other words, a more efficient algorithm (along with its efficient implementation) should generally lead to less energy consumption. This is because the more efficient algorithm will likely do one or more of the following: (1) execute fewer instructions and/or touch less data, (2) improve the hit rates in processor cache(s) and thus reduce data movement (including traffic on various interconnects), (3) reduce memory footprint and/or lay out data in memory in a way that results in more efficient accesses to memory, (4) fewer or more efficient data transfers over the network, (5) better layout and access to the disk to reduce I/O overhead, and (6) less contention for shared resources such as locks. All of these can directly reduce the power consumption by reducing the activity level in the computing infrastructure.

However, we need to be really careful and not equate shorter running time with less energy consumption. DVFS actually provides a direct contradiction to that idea – by running slower, and thereby taking longer, we save energy. One could argue (correctly), that DVFS changes the behavior of the hardware, rather than the software, and thus does not violate the idea of reducing energy consumption by making the program run faster. Nevertheless, even software restructuring to reduce the run-time may not always reduce energy consumption. This could happen because a more efficient algorithm puts more stress on the CPU and increases the energy consumption more than the amount by which it reduces the run-time. In general, the energy consumption and run-time of a program are affected differently by the choice of algorithm, data structure, data locality, caching behavior, etc. and the overall impact could be difficult to model and predict.

Enhancing Energy Efficiency Opportunities

It is often possible to increase energy efficiency of a program/service by enhancing opportunities for it to make use of energy saving techniques discussed above. This applies to both terminating programs (that take some inputs and then run until completion) and services (that receive an input or query, execute, and then wait for the next one). With idle power management, if we can increase the idle periods, the underlying hardware can use low power modes more effectively and thereby save energy. In a terminating program, this could be done by bunching together periods of IO, memory accesses to fetch data into the cache, and execution from the cache. In a service, this can be done by batching the queries together suitably so that multiple small idle periods turn into one larger idle period. With active power management, restructuring the workload so that the utilization of various devices can be more balanced allows energy saving by not having to change the active state too frequently.

As a concrete example, consider the case of network traffic flow coming into a switch or router. Suppose that we collect a batch of $n > 1$ packets and then forward them. In this case, we can put the switch/router port into low power mode while we are collecting the next batch. If this batching period is long enough that the overhead of transition into and out of low power state is a relatively small fraction of the total idle period, better energy efficiency is achieved than by forwarding the packets one by one. Of course, the cost is extra delay caused to the packets. Note that for this solution to work, we should be able to receive and queue up the packets any time; it is only the forwarding part that can go into the low power mode. This requirement may be difficult to satisfy or require additional hardware to capture packets arriving while the port is in low power mode. Another important issue is that the batching needs to happen for all of the traffic coming into a port, rather than for only some of the flows. For example, if the port is receiving several flows, and at least one of them is too latency sensitive to allow for batching, there may not be much advantage in batching the others.

Data Movement vs. Computation

Computation and communications are two key functions in all of the computing technology and both need to be fast and energy efficient. Unfortunately, communication (or data transfer) has not kept pace with computing on either front. This is true at all levels starting with on-chip wires. As the width of the wires shrinks, their resistance goes up, and is already measured in Mega-Ohms per inch. This makes data movement increasingly more costly in terms of power as compared with computation. While the transistor power decreases with smaller "feature size", the wire power does not necessarily scale down due to the wires becoming thinner.

Also, unless the wire length also decreases in the same proportion as the feature size (generally not true), the increased capacitance makes the wires slower. The net effect is that the energy consumed to move data on-chip by 1 cm is increasing while the energy required for computation (say, addition of two numbers in registers) has been going down. This is true for all interconnects/links including interconnect among cores, DRAM to memory controller, etc. The net result is that more attention must be paid to data movement to both on-chip and off-chip than has been done in the past. Even the system level interconnects such as PCI-E, Ethernet, Infiniband, etc. suffer from similar issues – their speed increase and energy efficiency has not kept pace with the computing.

There are several aspects to be considered in minimizing data movement. One key issue concerns data representation. For elementary data types, it helps to minimize their size, so that more "information" can be moved using the same number of bytes. For example, signed integers that are not expected to go beyond $2^{15} - 1$, are better represented as "short" rather than "int". Similarly, the floating point variables whose computation would be acceptable as single-precision (in terms of range and precision), should not be declared as double. A compact representation at larger granularity (e.g., arrays, structures, etc.) is also highly desirable provided it does not make the algorithm inefficient and thereby erase the advantages of compact representation.

Another key concept in minimizing data movement is *locality of access*. The basic idea is that if some data items are placed close together, the algorithm should be designed to access them together so that the data can be brought in larger chunks. Conversely, the data that is normally accessed together should be placed together. This applies to the algorithm design, memory allocation of variables by the compiler, placement and access of data on the disk, etc. Locality can be very difficult to achieve when the data to be accessed depends on external queries, and the query workload is highly variable.

The third key concept is a tradeoff between computation and data movement. In the past, computation was expensive and the algorithms emphasized the need to reuse what has already been computed. This is increasingly not true – it may be better to recompute the result locally yet again, instead of fetching it from elsewhere. For example, consider two nodes N_1 and N_2 in a multiprocessor system, each of which holds data items A and B in their caches. (Here A and B could be scalars or vectors.) Suppose that N_1 has already computed $A + B$, and N_2 needs it. It may be faster for N_2 to recompute it (in a separate variable) instead of requiring it to be transferred from N_1 (by accessing the variable holding $A + B$). This situation applies at higher levels as well. For an example, consider two joinable relational tables A and B stored on a disk accessible to two nodes N_1 and N_2. Suppose that both nodes have cached A and B in the memory and N_1 has computed $A \bowtie B$. Now if N_2 needs $A \bowtie B$, it may be more efficient for it to compute on its own, rather than asking N_1 to either send the result over the network or write it to the disk from where N_2 can read it.

Tradeoff Between Energy and Performance

For best performance, it is usually desirable to spread the load across all resource instances (e.g., servers in a cluster, cores in a CPU, all channels of DRAM, etc.) Such *load balancing* minimizes bottlenecks and hence leads to better performance. However, from the energy perspective, it is better to concentrate load on as few resource instances as possible so that the rest can be put in low power mode. The load balancing among the used resource is still important, although ideally this would be balancing of power consumption rather than load.

Restructuring of a program to enable better energy efficiency and actually exploiting this potential by power management technique itself has impact on the performance. For example, batching of requests means additional delays, which may be undesirable. Thus, the extent of permissible batching will be governed by the maximum tolerable delay. Also, any kind of power state transition causes delays, which again must be controlled. For example, placing the unused servers in S4 (hibernate) state may require 10's of seconds to resume operation, possibly followed by some task migration before normal operation can be resumed.

Given the different impact of various techniques on power consumption and performance, it is often desirable to speak of a metric that involves both performance and power. The simplest one is performance per watt, which may be reported as transactions/watt for transactional workload. Unfortunately, such a metric very much depends on the utilization level with or without power management. Without any power management, the metric will be the highest if the resource is 100% utilized and become very small as the utilization goes down to zero if the idle power consumption is significant. With aggressive power management, the maximum may be achieved at a very low performance level because at that level one could set the voltage to the lowest level and thereby gain a substantial reduction in active power. Thus performance per watt (or a similar work done per Joule of energy) needs to be interpreted carefully to be meaningful.

Parallelism vs. Energy Efficiency

Parallelism is a property of both hardware and software, and to the extent the available parallelism in the software can be mapped to that in the hardware, we can reduce the execution time and possibly the energy consumption. This section provides an overview of some of these issues.

Hardware vs. Software Parallelism

The hardware provides the following types of parallelisms

1. Instruction Level Parallelism (ILP), where the goal is to complete (or "retire") as many instructions as possible in one clock cycle. This is achieved by two techniques: (a) pipelining, divide the execution into a pipeline of stages, so that while stage i is working on nth instruction, stage $i - 1$ could be working on $(n + 1)$st instruction, etc., and (b) superscalar processing, where several independent instructions are processed in parallel by independent hardware units.
2. Data Level Parallelism (DLP), where multiple data elements (e.g., elements of a vector) are processed in parallel either by the same operation (SIMD) or different operations (MIMD). For example, two vectors can be added in the time it takes to add two elements.
3. Thread Level Parallelism (TLP), where multiple software threads execute in parallel on different cores or HW threads (usually followed by a synchronization point where the computed results are exchanged and prepared for the next phase). The well-known map reduce framework is a good example of this type of parallelism.

The ILP occurs naturally in software in that the sequence of instructions in a program need not be executed in that order – the only thing that matters is ordering with respect to the dependencies that result from one instruction using the result produced by another instruction. Branches are also problematic because they force the execution to move to another spot in the program. Program ILP is automatically exploited by the current architectures. For exploiting ILP, the instruction execution is divided into multiple "stages", such as instruction fetch, decode, operand fetch, execution, and result generation. These stages work in parallel in that while one instruction is in the nth stage, the next instruction could simultaneously be in (n-1)st stage. The hardware required to handle pipelining includes interface registers between stages and extensive logic to manage dependencies, forward results, handle branches and exceptions. A deeper pipeline requires more hardware and may need more complex logic. Although a non-pipelined CPU would not optimally use the hardware, it can be largely power gated and thus does not result in much of an energy burden. The power gating of many small stages with complex interface logic is more difficult and carries more overhead. The biggest problem with deep pipelines however is the handling of branches, which cause substantial inefficiency. Superscalar designs also suffer from inefficiency if there are not enough independent instructions to pack together. In general, simpler designs (e.g., Intel Atom vs. Core processor, or Arm vs. Intel) tend to be much more energy efficient.

Vector computations occur naturally in software and can exploit the hardware vector processing capabilities easily for faster and more energy efficient processing. The energy efficiency results from less per operation overhead of vector processor vs. scalar processor and the overall shorter processing time. The overhead of properly using the vector processing units largely occurs in terms of proper code generation by the compiler, rather than at runtime. Furthermore, unused units can be easily put in low power mode.

A similar situation occurs with TLP, since the "uncore" hardware is shared among all the cores (discussed in the next section) and the cores can work in parallel to finish the task much faster than a single core could. Exploiting the multi-core architecture generally requires parallelism at a much coarser granularity than provided by ILP or vector processing.

Application Level Parallelism vs. Energy Efficiency

Application level parallelism can be achieved by spreading the operation over multiple resources such as servers, storage devices, network interfaces, etc. For example, running the same application on multiple servers can scale up the overall query processing rate up to the point where some bottleneck develops due to software contention (e.g., locks), access to shared disks, limited network bandwidth, etc. Storage system bottlenecks can be relieved by duplicating read-intensive data across multiple storage devices, or partitioning and striping data intelligently across them. Similarly, network bottlnecks may be relieved by using multiple network interfaces.

There are two related perspectives on energy efficiency for this type of parallelism. One is that if the system is limited by some significant bottleneck, the less used resources will not be used efficiently. For example, if the throughput is limited by the storage system, the servers will experience stalls, and it may be possible to retain the same throughput by simply reducing the number of servers to the point where the storage bottleneck is relieved. The extra servers could then be shut down or put in deep sleep mode to conserve energy. The other perspective is of dynamic sizing of resources to match the needs. That is, if the resources (servers, disks, network interfaces) are not well utilized, better energy efficiency can be achieved by consolidating the workload on fewer resources (and shutting down the rest). Thus, the general principle is to size the resources such that all resources operate close to their bottleneck point, but not above it.

It is important to note that dynamic sizing may be very difficult in many cases and may itself have significant energy overhead. We will illustrate this with two examples. First, consider a cluster of web servers that are dynamically sized to match the web query load. Each time a web server is to be removed from service, we first need to direct all new queries to other servers, let it finish all existing queries and then put it to sleep. This involves energy overhead. Similarly, when a web-server is to be fired up, it needs to fetch all required content to its DRAM and processor caches, which again involves a significant energy overhead. Thus, the frequency of changes needs to be properly controlled. The second example concerns disks where dynamic movement of a lot of data is impractical. This is usually handled by staging the data on the largest set of disks needed and then varying the number of active disks. The problem here is that irrespective of whether a disk is in use or not, its data needs to be kept up to date. This requires that the disk be periodically spun up, and the data updated. The energy and reliability implications of repeated spin up/down of disks need to be considered in designing an algorithm for matching the demanded throughput with the supply.

Application level parallelism can also be used across data centers – for example, by running the application in multiple data centers in order to do the processing closer to the demand. A proper application of this idea can reduce delays, lower network bandwidth requirements, and also save energy by virtue of the increased locality and less data movement. It can also handle energy supply limitations – for example, by processing queries or tasks where the energy supply is plentiful.

Thread Level Parallelism

Because of the proliferation of multicore architectures, the topic of designing parallel algorithms has gained considerable importance. These algorithms are focused on completing the processing as quickly as possible on the available cores, but are limited by the available parallelism. A well formulated parallel algorithm should allow assignment of a substantial chunk of work to each core such that they can all work in parallel without any need for any synchronization or data sharing with other cores. A typical algorithm would then follow a map-reduce type of paradigm: execute in several rounds, where each round involves the following 3 steps: (a) data distribution to some subset of cores, (b) parallel computation on all these cores, and (c) shuffling or summarization of the data to prepare for the next round.

The key challenges in devising such an algorithm include: (a) coarse granularity of work assignment to the core so that a significant amount of parallel computation is done in each round, (b) ability to use all or most of the available cores, and (c) workload balancing across the cores so that the computation on all the cores finishes at about the same time. These objectives could be difficult to achieve in general, and the parallel algorithm formulation to achieve them could make it substantially different from the sequential algorithm. Often this amounts to an algorithm that would be less efficient in time and space than the original sequential algorithm if executed on a single core machine. This difference, along with the overheads of parallel execution could make the parallel algorithm inherently less energy efficient than the sequential algorithm, even though the parallel algorithm would most likely take much less time to execute than the sequential version.

Designing the parallel algorithm to engage all cores and equalize work for them is often not possible because the number of data partitions and the size of each partition is often data dependent. For example, with parallel merge sort, the number of lists to be merged varies at different levels of the merge tree. As another example, the pivoting in quicksort invariably results in unequal size lists. This results in achievable speedup of parallel processing significantly less than the number of processors, say N, and often decreases with N beyond a certain limit.

Comparing the energy consumption of parallel vs. sequential versions of a program could itself be a bit tricky. Consider, for simplicity, a situation where all of the k cores in a system are engaged. Suppose that the sequential algorithm runs for time τ on one core, and the parallel version is able to divide much of the work into k equal parts, one allocated to each core. Still, there will be some overhead of parallelization, and we can consider this as a fixed fraction of the code, as suggested by Amdahl's Law. Accordingly, let this sequential fraction be f_a.

Then the parallel algorithm will take $\tau(f_s + (1 - f)/k)$ time. Suppose that the power consumed for each core when executing the algorithm is $P_A^{(c)}$ and when idle $P_I^{(c)}$. (Note that the active power does include the idle part here.) Let $P_A^{(u)}$ and $P_I^{(u)}$ denote the active and idle power consumption of "uncore", or parts of the CPU other than cores (e.g., core interconnect, memory controller, etc.). Since the $(k - 1)$ cores must remain idle in the sequential program for the entire period τ, the energy consumption of the sequential program is given by:

$$E_S = [P_A^{(u)} + P_A^{(c)} + (k - 1) * P_I^{(c)}] \times \tau \tag{3}$$

Since all the cores are active simultaneously for a parallel algorithm, its energy consumption is given by:

$$E_P = [P_A^{(u)} + k \times P_A^{(c)}] \times \tau[f_s + (1 - f_s)/k] = [P_A^{(u)}/k + P_A^{(c)}] \times \tau[kf_s + (1 - f_s)] \tag{4}$$

Notice that in this comparison, we assume that the idle cores are not doing anything, and thus extra power is consumed even if they are put in a low power mode. If we instead assume that idle cores can be used for something else, their energy consumption should not be charged to the sequential program. With our assumptions, the comparison comes down to the following tradeoff: (a) additional "uncore" power spent by the sequential algorithm, and (b) additional overhead of the parallel implementation. Thus the parallel algorithm is better if

$$[P_A^{(u)}/k + P_A^{(c)}] \times [kf_s + (1 - f_s)] < [P_A^{(u)} + P_A^{(c)}] \tag{5}$$

which can be simplified to yield:

$$P_A^{(c)} < P_A^{(u)}(1/f_s - 1)/k \tag{6}$$

For example, if the core and uncore power are identical, this equation reduced to $k < (1/f_s - 1)$. This means that so long as the number of cores is not too large, the reduction in execution time will be more than the overhead of parallelism, and hence the power consumption of the parallel program will be lower. However, with a large number of cores, this is not true and the parallel program will not provide any power advantage. As an illustration, Fig. 4 shows the power consumption for the Cinebench 11.5 benchmark run on different number of cores. This figure is taken from [6], which talks about techniques for making software energy efficient. Notice the Amdahl's law in play – the execution time with 2 cores is more than 1/2 of execution time for 1 core, and similarly for 4 vs. 2 cores, and 8 vs. 4 cores. Also note that the 2 core power is less than twice that of single core power because not all the power consumed in the core.

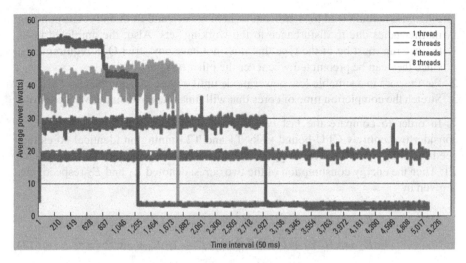

Fig. 4 Power consumption vs. number of cores. (Taken from [6])

In the more general situation where the different cores do different amount of work, the cores that finish earlier will remain idle, and will consume extra energy even if they are placed in a low power mode. This would make the parallel program less efficient.

Power Management of Parallel Computations

With both sequential and parallel programs, smart energy management can be exploited to reduce the energy consumption of the cores that are underutilized. In particular, cores that are idle can autonomously go into a suitable sleep mode using a suitable "runway". This also includes progressive algorithms where the core first enters a shallow sleep mode and then is promoted to a deeper sleep. However, such autonomous actions are not the most efficient, and an explicit control by the program (or a middleware that is aware of the program behavior) can do a much better job by properly scheduling tasks and taking energy management actions with a better knowledge of task schedules.

In particular, consider the scenario above where multiple cores do some computation in parallel and then "join" at the end. The cores may finish their work at different times either because the cores have nonhomogenous characteristics or the work given to them is unequal. In this case, the cores that can finish early be managed in the following three ways:

1. Schedule another unrelated task on the core and preempt this task when other cores reach the synchronization point. This approach is workable if the core has to wait for a long time before others reach the synchronization point,

otherwise it could lead to significant switching back and forth and corresponding inefficiencies due to disturbance to the working sets. Also, the unrelated tasks that we run must be of the type that does not have any strict QoS requirements, since they can be preempted whenever the other tasks are ready.

2. Put the core in a suitable low power mode until other cores complete.
3. Stretch the completion time of cores that will finish fast by using DVFS controls.

In order to compare the last two options with respect to energy efficiency, consider two entirely CPU-bound tasks T1 and T2 running on identical cores that need to "join" when finished. Suppose that task T2 takes only half the time as task T1. Then the energy consumption of the two cores, denoted E_1 and E_2 respectively, is given by

$$E_1 = (I_L.V_0 + 1/2.C.V_0^2.f_0)\tau$$

$$E_2 = (I_L.V_0 + 1/2.C.V_0^2.f_0)\tau/2 + I_L.V_0.\alpha_{lp}.\tau/2 \qquad (7)$$

where V_0 is the normal voltage, f_0 the normal frequency, and I_L is the leakage current. We assume that the leakage current reduces by a factor α_{lp} in the low power state. Now, if the core frequency for T2 is halved, i.e., core runs at $f_1 = f_0/2$, both tasks will finish at the same time. Let $V_1 < V_0$ denote the voltage compatible with the halved frequency. Then the energy consumption of T2 is given by

$$E_2' = (I_L.V_1 + 1/2.C.V_1^2.f_0/2)\tau \qquad (8)$$

Generally, we expect $E_2' < E_2$ if the frequency halving allows for significant voltage reduction. For instance, consider the earlier example where $f_0 = 2.0\,\text{GHz}$ with $V_0 = 1.2\,\text{V}$, $V_1 = 0.9\,\text{V}$, $I_L = 7.5\,\text{Amp}$, and effective Capacitance $(C) = 10\,\text{nanoFarad}$. Then, $E_1/t = 23.4\,\text{W}$ as before, but

$$E_2/\tau = (7.5 \times 1.2 + 5 \times 1.2^2 \times 2) \times 0.5 + 7.5 \times 1.2 \times 0.5 \times \alpha_{lp}$$

$$= 11.7 + 4.5\alpha_{lp}\,W$$

$$E_2'/\tau = (7.5 \times 0.9 + 5 \times 0.9^2 \times 1) = 10.8\,\text{W} \qquad (9)$$

It is seen that DVFS provides lower energy consumption (E_2') here irrespective of the value of α_{lp}. However, if the voltage reduction is limited to 1.1 V at the 1/2 frequency, we have $E_2'/\tau = 14.3\,\text{W}$. In this case, we need $\alpha_{lp} > 0.578$, for the DVFS control to beat the idle power control. In reality, we expect the α_{lp} to be much smaller than this threshold, and hence low power control may be preferred.

Energy Adaptation

Until now we have largely considered opportunistic reduction of energy consumption with as little impact on performance as possible. While this is very valuable, there are many situations where such an approach is inadequate. At the architectural level, the power and thermal densities continue to grow due to shrinking feature size and it is becoming necessary to limit the power consumption to ensure that the thermal limits are not exceeded. At the rack level, *power capping* may be necessary to ensure that the power circuits remain within their capacity. A typical reason for stress on power circuit capacity is that the physical infrastructure in a data center (e.g., racks, power distribution, cooling, etc.) is originally installed based on the requirements of the servers at that time. However, with increasing density of the servers, the power limits may be exceeded in the future, thereby requiring power capping. At the data center level, power capping may be necessary either due to use of renewable power supply (that naturally fluctuates) or due to the inadequacy of power and cooling infrastructure.

The worst case situation of power and cooling requirements in a data center will occur if all of the components of all the servers are simultaneously working at peak capacity. However, this is unrealistic and undesirable from the infrastructure cost perspective. Most data center servers run at a very low utilization typically, 10–20% range, and may hit 70–80% only a few times in a year. In fact, most data center operators will upgrade/expand their computing capacity much before it threatens to be a frequent bottleneck. Furthermore, even if the utilization of one component (say, CPU) goes to 100%, it is highly unlikely that the others (e.g., memory, network or the storage) could simultaneously be working at their maximum capacity. In particular, if the CPU is doing heavy IO or pulling in a lot of data from the memory, it will mostly be stalling, and not executing to the best of its capability.

In the so called "co-lo" (colocation) environments that are becoming popular, multiple companies lease and operate servers from a single server-farm operator. In such an environment, a simultaneous peak usage of all the servers belonging to various client companies is even less likely. Thus, it is highly desirable to design the power infrastructure in the data centers to a value that is substantially below the theoretical peak (often by a factor of 3 or more). The same goes for the cooling infrastructure – provisioning enough cooling capacity to handle the worst case heat generation situation is usually not sensible.

The underprovisioning of the power/cooling infrastructure does require the ability to handle those rare situations where the demand may exceed the supply. It is important to note here the relationship between power and cooling. If the cooling capacity is inadequate, the power consumption must be capped (even if there is no power constraint) so that no thermal emergencies are created. An intelligent power capping can adapt the system to the available power/cooling capacities without any significant impact on the performance. In fact, with an intelligent adaptation mechanism, it is possible and highly desirable to deliberately underdesign the power and cooling infrastructure so that a suitable balance between cost (both infrastructure and operational) and risk of violating quality of service (QoS) requirements is achieved. Such a tradeoff can result in huge cost savings with only a minor increase in the risk of degraded QoS.

The key difference between adaptation and the opportunistic energy management is that the former is a *mandatory reduction in energy consumption* and can only be done by compromising on some performance aspect such as delay or throughput. For this reason, it needs to be performed carefully, else the performance impact could be substantial. In particular, the energy deficit must be properly distributed among various resources with an eye towards the requirements and priorities of the applications using those resources. For example, consider a map-reduce application running on N servers. Such applications divide the work among all servers (map phase) and eventually collect/merge results (reduce phase). For the best performance, all servers should finish at about the same time, and under energy capping, it is necessary to allocate budgets so that this will be the case. A haphazard energy capping of the servers may result in significant imbalance and thus affect both the performance and the energy efficiency of the application.

A similar situation arises if multiple cores used to exploit thread level parallelism are energy capped – the energy budget of each core (and hence the DVFS controls used by it) should assure the balance. Finally, a balance is necessary across resources of various types as well. For example, running the memory or links at lower frequency while the cores are running at full frequency would result in CPU stalls and thus slower progress than if the energy budgets were balanced. Similarly, if the storage system in a data center is not power managed but the servers are, this would result in suboptimal performance, and hence less work done under energy limited situations than is possible.

It follows from the above discussion that there are two major issues in adaptation to energy limitations: (a) how to estimate suitable power budget for each system, subsystem and down to individual resources (CPU cores, links, DRAM channels, DRAMs, IO controllers, disks, etc.), and (b) how to apply power management in order to keep power consumption very close to the budget. For the first problem, we ideally want to assign power budgets so that it is possible to achieve the balance discussed above. Simple models that express power consumption as a linear equation $ax + b$ as a function of performance (x) are often used for this. (Here b represents the idle power and a is the power consumption per unit of work.) Although such an approach is often adequate, it is important to note that nonlinearities often arise when there is a resource bottleneck or contention for a shared resource. For example, while we may be able to use $ax + b$ type of expression for programs running individually on a machine, nonlinearities may arise when multiple such programs run concurrently on the same machine because of contention for cache or other resources.

An additional complication arises in energy adaptation due to variability in the workload itself. If the workload (e.g., rate or type of queries arriving at a server) varies, so will the energy consumption. Thus a budget computed initially cannot be kept constant even if the energy availability does not change; instead, it needs to vary with the workload as well. A too frequent a change is itself undesirable because a change in energy budget may require a change of the DVFS state, redirection of queries to other servers, or even migration of applications to other servers. A poorly designed control mechanism could result in ping-ponging, i.e., reduction of energy budget (and corresponding corrective actions) followed by an increase, followed by a reduction, etc. Such a behavior must be avoided.

Several of these issues have been explored in [7] which discusses the notion of "energy adaptive computing" in detail [8]. There are many other investigations related to energy capping; for example, Sharma et al. [9] propose a scheme to handle intermittent energy constraints by power cycling servers. The scheme is a purely power driven management scheme and independent of workload demands. An interesting aspect of this work is to pay attention to the energy surplus periods as well, when it may be possible to do additional computation, or speculatively prefetch data so that less data transfer activity takes place during the low energy periods.

Emerging Issues and Outlook

For a long time, the semiconductor industry thrived on the voltage reduction each time the technology moved to the next lower "feature size". This is known as "Dennard Scaling", introduced by Robert Dennard in 1974. The basic idea is as follows: Let's say we reduce the feature size by a factor of $\sqrt{2}$, which is typically how generations of semiconductor processes advance. An example of this is feature size going from 90 nm (nanometer) to 64 nm in late 1990s. Ideally, the reduction in the feature size reduces the capacitance C by the same factor and allows voltage reduction by $\sqrt{2}$ as well. Thus the overall dynamic power reduces by a factor of 2.0. Now, the feature size reduction allows twice as many transistors to be packed in the same chip, and we can raise the frequency by a factor of $\sqrt{2}$, which by itself would increase the power consumption by a factor of 2.0. But the two aspects combined result in *a new chip with same power consumption as before, but with twice as many transistors, each operating 1.414 times faster!*

Denard scaling started to break down in early 2000s and stopped around 2004. The reasons for this include the inability to lower voltage much, challenges in reducing transistor capacitance, difficulties in increasing frequency, and the difficult problem of increasing wire resistance and capacitance. This gave rise to growth in the "other dimension", i.e., the number of cores rather than performance of a single core. In fact, Moore's law was redefined to refer to the computing power of all the cores and thus continued unabated. Recently, even this trend has proved unsustainable. With the industry going for 7 nm technology next, the transistor widths are only a few tens of atoms wide (one nm = 3 Si atoms), which brings in numerous technological challenges including very high power densities (power dissipated per unit area). In fact, the technology is already at a point where we can easily put hundreds of CPU cores on a chip, but powering them all simultaneously is not possible.

Lowering the power density requires further lowering of the voltage. This is challenging since voltages are already close to the threshold required for reliable switching, and lowering them further introduces timing errors. Such errors must be detected and handled either by repeating the operation or by introducing additional delays to avoid the errors. This results in concerns of unreliability and residual "silent errors" in the computations – errors that somehow escape the mechanisms designed to catch them.

An added problem is that at such small feature sizes, it is extremely difficult to pattern each transister or wire accurately, which means that supposedly identical CPU cores, caches, logic units, etc. show considerable variability in their characteristics. This essentially makes the entire chip very heterogeneous, and poses challenges in using it optimally both in terms of its performance capabilities and power/thermal limits. In particular, some of the cores may run reliably only at much lower frequencies than the target frequency, while others may exceed their normal TDP (total dissipated power). In the past, such cores would be simply disabled, and the CPU sold with fewer cores, but this could reduce the yields considerably going forward.

We already discussed the need to minimize data movement for both performance and energy reasons, but effective techniques to simultaneously handle both data movement and computation remain unresolved. The traditional notions of algorithmic efficiency only concern computational complexity, rather than the data movement. A proper consideration of both requires new ways of designing algorithms and evaluating their complexity both in terms of performance and energy.

In short, the importance of power/energy considerations in both hardware and software design will continue to increase, and so will the additional challenges brought about by smaller feature sizes and low voltages. The main issues are high variability and increasing the chance of unreliability or "silent errors" which must be tackled along with the energy management.

References

1. G. Dhiman and T. S. Rosing, "Dynamic voltage frequency scaling for multi-tasking systems using online learning," in *ISLPED '07: Proceedings of the 2007 international symposium on Low power electronics and design*, 2007.
2. K. Kant, "A control scheme for batching dram requests to improve power efficiency," in *Proc. of ACM SIGMETRICS*, 2011, pp. 139–140.
3. ——, "Multi-state power management of communication links," in *Communication Systems and Networks (COMSNETS), 2011 Third International Conference on*. IEEE, 2011, pp. 1–10.
4. E. Pinheiro and R. Bianchini, "Energy conservation techniques for disk array-based servers," in *Proceedings of the 18th annual international conference on Supercomputing*, ser. ICS '04. New York, NY, USA: ACM, 2004.
5. D. Colarelli and D. Grunwald, "Massive arrays of idle disks for storage archives," in *Supercomputing '02: Proceedings of the 2002 ACM/IEEE conference on Supercomputing*. Los Alamitos, CA, USA: IEEE Computer Society Press, 2002, pp. 1–11.

6. M. Sabharwal, A. Agrawal, and G. Metri, "Enabling green it through energy-aware software," *IT Professional*, vol. 15, no. 1, pp. 19–27, Jan 2013.
7. K. Kant and M. Murugan and D. H. C. Du, "Willow: A Control System for Energy and Thermal Adaptive Computing," in *IPDPS '11*, 2011.
8. K. Kant, M. Murugan and D. H. C. Du, "Enhancing data center sustainability through energy adaptive computing," *ACM JETC (Special Issue)*, April 2012.
9. N. Sharma, S. Barker, D. Irwin, and P. Shenoy, "Blink: managing server clusters on intermittent power," in *Proceedings of the sixteenth international conference on Architectural support for programming languages and operating systems (ASPLOS)*, 2011.

Scheduling for Fault-Tolerance:
An Introduction

Guillaume Aupy and Yves Robert

Abstract Parallel execution time is expected to decrease as the number of processors increases. We show in this chapter that this is not as easy as it seems, even for perfectly parallel applications. In particular, processors are subject to faults. The more processors are available, the more likely faults will strike during execution. The main strategy to cope with faults in High Performance Computing is checkpointing. We introduce the reader to this approach, and explain how to determine the optimal checkpointing period through scheduling techniques. We also detail how to combine checkpointing with prediction and with replication.

Relevant core courses: Data Structures and Algorithms, Probabilities

Relevant PDC topics: Scalability in algorithms and architectures; Fault tolerance; Time

Context for use: Mid under-graduate curriculum. Having a minimal background in probabilities is better. The appendices are for students who are more advanced.

Learning outcomes: Comprehend that having access to more processors does not guarantee faster execution—introduce the notion of faults and easy algorithms to cope with faults

G. Aupy
Inria & Labri, University of Bordeaux, Bordeaux, France
e-mail: Guillaume.Aupy@inria.fr

Y. Robert (✉)
ENS Lyon, Lyon, France

University of Tennessee, Knoxville, TN, USA
e-mail: yves.robert@ens-lyon.fr; Yves.Robert@inria.fr

© Springer International Publishing AG, part of Springer Nature 2018   143
S. K. Prasad et al. (eds.), *Topics in Parallel and Distributed Computing*,
https://doi.org/10.1007/978-3-319-93109-8_6

Introduction

In this chapter, we present scheduling algorithms to cope with faults on large-scale parallel platforms. We study *checkpointing* and show how to derive the optimal checkpointing period. Then we explain how to combine checkpointing with *fault prediction*, and discuss how the optimal period is modified when this combination is used. And finally we follow the very same approach for the combination of checkpointing with *replication*. But wait. First, we have to help Alice out: she is having trouble with her laptop while writing her thesis.

Checkpointing on a Single Processor

Alice Needs Help

The most natural fault-tolerance technique when considering a fault-prone environment is to save your work periodically. This is what we (should) do in every-day's life. Alice is doing a very long and fastidious work: she is writing her PhD thesis, using an unreliable resource, namely a 4-year-old laptop. Because she is afraid of losing her precious work if the laptop crashes, she regularly saves her work on an external disk.

At first, because she knew that her laptop could not be trusted, Alice decided to save her work on the external disk every 3 h. Writing her file to disk takes approximatively 3 min. On the mid-afternoon of day 3, Alice's laptop crashed, she had to reboot it, and as a consequence she lost the last hour and a half of her work! Indeed, the crash happened right 90 min after her last saving on the external disk; she could have lost much more if, say, the crash had happened only 10 min before the next saving. Piqued, Alice decided that from now on, she would save her work on the external disk more frequently, every half hour of work instead of every 3 h. But after three additional days of work without further problem, she compared what she did during days 1, 2 and 3, and during days 4, 5 and 6. She noticed that she did less work on days 4, 5 and 6 than on days 1, 2 and 3 (even though she lost 90 min of work on the third day). Alice is puzzled now: what is the best frequency to save her work?

The technique of saving intermediate work is called *checkpointing*. Because Alice works for a constant amount of time between two checkpoints, her technique is called *periodic checkpointing*. In the following, we explain why she did more work during the three first days, and how she could find the best period between each checkpoint.

Modeling the Occurrence of Faults

Computing environments, such as Alice's laptop, are prone to faults. The first question is to quantify the rate or frequency at which these faults strike. To that purpose, one uses probability distributions, and more specifically, Exponential probability distributions. The definition of $Exp(\lambda)$, the Exponential distribution law of parameter λ, goes as follows:

- The probability density function is $f(t) = \lambda e^{-\lambda t} dt$ for $t \geq 0$;
- The cumulative distribution function is $F(t) = 1 - e^{-\lambda t}$ for $t \geq 0$;
- The mean is $\mu = \frac{1}{\lambda}$.

Consider a process executing in a fault-prone environment. The time-steps at which fault strike are non-deterministic, meaning that they vary from one execution to another. To model this, we use IID (Independent and Identically Distributed) random variables $X_1, X_2, X_3, \ldots$. Here X_1 is the delay until the first fault, X_2 is the delay between the first and second fault, X_3 is the delay between the second and third fault, and so on. All these random variables obey the same probability distribution $Exp(\lambda)$. We write $X_i \sim Exp(\lambda)$ to express that X_i obeys an Exponential distribution $Exp(\lambda)$.

Each random variable X_i has the same cumulative distribution function $F(t) = 1 - e^{-\lambda t}$: by definition, $F(t)$ gives the probability of the event $X_i < t$. In other words, $F(t) = \mathbb{P}(X_i < t)$ is the probability of having the next fault strike after a delay not larger than t. See Fig. 1 for the cumulative distribution function of $Exp(\frac{1}{6 \times 3,600})$. For simplicity, time is counted in hours in the figure, so that $\lambda = \frac{1}{6}$ and $\mu = 6$: in average, a fault will strike every 6 h. Reading the plot, we have $F(2) \approx 0.283$, which means that there is a 28% chance of having the next fault strike within 2 h.

We already observed that each random variable X_i has the same mean $\mathbb{E}(X_i) = \mu$. In average, a fault will strike every μ seconds. This is why μ is called the MTBF of the process, where MTBF stands for *Mean Time Between Faults*. The MTBF is

Fig. 1 Assuming $\lambda = 1/6$ (counting time in hours), the probability that a failure will strike within 2 h is $F(2) = \mathbb{P}(X < 2) = 1 - e^{-2/6} \approx 0.283$

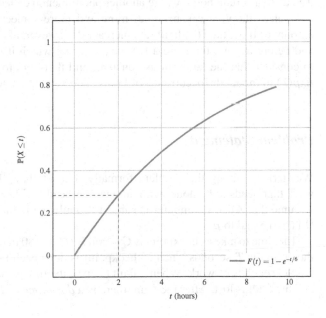

a key parameter to Alice's problem. One can show (see "Appendix 3: MTBF of a Platform with p Parallel Processors" for a proof) that the expected number of faults $N_{\text{faults}}(T)$ that will strike during T seconds is such that

$$\lim_{T \to \infty} \frac{N_{\text{faults}}(T)}{T} = \frac{1}{\mu} \tag{1}$$

Why are Exponential distribution laws so important? This is because of their *memoryless* property, which writes: if $X \sim Exp(\lambda)$, then $\mathbb{P}(X \geq t + s \mid X \geq s) = \mathbb{P}(X \geq t)$ for all $t, s \geq 0$. This equation means that at any instant, the delay until the next fault does not depend upon the time that has elapsed since the last fault. The memoryless property is equivalent to saying that the fault rate is constant. The fault rate at time t, RATE(t), is defined as the (instantaneous) rate of fault for the survivors to time t, during the next instant of time:

$$\text{RATE}(t) = \lim_{\Delta \to 0} \frac{F(t + \Delta) - F(t)}{\Delta} \times \frac{1}{1 - F(t)} = \frac{f(t)}{1 - F(t)} = \lambda = \frac{1}{\mu}$$

The fault rate is sometimes called a *conditional* fault rate since the denominator $1 - F(t)$ is the probability that no fault has occurred until time t, hence converts the expression into a conditional rate, given survival past time t.

We have discussed Exponential laws above, but other probability laws could be used. For instance, it may not be realistic to assume that the fault rate is constant: indeed, computers, like washing machines, suffer from a phenomenon called *infant mortality*: the probability of fault is higher in the first weeks than later on. In other words, the fault rate is not constant but instead decreasing with time. Well, this is true up to a certain point, where another phenomenon called *ageing* takes over: your computer, like your car, becomes more and more subject to faults after a certain amount of time: then the fault rate increases! However, after a few weeks of service and before ageing, there are a few years during which it is a good approximation to consider that the fault rate is constant, and therefore to use an Exponential law $Exp(\lambda)$ to model the occurrence of faults. The key parameter is the MTBF $\mu = \frac{1}{\lambda}$.

Problem Statement

We start by stating the problem formally. Let TIME$_{\text{base}}$ be the base time of the work that needs to be done, without any overhead (neither checkpoints nor faults). Assume that Alice's computer is subject to faults with a mean time between faults (MTBF) equal to μ.

The time to take a checkpoint is C seconds ($C = 180$ in the example). We say that the period is T seconds when a checkpoint is done each time Alice has completed $T - C$ seconds of work. When a fault occurs, the time between the last checkpoint and the fault is lost. After the fault, there is a *downtime* of D seconds to account for

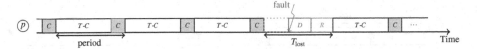

Fig. 2 An execution

the temporary unavailability (for example Alice's laptop is restarted, or the mouse is changed, or she now needs to use her brother Bob's laptop). Finally, in order to be able to resume the work, the content of the last checkpoint needs to be *recovered* which takes a time of R seconds (the external disk is connected and the checkpoint file is read). The sum of the time lost after the fault, of the downtime and of the recovery time is denoted T_{lost}. All these notations are depicted in Fig. 2.

Example

The difficulty of the problem is to trade-off between the time spent checkpointing, and the time lost in case of a fault. Consider an application such that $\text{TIME}_{\text{base}} = 30\,\text{min}$, and assume a checkpoint time of $C = 3\,\text{min}$, a downtime of $D = 1\,\text{min}$ and a recovery time of $R = 3\,\text{min}$.

We consider the following combinations:

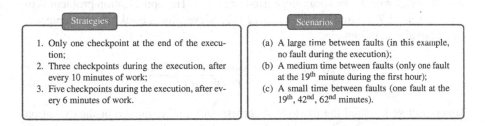

Strategies	Scenarios
1. Only one checkpoint at the end of the execution;	(a) A large time between faults (in this example, no fault during the execution);
2. Three checkpoints during the execution, after every 10 minutes of work;	(b) A medium time between faults (only one fault at the 19th minute during the first hour);
3. Five checkpoints during the execution, after every 6 minutes of work.	(c) A small time between faults (one fault at the 19th, 42nd, 62nd minutes).

In Fig. 3, we picture the execution of the application for the three different strategies, under the three different scenarios. This example shows that the lower the time between faults, the higher the frequency of checkpoints should be. However, the checkpointing strategy with the smallest period is not always the best one: sometimes, there are not enough faults to pay off the overhead of frequent checkpoints.

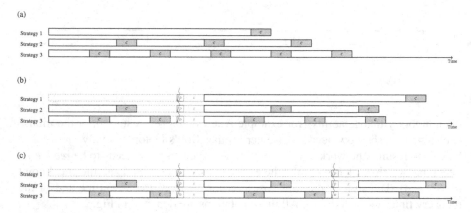

Fig. 3 The three strategies obtain different results depending upon the MTBF. (**a**) Large MTBF: there are no or very few faults. Checkpointing is too expensive. The first strategy wins. (**b**) Medium MTBF: there are more faults. It is good to checkpoint, but not too frequently, because of the corresponding overhead. The second strategy wins. (**c**) Small MTBF: there are many faults. The cost of the checkpoints is paid off because the time lost due to faults is dramatically reduced. The third strategy wins

Solution

Let $\text{TIME}_{\text{final}}(T)$ be the expectation of the total execution time of an application of size $\text{TIME}_{\text{base}}$ with a checkpointing period of size T. The optimization problem is to find the period T minimizing $\text{TIME}_{\text{final}}(T)$. However, for the sake of convenience, we rather aim at minimizing

$$\text{WASTE}(T) = \frac{\text{TIME}_{\text{final}}(T) - \text{TIME}_{\text{base}}}{\text{TIME}_{\text{final}}(T)}.$$

This objective is called the *waste* as it corresponds to the fraction of the execution time that does not contribute to the progress of the application (the time *wasted*). Of course minimizing the waste WASTE is equivalent to minimizing the total time $\text{TIME}_{\text{final}}$, because we have

$$(1 - \text{WASTE}(T))\,\text{TIME}_{\text{final}}(T) = \text{TIME}_{\text{base}},$$

but using the waste is more convenient. The waste varies between 0 and 1. When the waste is close to 0, it means that $\text{TIME}_{\text{final}}(T)$ is very close to $\text{TIME}_{\text{base}}$ (which is good), whereas, if the waste is close to 1, it means that $\text{TIME}_{\text{final}}(T)$ is very large compared to $\text{TIME}_{\text{base}}$ (which is bad).

First Source of Waste

Consider a *fault-free* execution of the application with periodic checkpointing. By definition, during each period of length T we take a checkpoint, which lasts for C time units, and only $T - C$ units of work are executed. Let TIME_{FF} be the execution time of the application in this setting. The fault-free execution time TIME_{FF} is equal to the time needed to execute the whole application, TIME_{base}, plus the time taken by the checkpoints:

$$\text{TIME}_{FF} = \text{TIME}_{base} + N_{ckpt}C,$$

where N_{ckpt} is the number of checkpoints taken. Additionally, we have

$$N_{ckpt} = \left\lceil \frac{\text{TIME}_{base}}{T - C} \right\rceil \approx \frac{\text{TIME}_{base}}{T - C}.$$

To discard the ceiling function, we assume that the execution time TIME_{base} is large with respect to the period or, equivalently, that there are many periods during the execution. Plugging back the (approximated) value $N_{ckpt} = \frac{\text{TIME}_{base}}{T-C}$, we derive that

$$\text{TIME}_{FF} = \frac{T}{T - C}\text{TIME}_{base}. \tag{2}$$

Similarly to the WASTE objective, the waste due to checkpointing in a fault-free execution, WASTE_{FF}, is defined as the fraction of the fault-free execution time that does not contribute to the progress of the application:

$$\text{WASTE}_{FF} = \frac{\text{TIME}_{FF} - \text{TIME}_{base}}{\text{TIME}_{FF}} \Leftrightarrow \left(1 - \text{WASTE}_{FF}\right)\text{TIME}_{FF} = \text{TIME}_{base}. \tag{3}$$

Combining Eqs. (2) and (3), we get:

$$\text{WASTE}_{FF} = \frac{C}{T}. \tag{4}$$

This result is quite intuitive: every T seconds, we waste C for checkpointing. This calls for a very large period in a fault-free execution (even an infinite period, meaning no checkpoint at all). However, a large period also implies that a large amount of work is lost whenever a fault strikes, as we discuss now.

Second Source of Waste

Consider the entire execution (with faults) of the application. Let TIME_{final} denote the expected execution time of the application in the presence of faults. This

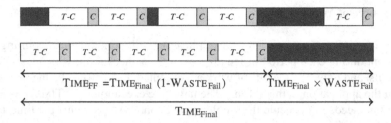

Fig. 4 An execution (top), and its re-ordering (bottom), to illustrate both sources of waste. Blackened intervals correspond to time lost due to faults: downtimes, recoveries, and re-execution of work that has been lost

execution time can be divided into two parts: (i) the execution of chunks of work of size $T - C$ followed by their checkpoint; and (ii) the time lost due to the faults. This decomposition is illustrated in Fig. 4. The first part of the execution time is equal to $\mathrm{TIME_{FF}}$. Let N_{faults} be the number of faults occurring during the execution, and let T_{lost} be the average time lost per fault. Then,

$$\mathrm{TIME_{final}} = \mathrm{TIME_{FF}} + N_{\mathrm{faults}} T_{\mathrm{lost}}.$$

In average, during a time $\mathrm{TIME_{final}}$, $N_{\mathrm{faults}} = \frac{\mathrm{TIME_{final}}}{\mu}$ faults happen (recall Eq. (1)). We need to estimate T_{lost} (see Fig. 2). A natural estimation for the moment when the fault strikes in the period is $\frac{T}{2}$. Intuitively, faults strike anywhere in the period, hence in average they strike in the middle of the period. Daly [6] give the proof of this result for Exponential distribution laws. We conclude that $T_{\mathrm{lost}} = \frac{T}{2} + D + R$, because after each fault there is a downtime and a recovery. This leads to:

$$\mathrm{TIME_{final}} = \mathrm{TIME_{FF}} + \frac{\mathrm{TIME_{final}}}{\mu}\left(D + R + \frac{T}{2}\right).$$

Let $\mathrm{WASTE_{fault}}$ be the fraction of the total execution time that is lost because of faults:

$$\mathrm{WASTE_{fault}} = \frac{\mathrm{TIME_{final}} - \mathrm{TIME_{FF}}}{\mathrm{TIME_{final}}} \Leftrightarrow (1 - \mathrm{WASTE_{fault}})\,\mathrm{TIME_{final}} = \mathrm{TIME_{FF}}$$

We derive:

$$\mathrm{WASTE_{fault}} = \frac{1}{\mu}\left(D + R + \frac{T}{2}\right). \tag{5}$$

Equations (4) and (5) show that each source of waste calls for a different period: a large period for $\mathrm{WASTE_{FF}}$, as already discussed, but a small period for $\mathrm{WASTE_{fault}}$, to decrease the amount of work to re-execute after each fault. Clearly, a trade-off is to be found. Here is how. By definition we have

$$\text{WASTE} = 1 - \frac{\text{TIME}_{\text{base}}}{\text{TIME}_{\text{final}}}$$

$$= 1 - \frac{\text{TIME}_{\text{base}}}{\text{TIME}_{\text{FF}}} \frac{\text{TIME}_{\text{FF}}}{\text{TIME}_{\text{final}}}$$

$$= 1 - (1 - \text{WASTE}_{\text{FF}})(1 - \text{WASTE}_{\text{fault}}).$$

Altogether, we derive the final result:

$$\text{WASTE} = \text{WASTE}_{\text{FF}} + \text{WASTE}_{\text{fault}} - \text{WASTE}_{\text{FF}}\text{WASTE}_{\text{fault}} \qquad (6)$$

$$= \frac{C}{T} + \left(1 - \frac{C}{T}\right)\frac{1}{\mu}\left(D + R + \frac{T}{2}\right). \qquad (7)$$

In Fig. 5, we plot WASTE as a function of the period T for a set of parameters. We obtain $\text{WASTE} = \frac{u}{T} + v + wT$, where $u = C\left(1 - \frac{D+R}{\mu}\right)$, $v = \frac{D+R-C/2}{\mu}$, and $w = \frac{1}{2\mu}$. It is easy to see that WASTE is minimized for $T = \sqrt{\frac{u}{w}}$. The First-Order (FO) formula for the optimal period is thus:

$$T_{\text{FO}} = \sqrt{2(\mu - (D + R))C}. \qquad (8)$$

and the optimal waste is $\text{WASTE}_{\text{FO}} = 2\sqrt{uw} + v$, therefore

$$\text{WASTE}_{\text{FO}} = \sqrt{\frac{2C}{\mu}\left(1 - \frac{D+R}{\mu}\right)} + \frac{D+R-C/2}{\mu}. \qquad (9)$$

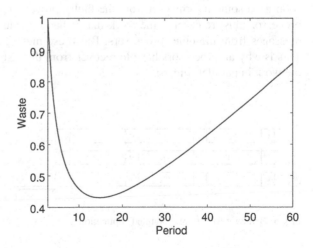

Fig. 5 Waste as a function of the period T, for $C = 3$, $D = 1$, $R = 3$ and $\mu = 40$. $T_{\text{FO}} \approx 14.7$. Shorter periods increase WASTE$_{\text{FF}}$ too much. Longer periods increase WASTE$_{\text{fault}}$ too much. T_{FO} achieves the best trade-off between both sources of waste

Finally, we show in "Appendix 1: First-Order Approximation of T_{FO} " why the computation above is a first order approximation.

In 1974, Young [18] obtained a different formula, namely $T_{\text{FO}} = \sqrt{2\mu C} + C$. Thirty years later, Daly [6] refined Young's formula and obtained $T_{\text{FO}} = \sqrt{2(\mu + R)C} + C$. Equation (8) is yet another variant of the formula, which we have obtained through the computation of the waste. There is no mystery, though. None of the three formulas is correct! They represent different first-order approximations, which collapse into the beautiful formula $T_{\text{FO}} = \sqrt{2\mu C}$ when μ is large in front of the resilience parameters D, C and R. This latter condition is the key to the accuracy of the approximation (see "Appendix 1: First-Order Approximation of T_{FO} "). Let us formulate our result as a theorem:

Theorem 1 *The optimal checkpointing period is* $T_{\text{FO}} = \sqrt{2\mu C} + o(\sqrt{\mu})$ *and the corresponding waste is* $\text{WASTE}_{\text{FO}} = \sqrt{\frac{2C}{\mu}} + o(\sqrt{\frac{1}{\mu}})$.

Theorem 1 has a wide range of applications. We discuss three of them in the following sections.

Checkpointing on a Parallel Platform

In this section, we deal with the problem of checkpointing a parallel application. We show how to reduce the optimization problem with p processors to the previous problem with only one processor. Most high performance applications are *tightly-coupled* applications, where each processor is frequently sending messages to, and receiving messages from the other processors. This implies that the execution can progress only when all processors are up and running. This also implies that when a fault strikes one processor, the whole application must be restarted from the last checkpoint. Indeed, even though the other processors are still alive, they will very soon need some information from the faulty processor. But to catch up, the faulty processor must re-execute the work that it has lost, during which it had received messages from the other processors. But these messages are no longer available. This is why all processors have to recover from the last checkpoint and re-execute the work in parallel (Fig. 6).

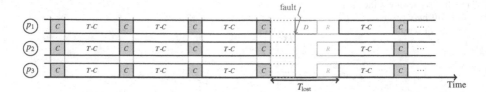

Fig. 6 Behavior for a tightly coupled application

(a)   (b)

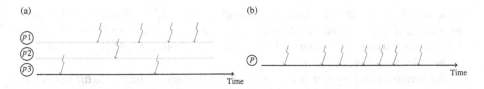

Fig. 7 Platform model: the super-processor replaces $p = 3$ processors. (a) Three faulty processors.... (b) ...make up for an equivalent even more faulty processor!

(a)

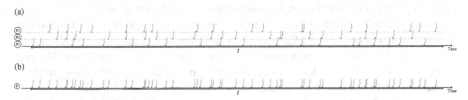

(b)

Fig. 8 Intuition of the proof of Proposition 1. (a) If three processors have around 20 faults during a time t ($\mu_{\text{ind}} = \frac{t}{20}$).... (b) ...during the same time, the equivalent processor has around 60 faults ($\mu = \frac{t}{60}$)

Let us recap. Each time a fault strikes somewhere on the platform, the application stops, all processors perform a downtime and a recovery, and they re-execute the work during a time T_{lost}. This sounds familiar. We can see the whole platform as a single *super-processor*, very powerful (its speed is p times that of individual processors) but also very prone to faults: all the faults strike this poor super-processor! See Fig. 7 for an illustration.

We can apply Theorem 1 to the super-processor and determine the optimal checkpointing period as $T_{\text{FO}} = \sqrt{2\mu C} + o(\sqrt{\mu})$, where μ now is the MTBF of the super-processor. How can we compute this MTBF? Have a look at Fig. 8. We see that the super-processor is hit by faults p times more frequently than the individual processors. We should then conclude that its MTBF is p times smaller than that of each processor. We state this result formally:

Proposition 1 *Consider a platform with p identical processors, each with MTBF μ_{ind}. Let μ be the MTBF of the platform. Then*

$$\mu = \frac{\mu_{ind}}{p} \tag{10}$$

Proof If the inter-arrival times of the faults on each individual processor are IID random variables (recall that IID means Independent and Identically Distributed) with probability distribution $Exp(\lambda)$ (where $\lambda = \frac{1}{\mu_{\text{ind}}}$), then the inter-arrival times of the faults on the super-processor are IID random variables with probability distribution $Exp(p\lambda)$, which will prove the result.

The arrival time of the first fault on the super-processor is a random variable $Y_1 \sim Exp(\lambda)$. This is because Y_1 is the minimum of $X_1^{(1)}, X_1^{(2)} \ldots, X_1^{(p)}$, where $X_1^{(i)}$

is the arrival time of the first fault on processor P_i. But $X_1^{(i)} \sim Exp(\lambda)$ for all i, and the minimum of p random variables following an Exponential distribution $Exp(\lambda_i)$ is a random variable following an Exponential distribution $Exp(\sum_{i=1}^{p} \lambda_i)$ (see the textbook by Ross [16, p. 288]).

The memoryless property of Exponential distributions is the key to the result for the delay between the first and second fault on the super-processor. Knowing that first fault occurred on processor P_1 at time t, what is the (conditional) probability distribution of a random variable for the occurrence of the first fault on processor P_2? This probability distribution is conditioned on the information that P_2 has been alive for t seconds. The memoryless property states that the probability distribution of the arrival time of the first fault on P_2 is not changed at all by when given this information! It is still an Exponential distribution $Exp(\lambda)$. Of course this holds true not only for P_2, but for each processor. And we can use the same minimum trick as for the first fault. Finally, the reasoning is the same for the third fault, and so on.

This concludes the proof. We refer the reader to "Appendix 3: MTBF of a Platform with p Parallel Processors" for another proof, where we also prove Eq. (1).

Proposition 1 shows that scale is the enemy of fault-tolerance. If we double up the number of components in the platform, we divide the MTBF by 2, and the minimum waste automatically increases by a factor $\sqrt{2} \approx 1.4$ (see Eq. (9)). And this assumes that the checkpoint time C remains constant. With twice as many processors, there is twice more data to write onto stable storage, hence the aggregated I/O bandwidth of the platform must be doubled to match this latter requirement.

Fault Prediction

A possible way to cope with the numerous faults and their impact on the execution time is to try and predict them. In this section we do not explain how this is done, although Gainaru et al. [10], Yu et al. [19] and Zheng et al. [21] provide more details for the interested reader.

A *fault predictor* (or simply a predictor) is a mechanism that warns the user about upcoming faults on the platform. More specifically, a predictor is characterized by two key parameters, its recall r, which is the fraction of faults that are indeed predicted, and its precision pr, which is the fraction of predictions that are correct (i.e., correspond to actual faults). In this section, we discuss how to combine checkpointing and prediction to decrease the platform waste.

We start with a few definitions. Let μ_{Pr} be the mean time between predicted events (both true positive and false positive), and μ_{NPr} the mean time between unpredicted faults (false negative). The relations between μ_{Pr}, μ_{NPr}, μ, r and pr are as follows:

- Rate of unpredicted faults: $\frac{1}{\mu_{NPr}} = \frac{1-r}{\mu}$, since $1 - r$ is the fraction of faults that are unpredicted;

- Rate of predicted faults: $\frac{r}{\mu} = \frac{pr}{\mu_{\text{Pr}}}$, since r is the fraction of faults that are predicted, and pr is the fraction of fault predictions that are correct.

To illustrate all these definitions, consider the time interval below and the different events occurring:

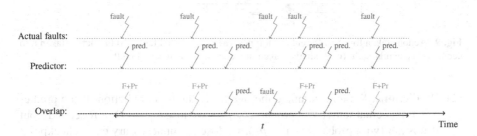

During this time interval of length t, the predictor predicts six faults, and there were five actual faults. One fault was not predicted. This gives approximatively: $\mu = \frac{t}{5}$, $\mu_{\text{Pr}} = \frac{t}{6}$, and $\mu_{\text{NPr}} = t$. For this predictor, the recall is $r = \frac{4}{5}$ (green "F+Pr" arrows over red "fault" arrows), and its precision is $pr = \frac{4}{6}$ (green "F+Pr" arrows over blue "pred." arrows).

Now, given a fault predictor of parameters pr and r, can we improve the waste? More specifically, how to modify the periodic checkpointing algorithm to get better results? In order to answer this question, we introduce *proactive checkpointing*: when there is a prediction, we assume that the prediction is given early enough so that we have time for a checkpoint of size C_{pr} (which can be different from C). We consider the following simple algorithm:

- While no fault prediction is available, checkpoints are taken periodically with period T;
- When a fault is predicted, we take a proactive checkpoint (of length C_{pr}) as late as possible, so that it completes right at the time when the fault is predicted to strike. After this checkpoint, we complete the execution of the period (see Fig. 9b, c);

We compute the expected waste as before. We reproduce Eq. (6) below:

$$\text{WASTE} = \text{WASTE}_{\text{FF}} + \text{WASTE}_{\text{fault}} - \text{WASTE}_{\text{FF}}\text{WASTE}_{\text{fault}} \qquad (11)$$

While the value of WASTE_{FF} is unchanged ($\text{WASTE}_{\text{FF}} = \frac{C}{T}$), the value of $\text{WASTE}_{\text{fault}}$ is modified because of predictions. As illustrated in Fig. 9, there are different scenarios that contribute to $\text{WASTE}_{\text{fault}}$. We classify them as follows:

(1) **Unpredicted faults:** This overhead occurs each time an unpredicted fault strikes, that is, on average, once every μ_{NPr} seconds. Just as in Eq. (5), the corresponding waste is $\frac{1}{\mu_{\text{NPr}}} \left[\frac{T}{2} + D + R \right]$.

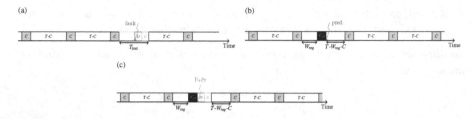

Fig. 9 Actions taken for the different event types. (**a**) Unpredicted fault. (**b**) Prediction taken into account – no actual fault. (**c**) Prediction taken into account – with actual fault

(2) Predictions: We now compute the overhead due to a prediction. If the prediction is an actual fault (with probability pr), we lose $C_{pr} + D + R$ seconds, but if it is not (with probability $1 - pr$), we lose the unnecessary extra checkpoint time C_{pr}. Hence

$$T_{\text{lost}} = pr(C_{pr} + D + R) + (1 - pr)C_{pr} = C_{pr} + pr(D + R)$$

We derive the final value of $\text{WASTE}_{\text{fault}}$:

$$
\begin{aligned}
\text{WASTE}_{\text{fault}} &= \frac{1}{\mu_{\text{NPr}}}\left(\frac{T}{2} + D + R\right) + \frac{1}{\mu_{\text{Pr}}}\left(C_{pr} + pr(D + R)\right) \\
&= \frac{1 - r}{\mu}\left(\frac{T}{2} + D + R\right) + \frac{r}{pr\,\mu}\left(C_{pr} + pr(D + R)\right) \\
&= \frac{1}{\mu}\left((1 - r)\frac{T}{2} + D + R + \frac{rC_{pr}}{pr}\right)
\end{aligned}
$$

We can now plug this expression back into Eq. (11):

$$
\begin{aligned}
\text{WASTE} &= \text{WASTE}_{\text{FF}} + \text{WASTE}_{\text{fault}} - \text{WASTE}_{\text{FF}}\text{WASTE}_{\text{fault}} \\
&= \frac{C}{T} + \left(1 - \frac{C}{T}\right)\frac{1}{\mu}\left(D + R + \frac{rC_{pr}}{pr} + \frac{(1 - r)T}{2}\right).
\end{aligned}
$$

To compute the value of T_{FO}^{pr}, the period that minimizes the total waste, we use the same reasoning as in section "Solution" and obtain:

$$T_{\text{FO}}^{pr} = \sqrt{\frac{2\left(\mu - \left(D + R + \frac{rC_{pr}}{pr}\right)\right)C}{1 - r}}.$$

We observe the similarity of this result with the value of T_{FO} from Eq. (8). If μ is large in front of the resilience parameters, we derive that $T_{\text{FO}}^{pr} = \sqrt{\frac{2\mu C}{1 - r}}$. This tells us that the recall is more important than the precision. If the predictor is capable of predicting, say, 84% of the faults, then $r = 0.84$ and $\sqrt{1-r} = 0.4$. The optimal period gets 2.5 times larger, and the waste is decreased by 60%. Prediction can help! See "Appendix 4: Going Further with Prediction" for further information.

Replication

Another possible way to cope with the numerous faults and their impact on the execution time is to use replication. Replication consists in duplicating all computations. Processors are grouped by pairs, such that each processor has a *buddy* (another processor performing exactly the same computations, receiving the same messages, etc.). See Fig. 10 for an illustration. We say that the two processes in a given pair are *replicas*. When a processor is hit by a fault, its buddy is not impacted. The execution of the application can still progress, until the buddy itself is hit by a fault later on. This sounds quite expensive: by definition, half of the resources are wasted (and this does not include the overhead of maintaining a consistent state between the two processors of each pair). At first sight, the idea of using replication on a large parallel platform is puzzling: who is ready to waste half of these expensive supercomputers?

In this section, we explain how replication can be used in conjunction with checkpointing and under which conditions it becomes profitable. In order to do this, we compare the checkpointing technique introduced earlier to the replication technique.

A *perfectly parallel application* is an application such that in a fault-free, checkpoint-free environment, the time to execute the application (TIME$_{\text{Base}}$) decreases linearly with the number of processors. More precisely:

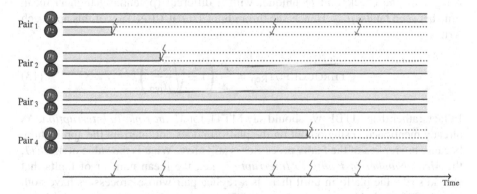

Fig. 10 Processor pairs for replication: each blue processor is paired with a red processor. In each pair, both processors do the same work

$$\text{TIME}_{\text{base}}(p) = \frac{\text{TIME}_{\text{base}}(1)}{p}.$$

Consider the execution of a perfectly parallel application on a platform with $p = 2P$ processors, each with individual MTBF μ_{ind}. As in the previous sections, the optimization problem is to find the strategy minimizing $\text{TIME}_{\text{final}}$. Because we compare two approaches using a different number of processors, we introduce the THROUGHPUT, which is defined as the total number of useful flops per second:

$$\text{THROUGHPUT} = \frac{\text{TIME}_{\text{base}}(1)}{\text{TIME}_{\text{final}}}$$

Note that for an application executing on p processors, THROUGHPUT $=$ $p\,(1 - \text{WASTE})$.

The *standard* approach, as seen before, is to use all $2P$ processors to fully parallelize the execution of the application on the platform. This would be optimal in a fault-free environment, but we are required to checkpoint frequently because faults repeatedly strike the p processors. According to Proposition 1, the platform MTBF is $\mu = \frac{\mu_{\text{ind}}}{p}$. According to Theorem 1, the waste is (approximately) WASTE $=$ $\sqrt{\frac{2C}{\mu}} = \sqrt{\frac{2Cp}{\mu_{\text{ind}}}}$. We have:

$$\text{THROUGHPUT}_{\text{Std}} = p \left(1 - \sqrt{\frac{2Cp}{\mu_{\text{ind}}}} \right) \tag{12}$$

The second approach uses *replication*. There are P pairs of processors, all computations are executed twice, hence only half the processors produce useful flops. One way to see the replication technique is as if there were half the processors using only the checkpoint technique, with a different (potentially higher) mean time between faults, μ_{rep}. Hence, the throughput THROUGHPUT$_{\text{Rep}}$ of this approach writes:

$$\text{THROUGHPUT}_{\text{Rep}} = \frac{P}{2} \left(1 - \sqrt{\frac{2C}{\mu_{\text{rep}}}} \right) \tag{13}$$

In fact, rather than MTBF, we should say MTTI, for *Mean Time To Interruption*. As already mentioned, a single fault on the platform does not interrupt the application, because the replica of the faulty processor is still alive. What is the value of *MNFTI*, the *Mean Number of Faults To Interruption*, i.e., the mean number of faults that should strike the platform until there is a replica pair whose processors have both been hit? If we find how to compute *MNFTI*, we are done, because we know that

$$\mu_{\text{rep}} = \mathit{MNFTI} \times \mu = \mathit{MNFTI} \times \frac{\mu_{\text{ind}}}{p}$$

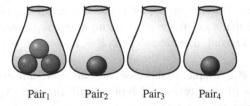

Pair₁   Pair₂   Pair₃   Pair₄

Fig. 11 Modeling the state of the platform of Fig. 10 as a balls-into-bins problem. We put a red ball in bin Pair$_i$ when there is a fault on its red processor p_1, and a blue ball when there is a fault on its blue processor p_2. As long as no bin has received a ball of each color, the game is on

We make an analogy with a balls-into-bins problem to compute *MNFTI*. The classical problem is the following: what is the expected number of balls that you will need, if you throw these balls randomly into P bins, until one bins gets two balls? The answer to this question is given by Ramanujan's Q-Function (see Flajolet [9]), and is equal to $\lceil q(P) \rceil$ where $q(P) = \frac{2}{3} + \sqrt{\frac{\pi P}{2}} + \sqrt{\frac{\pi}{288P}} - \frac{4}{135P} + \dots$. When $P = 365$, this is the birthday problem where balls are persons and bins are calendar dates; in the best case, one needs two persons; in the worst case, one needs $P + 1 = 366$ persons; on average, one needs $\lceil q(P) \rceil = 25$ persons.[1]

In the replication problem, the bins are the processor pairs, and the balls are the faults. However, the analogy stops here. The problem is more complicated, see Fig. 11 to see why. Each processor pair is composed of a blue processor and of a red processor. Faults are (randomly) colored blue or red too. When a fault strikes a processor pair, we need to know which processor inside that pair: we decide that it is the one of the same color as the fault. Blue faults strike blue processors, and red faults strike red processors. We now understand that we may need more than two faults hitting the same pair to interrupt the application: we need one fault of each color. The balls-and-bins problem to compute *MNFTI* is now clear: what is the expected number of red and blue balls that you will need, if you throw these balls randomly into P bins, until one bins gets both one red ball and one blue ball? To the best of our knowledge, there is no closed-form solution to answer this question, but a recursive computation does the job:

Proposition 2 *MNFTI* $= \mathbb{E}(NFTI|0)$ *where*

$$\mathbb{E}(NFTI|n_f) = \begin{cases} 2 & \text{if } n_f = P, \\ \frac{2P}{2P - n_f} + \frac{2P - 2n_f}{2P - n_f} \mathbb{E}\left(NFTI|n_f + 1\right) & \text{otherwise.} \end{cases}$$

Proof Let $\mathbb{E}(NFTI|n_f)$ be the expectation of the number of faults needed to interrupt the application, knowing that the application is still running and that faults

---

[1] As a side note, one needs only 23 persons for the probability of a common birthday to reach 0.5 (a question often asked in geek evenings).

have already hit n_f different processor pairs. Because each pair initially has 2 replicas, this means that n_f different pairs are no longer replicated, and that $P - n_f$ are still replicated. Overall, there are $n_f + 2(P - n_f) = 2P - n_f$ processors still running.

The case $n_f = P$ is simple. In this case, all pairs have already been hit, and all pairs have only one of their two initial replicas still running. A new fault will hit such a pair. Two cases are then possible:

1. The fault hits the running processor. This leads to an application interruption, and in this case $\mathbb{E}(NFTI|P) = 1$.
2. The fault hits the processor that has already been hit. Then the fault has no impact on the application. The *MNFTI* of this case is then: $\mathbb{E}(NFTI|P) = 1 + \mathbb{E}(NFTI|P)$.

The probability of fault is uniformly distributed between the two replicas, and thus between these two cases. Weighting the values by their probabilities of occurrence yields:

$$\mathbb{E}(NFTI|P) = \frac{1}{2} \times 1 + \frac{1}{2} \times (1 + \mathbb{E}(NFTI|P)),$$

hence $\mathbb{E}(NFTI|P) = 2$.

For the general case $0 \le n_f \le P - 1$, either the next fault hits a new pair, i.e., a pair whose 2 processors are still running, or it hits a pair that has already been hit, hence with a single processor running. The latter case leads to the same sub-cases as the $n_f = P$ case studied above. The fault probability is uniformly distributed among the $2P$ processors, including the ones already hit. Hence the probability that the next fault hits a new pair is $\frac{2P-2n_f}{2P}$. In this case, the expected number of faults needed to interrupt the application fail is one (the considered fault) plus $\mathbb{E}(NFTI|n_f + 1)$. Altogether we have:

$$\mathbb{E}(NFTI|n_f) = \frac{2P - 2n_f}{2P} \times (1 + \mathbb{E}(NFTI|n_f + 1))$$
$$+ \frac{2n_f}{2P} \times \left(\frac{1}{2} \times 1 + \frac{1}{2}(1 + \mathbb{E}(NFTI|n_f)) \right).$$

Therefore,

$$\mathbb{E}(NFTI|n_f) = \frac{2P}{2P - n_f} + \frac{2P - 2n_f}{2P - n_f} \mathbb{E}(NFTI|n_f + 1).$$

Let us compare the throughput of each approach with an example. From Eqs. (12) and (13), we have

$$\text{THROUGHPUT}_{\text{Rep}} \geq \text{THROUGHPUT}_{\text{Std}} \Leftrightarrow (1 - \sqrt{\frac{2Cp}{MNFTI\ \mu_{\text{ind}}}}) \geq 2(1 - \sqrt{\frac{2Cp}{\mu_{\text{ind}}}})$$

which we rewrite into

$$C \geq \frac{\mu_{\text{ind}}}{2p} \frac{1}{(2 - \frac{1}{\sqrt{MNFTI}})^2} \tag{14}$$

Take a parallel machine with $p = 2^{20}$ processors. This is a little more than one million processors, but this corresponds to the size of the largest platforms today. Using Proposition 2, we compute $MNFTI = 1284.4$. Assume that the individual MTBF is 10 years, or in seconds $\mu_{\text{ind}} = 10 \times 365 \times 24 \times 3600$. After some painful computations, we derive that replication is more efficient if the checkpoint time is greater than 293 s (around 6 min). This sets a target both for architects and checkpoint protocol designers.

Maybe you can say that $\mu_{\text{ind}} = 10$ years is pessimistic, because one would observe that $\mu_{\text{ind}} = 100$ years in current supercomputers. Because $\mu_{\text{ind}} = 100$ years allows to checkpoint up to 1 h, you would decide that replication is not worth it. But maybe you can also say that $\mu_{\text{ind}} = 10$ years is optimistic for processors equipped with thousands of cores and rather take $\mu_{\text{ind}} = 1$ year. In that case, unless you checkpoint in less than 30 s, better be prepared for replication. The beauty of performance models is that you can decide which approach is better *without bias nor a-priori*, simply by plugging your own parameters into Eq. (14).

Conclusion

In this chapter, we have dealt with fail-stop faults, i.e. faults that cause the application to crash and require to repair the resource or to find a spare one, and to re-execute work from some state of the application that had been previously saved. Other techniques involve to reconstruct the data lost by the failing processor from redundant information (e.g., checksums) maintained by the other processors. While unrecoverable, a fail-stop error has the nice characteristic that it can be detected immediately. On the contrary, a *silent error*, a.k.a. *silent data corruption*, gets unnoticed until it manifests after some random delay, e.g. because corrupted data is activated. Silent errors come from many sources, from errors in the arithmetic unit (due to low voltages) to bit flips in cache (due to cosmic radiation). Silent errors are difficult to detect, and because of the detection latency, they are even more difficult to correct. We refer the interested reader to studies such as Cappello et al. [4] or Gainaru et al. [11] to know more about the fascinating problems and solution techniques in the area of fault-tolerant computing at very large scale. See also the monograph [13] for a recent survey of fault-tolerant techniques for High Performance Computing.

Exascale computing (10^{18} operations per second, which require either one million processors, each with one thousand cores, or one hundred thousand processors, each with ten thousand cores) is a very large scale, but it is the scale of future-generation machines that will be with us in less than 10 years. Thus the area of resilience at scale is extremely important, and clever scheduling techniques are needed to help solve all the problems. Alice needs more help.[2]

Appendix 1: First-Order Approximation of T_{FO}

It is interesting to point out why the value of T_{FO} given by Eq. (8) is a first-order approximation, even for large jobs. Indeed, there are several restrictions for the approach to be valid:

- We have stated that the expected number of faults during execution is $N_{faults} = \frac{\text{TIME}_{final}}{\mu}$, and that the expected time lost due to a fault is $T_{lost} = \frac{T}{2} + D + R$. Both statements are true individually, but the expectation of a product is the product of the expectations only if the random variables are independent, which is not the case here because TIME_{final} depends upon the fault inter-arrival times.
- In Eq. (4), we have to enforce $C \leq T$ in order to have $\text{WASTE}_{EFF} \leq 1$.
- In Eq. (5), we have to enforce $D + R + \frac{T}{2} \leq \mu$ in order to have $\text{WASTE}_{fault} \leq 1$. We must cap the period to enforce this latter constraint. Intuitively, we need μ to be large enough for Eq. (5) to make sense. However, for large-scale platforms, regardless of the value of the individual MTBF μ_{ind}, there is always a threshold in the number of components p above which the platform MTBF, $\mu = \frac{\mu_{ind}}{p}$, becomes too small for Eq. (5) to be valid.
- Equation (5) is accurate only when two or more faults do not take place within the same period. Although unlikely when μ is large in front of T, the possible occurrence of many faults during the same period cannot be eliminated.

To ensure that the condition of having at most a single fault per period is met with a high probability, we cap the length of the period: we enforce the condition $T \leq \alpha\mu$, where α is some tuning parameter chosen as follows. The number of faults during a period of length T can be modeled as a Poisson process of parameter $\beta = \frac{T}{\mu}$. The probability of having $k \geq 0$ faults is $P(X = k) = \frac{\beta^k}{k!}e^{-\beta}$, where X is the random variable showing the number of faults. Hence the probability of having two or more faults is $\pi = P(X \geq 2) = 1 - (P(X = 0) + P(X = 1)) = 1 - (1+\beta)e^{-\beta}$. If we assume $\alpha = 0.27$ then $\pi \leq 0.03$, hence a valid approximation when bounding the period range accordingly. Indeed, with such a conservative value for α, we have

---

[2]By the way, there is a nice little exercise in "Appendix 6: Scheduling a Linear Chain of Tasks" if you are motivated to help.

overlapping faults for only 3% of the checkpointing segments in average, so that the model is quite reliable. For consistency, we also enforce the same type of bound on the checkpoint time, and on the downtime and recovery: $C \leq \alpha\mu$ and $D + R \leq \alpha\mu$. However, enforcing these constraints may lead to use a sub-optimal period: it may well be the case that the optimal period $\sqrt{2(\mu - (D + R))C}$ of Eq. (8) does not belong to the admissible interval $[C, \alpha\mu]$. In that case, the waste is minimized for one of the bounds of the admissible interval. This is because, as seen from Eq. (7), the waste is a convex function of the period.

We conclude this discussion on a positive note. While capping the period, and enforcing a lower bound on the MTBF, is mandatory for mathematical rigor, simulations in Aupy et al. [2] show that actual job executions can always use the value from Eq. (8), accounting for multiple faults whenever they occur by re-executing the work until success. The first-order model turns out to be surprisingly robust!

Appendix 2: Optimal Value of T_{FO}

There is a beautiful method to compute the optimal value of T_{FO} accurately. First we show how to compute the expected time $\mathbb{E}(\mathrm{TIME}(T - C, C, D, R, \lambda))$ to execute a work of duration $T - C$ followed by a checkpoint of duration C, given the values of C, D, and R, and a fault distribution $Exp(\lambda)$. If a fault interrupts a given trial before success, there is a downtime of duration D followed by a recovery of length R. We assume that faults can strike during checkpoint and recovery, but not during downtime.

Proposition 3

$$\mathbb{E}(\mathrm{TIME}(T - C, C, D, R, \lambda)) = e^{\lambda R}\left(\frac{1}{\lambda} + D\right)(e^{\lambda T} - 1).$$

Proof For simplification, we write TIME instead of $\mathrm{TIME}(T - C, C, D, R, \lambda)$ in the proof below. Consider the following two cases:

(i) Either there is no fault during the execution of the period, then the time needed is exactly T;

(ii) Or there is one fault before successfully completing the period, then some additional delays are incurred. More specifically, as seen for the first order approximation, there are two sources of delays: the time spent computing by the processors before the fault (accounted for by variable $\mathrm{TIME}_{\mathrm{lost}}$), and the time spent for downtime and recovery (accounted for by variable $\mathrm{TIME}_{\mathrm{rec}}$). Once a successful recovery has been completed, there still remain $T - C$ units of work to execute.

Thus TIME obeys the following recursive equation:

$$\text{TIME} = \begin{cases} T & \text{if there is no fault} \\ \text{TIME}_{\text{lost}} + \text{TIME}_{\text{rec}} + \text{TIME} & \text{otherwise} \end{cases} \tag{15}$$

$\text{TIME}_{\text{lost}}$ denotes the amount of time spent by the processors before the first fault, knowing that this fault occurs within the next T units of time. In other terms, it is the time that is wasted because computation and checkpoint were not successfully completed (the corresponding value in Fig. 2 is $T_{\text{lost}} - D - R$).
TIME_{rec} represents the amount of time needed by the system to recover from the fault (the corresponding value in Fig. 2 is $D + R$).

The expectation of TIME can be computed from Eq. (15) by weighting each case by its probability to occur:

$$\mathbb{E}(\text{TIME}) = \mathbb{P}\,(\text{no fault}) \cdot T + \mathbb{P}\,(\text{a fault strikes}) \cdot \mathbb{E}\,(\text{TIME}_{\text{lost}} + \text{TIME}_{\text{rec}} + \text{TIME})$$

$$= e^{-\lambda T} T + (1 - e^{-\lambda T}) \left(\mathbb{E}(\text{TIME}_{\text{lost}}) + \mathbb{E}(\text{TIME}_{\text{rec}}) + \mathbb{E}(\text{TIME})\right) ,$$

which simplifies into:

$$\mathbb{E}(\text{TIME}) = T + (e^{\lambda T} - 1) \left(E(\text{TIME}_{\text{lost}}) + E(\text{TIME}_{\text{rec}})\right) \tag{16}$$

We have $\mathbb{E}(\text{TIME}_{\text{lost}}) = \int_0^\infty x\mathbb{P}(X = x | X < T)dx = \frac{1}{\mathbb{P}(X<T)} \int_0^T xe^{-\lambda x}dx$, and $\mathbb{P}(X < T) = 1 - e^{-\lambda T}$. Integrating by parts, we derive that

$$\mathbb{E}(\text{TIME}_{\text{lost}}) = \frac{1}{\lambda} - \frac{T}{e^{\lambda T} - 1} \tag{17}$$

Next, the reasoning to compute $\mathbb{E}(\text{TIME}_{\text{rec}})$, is very similar to $\mathbb{E}(\text{TIME})$ (note that there can be no fault during D but there can be some during R):

$$\mathbb{E}(\text{TIME}_{\text{rec}}) = e^{-\lambda R}(D + R) + (1 - e^{-\lambda R})(D + \mathbb{E}(R_{lost}) + \mathbb{E}(\text{TIME}_{\text{rec}}))$$

Here, R_{lost} is the amount of time lost to executing the recovery before a fault happens, knowing that this fault occurs within the next R units of time. Replacing T by R in Eq. (17), we obtain $\mathbb{E}(R_{lost}) = \frac{1}{\lambda} - \frac{R}{e^{\lambda R}-1}$. The expression for $\mathbb{E}(\text{TIME}_{\text{rec}})$ simplifies to

$$\mathbb{E}(\text{TIME}_{\text{rec}}) = De^{\lambda R} + \frac{1}{\lambda}(e^{\lambda R} - 1)$$

Plugging the values of $\mathbb{E}(\text{TIME}_{\text{lost}})$ and $\mathbb{E}(\text{TIME}_{\text{rec}})$ into Eq. (16) leads to the desired value:

$$\mathbb{E}(\text{TIME}(T - C, C, D, R, \lambda)) = e^{\lambda R} \left(\frac{1}{\lambda} + D\right)(e^{\lambda T} - 1)$$

Proposition 3 is the key to proving that the optimal checkpointing strategy is periodic. Indeed, consider an application of duration TIME$_{\text{base}}$, and divide the execution into periods of different lengths T_i, each with a checkpoint as the end. The expectation of the total execution time is the sum of the expectations of the time needed for each period. Proposition 3 shows that the expected time for a period is a convex function of its length, hence all periods must be equal and $T_i = T$ for all i.

There remains to find the best number of periods, or equivalently, the size of each work chunk before checkpointing. With k periods of length $T = \frac{\text{TIME}_{\text{base}}}{k}$, we have to minimize a function that depends on k. This is easy for a skilled mathematician who knows the Lambert function $\mathbb{L}$ (defined as $\mathbb{L}(z)e^{\mathbb{L}(z)} = z$). She would find the optimal rational value k_{opt} of k by differentiation, prove that the objective function is convex, and conclude that the optimal value is either $\lfloor k_{opt} \rfloor$ or $\lceil k_{opt} \rceil$, thereby determining the optimal period T_{opt}. What if you are not a skilled mathematician? No problem, simply use T_{FO} as a first-order approximation, and be comforted that the first-order terms in the Taylor expansion of T_{opt} is ... T_{FO}! See Bougeret et al. [3] for all details.

Appendix 3: MTBF of a Platform with p Parallel Processors

In this section we give another proof of Proposition 1. Interestingly, it applies to any continuous probability distribution with bounded (nonzero) expectation, not just Exponential laws.

First we prove that Eq. (1) does hold true. Consider a single processor, say processor p_q. Let $X_i, i \geq 0$ denote the IID (independent and identically distributed) random variables for the fault inter-arrival times on p_q, and assume that $X_i \sim D_X$, where D_X is a continuous probability distribution with bounded (nonzero) expectation μ_{ind}. In particular, $\mathbb{E}(X_i) = \mu_{\text{ind}}$ for all i. Consider a fixed time bound F. Let $n_q(F)$ be the number of faults on p_q until time F. More precisely, the $(n_q(F))$-th fault is the last one to happen before time F or at time F, and the $(n_q(F) + 1)$-st fault is the first to happen after time F. By definition of $n_q(F)$, we have

$$\sum_{i=1}^{n_q(F)} X_i \leq F < \sum_{i=1}^{n_q(F)+1} X_i.$$

Using Wald's equation (see the textbook of Ross [16, p. 420]), with $n_q(F)$ as a stopping criterion, we derive:

$$\mathbb{E}\left(n_q(F)\right)\mu_{\text{ind}} \leq F \leq (\mathbb{E}\left(n_q(F)\right) + 1)\mu_{\text{ind}},$$

and we obtain:

$$\lim_{F \to +\infty} \frac{\mathbb{E}\left(n_q(F)\right)}{F} = \frac{1}{\mu_{\text{ind}}}. \tag{18}$$

As promised, Eq. (18) is exactly Eq. (1).

Now consider a platform with p identical processors, whose fault inter-arrival times are IID random variables that follow the distribution D_X. Unfortunately, if D_X is not an Exponential law, then the inter-arrival times of the faults of the whole platform, i.e., of the super-processor of section "Checkpointing on a Parallel Platform", are no longer IID. The minimum trick used in the proof of Proposition 1 works only for the first fault. For the following ones, we need to remember the history of the previous faults, and things get too complicated. However, we could still define the MTBF μ of the super-processor. Using Eq. (18), μ must satisfy:

$$\lim_{F \to +\infty} \frac{\mathbb{E}\left(n(F)\right)}{F} = \frac{1}{\mu},$$

where $n(F)$ is the number of faults on the super-processor until time F. But does the limit always exist? and if yes, what is its value?

The answer to both questions is not difficult. Consider a fixed time bound F as before. Let $n(F)$ be the number of faults on the whole platform until time F, and let $m_q(F)$ be the number of these faults that strike component number q. Of course we have $n(F) = \sum_{q=1}^{p} m_q(F)$. By definition, $m_q(F)$ is the number of faults on component q until time F. From Eq. (18) again, we have for each component q:

$$\lim_{F \to +\infty} \frac{\mathbb{E}\left(m_q(F)\right)}{F} = \frac{1}{\mu_{\text{ind}}}.$$

Since $n(F) = \sum_{q=1}^{p} m_q(F)$, we also have:

$$\lim_{F \to +\infty} \frac{\mathbb{E}\left(n(F)\right)}{F} = \frac{p}{\mu_{\text{ind}}} \tag{19}$$

which answers both questions at the same time!

The curious reader may ask how to extend Eq. (19) when processors have different fault-rates. Let $X_i^{(q)}$, $i \geq 0$ denote the IID random variables for the fault inter-arrival times on p_q, and assume that $X_i^{(q)} \sim D_X^{(q)}$, where $D_X^{(q)}$ is a continuous probability distribution with bounded (nonzero) expectation $\mu^{(q)}$. For instance if $\mu^{(2)} = 3\,\mu^{(1)}$, then (in expectation) processor 1 experiences three times more failures than processor 2. As before, consider a fixed time bound F, and let $n_q(F)$ be the number of faults on p_q until time F. Equation (18) now writes

$\lim_{F \to +\infty} \frac{\mathbb{E}(m_q(F))}{F} = \frac{1}{\mu^{(q)}}$. Now let $n(F)$ be the total number of faults on the whole platform until time F. The same proof as above leads to

$$\lim_{F \to +\infty} \frac{\mathbb{E}(n(F))}{F} = \sum_{q=1}^{p} \frac{1}{\mu^{(q)}} \tag{20}$$

Kella and Stadje [14] provide more results on the superposition of renewal processes (which is the actual mathematical name of the topic discussed here!).

Appendix 4: Going Further with Prediction

The discussion on predictions in section "Fault Prediction" has been kept overly simple. For instance when a fault is predicted, sometimes there is not enough time to take proactive actions, because we are already checkpointing. In this case, there is no other choice than ignoring the prediction.

Furthermore, a better strategy should take into account at what point in the period does the prediction occur. After all, there is no reason to always trust the predictor, in particular if it has a bad precision. Intuitively, the later the prediction takes place in the period, the more likely we are inclined to trust the predictor and take proactive actions. This is because the amount of work that we could lose gets larger and larger. On the contrary, if the prediction happens in the beginning of the period, we have to trade-off the probability that the proactive checkpoint may be useless (if we take a proactive action) with the small amount of work that may be lost in the case where a fault would actually happen (if we do not trust the predictor). The optimal approach is to never trust the predictor in the beginning of a period, and to always trust it in the end; the cross-over point $\frac{C_{pr}}{pr}$ depends on the time to take a proactive checkpoint and on the precision of the predictor. Details are provided by Aupy et al. [2] for details.

Finally, it is more realistic to assume that the predictor cannot give the exact moment where the fault is going to strike, but rather will provide an interval of time, a.k.a. a prediction window. Aupy et al. [1] provide more information.

Appendix 5: Going Further with Replication

In the context of replication, there are two natural options for "counting" faults. The option chosen in section "Replication" is to allow new faults to hit processors that have already been hit. This is the option chosen by Ferreira et al. [8], who introduced the problem. Another option is to count only faults that hit *running processors*, and thus effectively kill replica pairs and interrupt the application. This second option may seem more natural as the running processors are the only ones that are important for the application execution. It turns out that both options are almost equivalent, the values of their *MNFTI* only differ by one, as shown by Casanova et al. [5].

Speaking of faults, an important question is: why don't we repair (or rejuvenate) processors on the fly, instead of doing so only when the whole application is forced to stop, recover from the last checkpoint, and restart execution? The answer is technical: current HPC resource management systems assign the user a fixed set of resources for the execution, and do not allow new resources (such as spare nodes) to be dynamically added during the execution. In fact, frequently, a new configuration is assigned to the user at restart time. But nothing prevents us from enhancing the tools! It should then be possible to reserve a few additional nodes in addition to the computing nodes. These nodes would be used to migrate the system image of a replica node as soon as its buddy fails, in order to re-create the failed node on the fly. Of course the surviving node must be isolated from the application while the migration is taking place, in order to maintain a coherent view of both nodes, and this induces some overhead. It would be quite interesting to explore such strategies.

Here a few bibliographical notes about replication. Replication has long been used as a fault-tolerance mechanism in distributed systems (see the survey of Gartner [12]), and in the context of volunteer computing (see the work of Kondo et al. [15]). Replication has recently received attention in the context of HPC (High Performance Computing) applications. Representative papers are those by Schroeder and Gibson [17], Zheng and Lan [20], Engelmann, Ong, and Scorr [7], and Ferreira et al. [8]. While replicating all processors is very expensive, replicating only critical processes, or only a fraction of all processes, is a direction being currently explored under the name *partial replication*.

Speaking of critical processes, we make a final digression. The de-facto standard to enforce fault-tolerance in critical or embedded systems is *Triple Modular Redundancy*, or TMR. Computations are triplicated on three different processors, and if their results differ, a voting mechanism is called. TMR is not used to protect from fail-stop faults, but rather to detect and correct errors in the execution of the application. While we all like, say, safe planes protected by TMR, the cost is tremendous: by definition, two thirds of the resources are wasted (and this does not include the overhead of voting when an error is identified).

Appendix 6: Scheduling a Linear Chain of Tasks

In this exercise you are asked to help Alice (again). She is still writing her thesis but she does not want to checkpoint at given periods of time. She hates being interrupted in the middle of something because she loses concentration. She now wants to checkpoint only at the end of a chapter. She still has to decide after which chapters it is best to checkpoint.

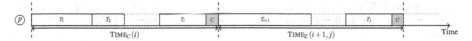

Fig. 12 Hint for the exercise

The difference with the original problem is that the checkpoints can only be taken at given time-steps. If we formulate the problem in a abstract way, we have a linear chain of n tasks (the n chapters in Alice's thesis), $T_1, T_2, \ldots, T_n$. Each task T_i has weight w_i (the time it takes to write that chapter). The cost to checkpoint after T_i is C_i. The time to recover from a fault depends upon where the last checkpoint was taken. For example, assume that T_i was checkpointed, and that T_{i+1}, T_{i+2} were not. If a fault strikes during the execution of T_{i+3}, we need to roll back and read the checkpoint of T_i from stable storage, which costs R_i. Then we start re-executing T_{i+1} and the following tasks. Note that the costs C_i and R_i are likely proportional to the chapter length).

As before, the inter-arrival times of the faults are IID random variables following the Exponential law $Exp(\lambda)$. We must decide after which tasks to checkpoint, in order to minimize the expectation of the total time. Figure 12 gives you a hint. $\text{TIME}_C(i)$ is the optimal solution for the execution of tasks $T_1, T_2, \ldots, T_i$. The solution to the problem is $\text{TIME}_C(n)$, and we use a dynamic programming algorithm to compute it. In the algorithm, we need to know $\text{TIME}_Z(i+1,j)$, the expected time to compute a segment of tasks $[T_{i+1} \ldots T_j]$ and to checkpoint the last one T_j, knowing that there is a checkpoint before the first one (hence after T_i) and that no intermediate checkpoint is taken. TIME_Z stands for *Zero intermediate checkpoint*. It turns out that we already know the value of $\text{TIME}_Z(i + 1, j)$: check that we have

$$\text{TIME}_Z(i + 1, j) = \mathbb{E}\left(\text{TIME}\left(\sum_{k=i+1}^{j} w_k, C_j, D, R_i, \lambda\right)\right)$$

and use Proposition 3.

References

1. G. Aupy, Y. Robert, F. Vivien, and D. Zaidouni. Checkpointing strategies with prediction windows. In *Dependable Computing (PRDC), 2013 IEEE 19th Pacific Rim International Symposium on*, pages 1–10. IEEE, 2013.
2. G. Aupy, Y. Robert, F. Vivien, and D. Zaidouni. Checkpointing algorithms and fault prediction. *Journal of Parallel and Distributed Computing*, 74(2):2048–2064, 2014.
3. M. Bougeret, H. Casanova, M. Rabie, Y. Robert, and F. Vivien. Checkpointing strategies for parallel jobs. In *Proceedings of SC'11*, 2011.
4. F. Cappello, A. Geist, B. Gropp, L. V. Kalé, B. Kramer, and M. Snir. Toward Exascale Resilience. *Int. Journal of High Performance Computing Applications*, 23(4):374–388, 2009.

5. H. Casanova, Y. Robert, F. Vivien, and D. Zaidouni. Combining process replication and checkpointing for resilience on exascale systems. Research report RR-7951, INRIA, May 2012.
6. J. T. Daly. A higher order estimate of the optimum checkpoint interval for restart dumps. *FGCS*, 22(3):303–312, 2004.
7. C. Engelmann, H. H. Ong, and S. L. Scorr. The case for modular redundancy in large-scale highh performance computing systems. In *Proc. of the 8th IASTED Infernational Conference on Parallel and Distributed Computing and Networks (PDCN)*, pages 189–194, 2009.
8. K. Ferreira, J. Stearley, J. H. I. Laros, R. Oldfield, K. Pedretti, R. Brightwell, R. Riesen, P. G. Bridges, and D. Arnold. Evaluating the Viability of Process Replication Reliability for Exascale Systems. In *Proc. of the ACM/IEEE SC Conf.*, 2011.
9. P. Flajolet, P. J. Grabner, P. Kirschenhofer, and H. Prodinger. On Ramanujan's Q-Function. *J. Computational and Applied Mathematics*, 58:103–116, 1995.
10. A. Gainaru, F. Cappello, and W. Kramer. Taming of the shrew: Modeling the normal and faulty behavior of large-scale hpc systems. In *Proc. IPDPS'12*, 2012.
11. A. Gainaru, F. Cappello, M. Snir, and W. Kramer. Failure prediction for hpc systems and applications: Current situation and open issues. *Int. J. High Perform. Comput. Appl.*, 27(3):273–282, 2013.
12. F. Gärtner. Fundamentals of fault-tolerant distributed computing in asynchronous environments. *ACM Computing Surveys*, 31(1), 1999.
13. T. Hérault and Y. Robert, editors. *Fault-Tolerance Techniques for High-Performance Computing*, Computer Communications and Networks. Springer Verlag, 2015.
14. O. Kella and W. Stadje. Superposition of renewal processes and an application to multi-server queues. *Statistics & probability letters*, 76(17):1914–1924, 2006.
15. D. Kondo, A. Chien, and H. Casanova. Scheduling Task Parallel Applications for Rapid Application Turnaround on Enterprise Desktop Grids. *J. Grid Computing*, 5(4):379–405, 2007.
16. S. M. Ross. *Introduction to Probability Models, Eleventh Edition*. Academic Press, 2009.
17. B. Schroeder and G. Gibson. Understanding failures in petascale computers. *Journal of Physics: Conference Series*, 78(1), 2007.
18. J. W. Young. A first order approximation to the optimum checkpoint interval. *Comm. of the ACM*, 17(9):530–531, 1974.
19. L. Yu, Z. Zheng, Z. Lan, and S. Coghlan. Practical online failure prediction for blue gene/p: Period-based vs event-driven. In *Dependable Systems and Networks Workshops (DSN-W)*, pages 259–264, 2011.
20. Z. Zheng and Z. Lan. Reliability-aware scalability models for high performance computing. In *Proc. of the IEEE Conference on Cluster Computing*, 2009.
21. Z. Zheng, Z. Lan, R. Gupta, S. Coghlan, and P. Beckman. A practical failure prediction with location and lead time for blue gene/p. In *Dependable Systems and Networks Workshops (DSN-W)*, pages 15–22, 2010.

Part II
For Students

MapReduce – The Scalable Distributed Data Processing Solution

Bushra Anjum

Abstract MapReduce is a programming paradigm used for processing massive data sets with a scalable and parallel approach on a cluster of distributed compute nodes. In this chapter we aim to provide background on the MapReduce programming paradigm and framework, highlighting its significance and usage for data crunching in today's scenario. Alongside, students will be introduced to important concepts such as Big Data, scalability, parallelization and divide & conquer. The chapter provides ample examples, both beginner level and advanced, for students to become proficient in recognizing problems suitable for a MapReduce solution and to define efficient Map and Reduce functions for those data sets.

Relevant core courses DS/A (Data Structures and Algorithms), CS2 (Second Programming Course in the Introductory Sequence)

Relevant PDC topics Why and what is parallel/distributed computing? (A), Concurrency (K/C), Cluster Computing (A), Scalability in algorithms and architectures (A), Speedup (C), Divide & conquer (parallel aspects) (A), Recursion (parallel aspects) (A), Scan (parallel-prefix) (K/C), Reduction (map-reduce) (K/A), Time (C/A), Sorting (K)

Context for use This chapter is a student centric resource and intended as supplementary material to any course focused on distributed systems and parallel algorithms. Students are expected to have a CS1 level basic knowledge going into this chapter.

Learning Outcomes After finishing this chapter, the student will be able to:

1. Recognize the growing need for scalable and distributed data processing solutions and how this need is addressed by the MapReduce paradigm.
2. Describe the MapReduce programming abstraction and runtime environment.

B. Anjum (✉)
Technical Lead and Senior Software Engineer Amazon, San Luis Obispo, CA, USA
e-mail: bushra.anjum@gmail.com

© Springer International Publishing AG, part of Springer Nature 2018   173
S. K. Prasad et al. (eds.), *Topics in Parallel and Distributed Computing*,
https://doi.org/10.1007/978-3-319-93109-8_7

3. Analyze the strengths and limitations of the MapReduce platform.
4. Define the problem characteristics that make it a candidate for a MapReduce solution.
5. Evaluate a given problem for its suitability to be solved using a MapReduce approach.
6. Transform a candidate problem into the map and reduce computing phases.
7. Outline the growing system of open source components, around the MapReduce framework, for large-scale data processing.

Background and Introduction

Professor Patrick Wolfe, executive director of the University College of London's Big Data Institute, has recently said in an interview with Business Insider [1], "The rate at which we're generating data is rapidly outpacing our ability to analyze it. The trick here is to turn these massive data streams from a liability into a strength."

Indeed digital data has been growing at an exponential rate, doubling every 2 years, and it is predicted [2] that by 2020 the digital universe will contain nearly as many bits as there are stars in the physical universe!

The increase of data opens up huge learning and analysis opportunities. However, it also comes with its unique set of challenge, e.g., How to store the data? How to process it economically in an acceptable amount of time? The complexity is not limited to the enormous size of the data, but also includes other dimensions such as the speed at which the data is produced, the various formats it comes in, etc. The popular term used to describe such datasets is "Big Data". Big Data is defined by the 'three Vs' of data: volume, variety, and velocity [3]. *Volume* refers to the size of the data, *Variety* refers to the fact that the data is often coming from a variety of different sources and in many different formats, and *Velocity* relates to the speed at which the data is being generated. Storing and processing these ever growing datasets require *scalable* storage and computing solutions. Let's define scalability next.

Scalability has been classically defined as the ability to process data even when it is larger than the available capacity of a server machine. One way to achieve this is to process the dataset piece by piece, i.e., to take the first piece of data, operate on it, write the results back then take the next piece of data, operate on it, write the results back and so on.

As an example, let's assume that we are given an extensive weather dataset consisting of millions of temperature, humidity, and air pressure readings. We would like to calculate the average temperature, average humidity, and average air pressure values over all the records in the dataset. Since the dataset is large and cannot fit inside a single server, we will analyze it by first breaking it down into smaller blocks and then working on it one block at a time. So we bring the first block of data in and calculate the three statistics (average temperature, average humidity, and average air pressure) on it. Then we bring the second block and calculate and update the three statistics to reflect the data present in both blocks. Then bring the third block,

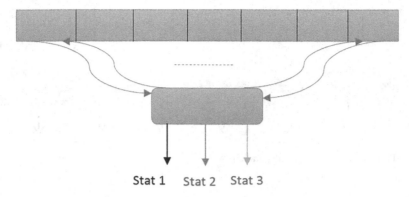

Fig. 1 Distributed processing without parallelism

and so on. After bringing in and processing the last block, we will have the average temperature, average humidity, and average air pressure statistics calculated over the entire dataset. This approach is shown in Fig. 1.

This block by block processing will give us the required results. Even so, a potential problem with this method is that as the size of the dataset increases, so does the overall processing time, as this approach is **serial** in nature. One way to scale the solution is to make this server machine better with a faster processor, more memory, etc. This approach is known as **"Scaling up"** (or vertical scaling), and it takes place through an improvement in the specification of a resource (e.g., upgrading a server with more main memory, larger hard drive, or a faster CPU, etc.). Still, there are limits to the upgrades that can be done to a computer system governed by the law of diminishing returns, price and power considerations, limiting Input/Output latencies, etc.

The alternate approach is to **parallelize** the solution. With this approach we could use multiple servers where each server works on an individual block of data, while all of them operate in parallel. This approach is called **"Scaling out"** (or horizontal scaling), and it takes place through an increase in the number of resources available to solve a problem (e.g., adding more hard drives to a storage system or adding more servers to support an application, etc.).

Scaling out is an excellent way to build Big Data applications, as it allows distribution of workloads to multiple servers operating in parallel. Thus by scaling out, we can use hundreds or even thousands of commodity servers and apply them all to the same problem. Such a collection of server machines, connected via hardware, networks, and software, and working in parallel on the same problem, is called a cluster.

Scaling out and parallelizing the solution is an attractive option, however, as a result of processing different data blocks on different server machines, now we have results distributed all over the cluster. In the context of our example, this means that we have an average temperature, average humidity, and average air pressure reading calculated over the first block on one server, an average temperature,

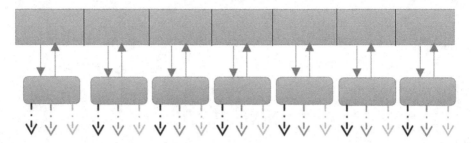

Fig. 2 Distributed processing with parallelism

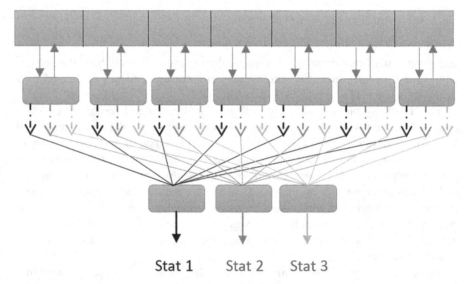

Stat 1   Stat 2   Stat 3

Fig. 3 Multi-stage distributed processing with parallelism

average humidity, and average air pressure reading calculated over the second block
on another server and so on. Whereas, we are interested in consolidated results
calculated over the entire dataset. This situation is depicted in Fig. 2 above.

Here is an interesting insight though, the key insight that led to the development
of the MapReduce paradigm. We already have individual statistics calculated per
block of data. Now, we can use more servers to parallelize the consolidation task
too! That is, take all those average temperatures, average humidity, and average
air pressure statistics calculated per block and combine them on additional server
machines, in parallel if possible. This added step is demonstrated in Fig. 3 above.

Hence scaling out is part of the solution, the other part is having distributed
algorithms that can run on these clusters and produce desired results. *MapReduce*
is one category of such distributed algorithms. It started when Google developed
and published on Google File System (GFS), a scalable and distributed data storage

solution in 2003 to store the large corpus of web crawling data [4]. Then in 2004 Google further presented the MapReduce framework to help search and find the insights from the data stored in the GFS [5]. Inspired by Google's proprietary GFS and MapReduce, their open source equivalents were developed by the Apache Software Foundation and became the Hadoop project [6]. Much like Google's MapReduce is layered on top of GFS, in Hadoop, MapReduce framework is layered on top of Hadoop Distributed File System (HDFS), a distributed fault tolerant storage facility. Hadoop and its variants are currently in use at Yahoo!, Facebook, Amazon and Google-IBM NSF clusters, to name a few.

We have almost described the MapReduce paradigm here. Let's look at it formally in the next section.

MapReduce

MapReduce is a scalable distributed data processing solution that works in collaboration with a massively scalable distributed file system, such as HDFS. HDFS is responsible for taking large datasets, dividing them into smaller blocks and storing them on individual nodes of a cluster while providing additional services like availability, fault tolerance, replication, persistence, etc. MapReduce framework, which is layered on top of HDFS, is responsible for bringing the computation to the data stored in these nodes and running it in parallel.

MapReduce consists of *two separate and distinct computation phases,* the Map phase and the Reduce phase. During the first phase, the framework runs a map function (also called a mapper) in parallel on the entire dataset stored in the HDFS. In the second phase, the framework runs a reduce function (also known as a reducer) on all the data produced by the mappers during the Map phase. The output from the reducers, which is the final result of the data processing job, is written back to the HDFS cluster. As the sequence of the name MapReduce implies, the reduce task is always performed after the map task is finished.

A map function is executed in parallel on each node in the HDFS cluster that is storing a block of the input data. The mapper reads the block of data one record or one line at a time, depending on the type of data. The data is read record by record if it is structured in nature such as originating from a database, or it may be read line by line, if the data is unstructured such as log files, social media streams, etc. The mapper then processes this record or line of data and outputs the results of its processing in a specialized format. The format is essentially a list consisting of key-value pairs, i.e., $\{(key_1, value_1), (key_2, value_2), \ldots, (key_n, value_n)\}$. For example, for our weather dataset, the mapper may emit the following three key-value pairs for each record of input read: **{('temperature': value_of_temperature_in_current_record), ('humidity': value_of_humidity_in_current_record), ('air pressure': value_of_airpressure_ in_ current_record)}**. Then the Reduce phase begins. Here several reducers work in parallel, each taking as input a 'subset' of key-value pairs produced by the map function and combining those into a final result. For weather data example, three reducers may work in parallel. The first reducer may combine all the key-value pairs that have the key 'temperature' into a single result (**'temperature':** **avg_value_of_temperature_of_all_records**). The second reducer may combine all the key-value pairs with the key 'humidity' into a single result (**humidity:** **avg_value_of_humidity_of_all_records**), and a third reducer may combine all the key-value pairs with the key 'air pressure' into a single result (**air pressure:** **avg_value_of_airpressure_of_all_records**). If the reader recalls, this is what we suggested in Fig. 3 above.

In general, a map and a reduce function can be defined by the following mappings:

$$map : value_{input} \rightarrow \left(key_{output}, value_{intermediate}\right)$$

$$reduce : (key_{output}, \{list\,(value_{intermediate})\}) \rightarrow value_{output}$$

The reader may be wondering how all the key-value pairs associated with a unique key end up at a single reducer? This functionality is provided by the "shuffle" phase of the underlying MapReduce framework.

After the mapper, and before the reducer, a background shuffle phase comes into play. It involves sorting the mapper outputs, combining all the key-value pairs with the same key into a list format {key, list(values)} and deciding on which reducers to send the list to for further processing. The shuffle phase assures that every key-value pair with the same key goes to the same reducer. It is important to mention here that a single reducer may process more than one list but, a list corresponding to a unique key is only handled by a single reducer.

Now is a good place to call out the difference between the MapReduce abstraction (also called the programming paradigm) and the MapReduce framework (also called the runtime system). As a user of MapReduce, we load the data into the HDFS and write the MapReduce abstractions, i.e., a 'serial' map function and a 'serial' reduce function, to process the data. The MapReduce system then takes care of everything else such as taking the map function and applying it in parallel to all the input blocks, shuffling intermediate results produced by the mappers and re-routing them to the appropriate reducers, running the reducers in parallel and writing the final output back to the HDFS. The MapReduce framework also provides distributed processing services such as scheduling, synchronization, parallelization, maintaining data and code locality, monitoring, failure recovery, etc. As far as the user is concerned, all this happens automatically. Therefore one of the strengths of MapReduce, and main contributor to its widespread popularity is the ability of the framework to separate the 'what' of distributed processing from the 'how'. The user focuses on the data problem they are trying to solve, and all the required aspects of distributed code execution are transparently handled for them by the framework.

Let's spend a little more time discussing the framework. The MapReduce framework uses the master-worker architecture. The master process is responsible for task scheduling, overall resource management, monitoring, and failure recovery.

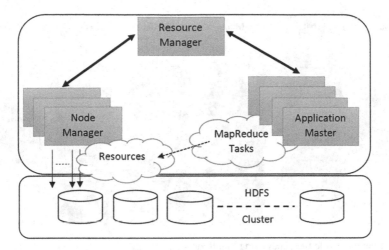

Fig. 4 YARN execution framework for Hadoop 2.0

The worker processes are responsible for managing per node resources and job executions. As an example, we will briefly discuss YARN here, the MapReduce execution framework for Hadoop v2 [7]. YARN, which stands for 'Yet Another Resource Negotiator,' is built on top of HDFS and has three essential elements, as shown in Fig. 4 above:

1. A singleton master process called the 'Resource Manager' (RM). The RM keeps track of the worker processes; which cluster node they are running at, how many resources they have available and how to assign those resources to the MapReduce tasks. RM accepts MapReduce job requests, allocate resources to the job and schedules the execution.
2. An 'Application Master' (AM) is spawned by the RM for every accepted MapReduce job request. AM has the responsibility of negotiating appropriate resources from the RM, starting the map and reduce tasks on the assigned resources and monitoring for progress.
3. A per node (or per group of nodes) worker process called the 'Node Manager' (NM) is responsible for announcing itself to the RM along with its available resources (memory, cores) and sending periodic updates.

We will shift our focus back to the MapReduce programming paradigm, which is the original intent of this chapter. Let's take the classic example of counting word frequency and see how it can efficiently be solved using the MapReduce approach.

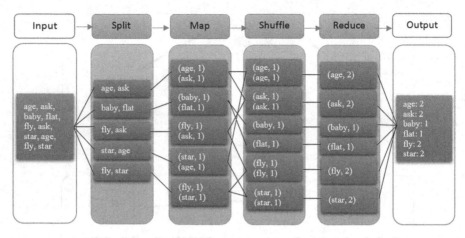

Fig. 5 Counting world frequency using MapReduce

Example: Counting Word Frequency

The 'Word Count' example is the 'Hello World' equivalent of the MapReduce paradigm. In this example, we count the number of occurrences of each unique word in an input dataset, possibly a huge dataset, which consists of text files.

The first step is to split the input files into smaller blocks and to store each block on a distinct node in a distributed cluster with the help of an HDFS. The mapper then looks at the block of text, one line at a time, and emits each word with a count of 1, i.e., the map function output is the key-value pair (word$_i$, 1). It is primarily marking the word as being seen once. All the mappers operate in parallel on the cluster nodes and emit similar key-value pairs. Next, the shuffle phase collects all the words emitted by the mappers, sorts them alphabetically, makes a list for each unique word and sends each list to a reducer. The output of the shuffle phase looks like:{(word$_1$, 1), (word$_1$, 1), ..., (word$_1$, 1)}, {(word$_2$, 1), (word$_2$, 1), ..., (word$_2$, 1)}, ..., {(word$_n$, 1), (word$_n$, 1), ..., (word$_n$, 1)}. The reducer then sums the number of occurrences in the input list and emits that value as the final result. The MapReduce mappings of this problem is given below and a pictorial depiction of the various phases in Fig. 5.

$$map : word_1 word_2 \ldots word_n \rightarrow \{(word_1, 1), (word_2, 1), \ldots (word_n, 1)\}$$

$$reduce : (word_i, \{1, 1, \ldots 1\}) \rightarrow word_i : \sum\nolimits_{All} 1$$

Let's look at another example, where we combine dictionaries using the MapReduce distributed programming paradigm.

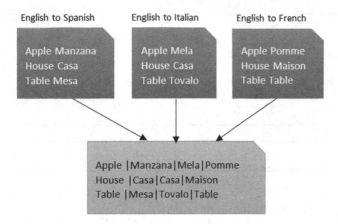

Fig. 6 Combining dictionaries using MapReduce

Example: Combining Dictionaries

In this example, we will take a set of translation dictionaries, English-Spanish, English-Italian, English-French, and create a dictionary file that has the English word followed by all the different translations separated by the pipe (|) character. For example, looking at Fig. 6, if the input files are as shown in the top boxes, we expect the final output as shown in the box below. This example is a modified version of the dictionary example discussed at the DZone blog [8].

In this example, each dictionary will be parsed by a mapper (or a set of mappers) line by line, emitting each English word and its corresponding translation as the output of the map function. The reducer, with the help of the shuffle phase, will then receive all the translations related to a particular word and combine them into the final output. We present these mappings below.

$$map : word\ translation \rightarrow (word, translation)$$

$$reduce : (word_i, \{list\ (translation)\}) \rightarrow word_i|translation_1| \dots |translation_n$$

For this example, we also present simplified code excerpts of a Java-based implementation of the map and reduce functions. The complete code, along with running instructions can be found at [7].

```
public void map(String key_word, String value
  _translation, Context context)
{
   context.write(key_word, value_translation);
}
```

```
public void reduce(String key_word, Iterable<String>
                   values, Context context)
{
   String translations = "";
   for (String value_translation : values)
   {
      translations += "|" + value_translation;
   }
   context.write(key_word, translations);
}
```

We have provided several other examples at the end of this chapter.

Strengths and Limitations of MapReduce

MapReduce is a programming model (and execution framework) for processing large datasets distributed across a cluster with a parallel, distributed algorithm. It has found merits in many applications, such as recommendation systems, processing of logs, marketing analysis, warehousing of data, fraud detection etc.

MapReduce is ideal for running batch computations over large datasets as it can easily scale to hundreds, even thousands, of server nodes. The framework takes the computation to the data rather than bringing the data from various cluster nodes to a centralized processing location. Running mappers on the same node as the data block achieves data locality, consequently conserving precious network bandwidth and allowing for faster processing [9]. The framework is designed to both take advantage of massive parallelism while at the same time hiding messy internal details (parallelization, synchronization, failure recovery, etc.) from the programmer.

Another advantage of the MapReduce programming paradigm is its flexibility. MapReduce programming has the capability to operate on different types and sources of data, whether they are structured (database records) or unstructured (from social media, email, or clickstream, etc.). MapReduce can work on all of them with the help of the various input processing libraries available with the framework.

MapReduce framework is built to be both available and resilient. The underlying HDFS ensures that when data is sent to an individual node in the entire cluster, the very same set of data is replicated at other numerous nodes that make up the cluster. Thus, if there is any failure that affects a particular node, there are always other copies that can still be accessed whenever the need may arise. This replication always assures the availability of data. On top of that MapReduce framework has baked in fault tolerance. In a distributed system, failures are a norm. Anything from a processing node to a network connection to a storage disk may fail at any time. In fact this is one of the prime responsibilities of the Master process. The framework can quickly recognize failures that occur and then apply a quick and automatic recovery solution.

MapReduce works well in its domain, offline batch processing, however, it is less effective outside of it. For example, MapReduce is not an ideal solution for tasks that need a shared state or global coordination. MapReduce does not support shared mutable state. The technique is, in general, *embarrassingly parallel*. There is only a single opportunity for a global synchronization in MapReduce which is after the map phase ends and before the reduce phase begins [10].

Also, as it is designed for large and distributed datasets, the performance is not ideal when it operates on small datasets or individual records. The MapReduce framework has considerable startup and execution costs such as setting up the parallel environment, task creation, communication, synchronization, etc. These overheads are usually negligible as the framework is optimized to conduct batch operations over a large amount of data. However, for smaller problems, it is probably going to be faster to process the data serially on a single fast processor than use a distributed MapReduce system.

MapReduce is not ideal for real-time processing of data, or iterative and interactive applications either. This is because both Iterative and Interactive applications require faster data sharing across parallel jobs. Unfortunately, in most current frameworks, the only way to reuse data between computations (Ex – between two MapReduce jobs) is to write it to an external stable storage system. A framework like YARN can enable such applications by scheduling multiple tasks, however, data sharing remains slow in MapReduce due to replication, serialization, and disk IO. Apache Software Foundation introduced Spark for speeding up the Hadoop computational computing software process [11]. The main feature of Spark is its in-memory cluster computing that increases the processing speed of an application. It is used to model efficiently several other types of computations such as Interactive Queries and Stream Processing.

The Hadoop-MapReduce Ecosystem

The Hadoop platform primarily consists of two essential services: a reliable, distributed file system called HDFS and the high-performance parallel data processing engine called MapReduce. Though they form the core of the Hadoop project, they are just two parts of a growing system of open source components for large-scale data processing. Below we briefly discuss some of the related technologies. The reader is encouraged to look at the references for more details.

Hive [12] was originally developed at Facebook for business analysts to be able to access data on Hadoop using an SQL-like engine. Hive offers techniques to map a tabular structure onto a distributed file system like HDFS, and also allows querying of the data from this mapped tabular structure using an SQL dialect known as HiveQL. HiveQL queries are executed via MapReduce, i.e., when a HiveQL query is issued, it triggers Map and Reduce tasks to perform the operation specified in the query.

Pig [13], developed at Yahoo, is a platform for constructing data flows for extract, transform, and load (ETL) processing and analysis of large datasets. Pig uses a high-level scripting language called Pig Latin. Pig Latin queries and commands are compiled into one or more MapReduce tasks and then executed on a Hadoop cluster.

Where Hive was developed to process completely structured data, Pig can be used for both structured as well as unstructured data (a pig will eat anything!). Both Pig and Hive queries get converted into MapReduce tasks under the hood.

The MapReduce framework is at its best when the data is huge, and we want to batch process it offline. However, it is not suitable for real-time processing or random read and write accesses. It led to the development of Apache HBase [14], the distributed, scalable, NoSQL database for Hadoop, built on top of HDFS, that is great for quick updates and low latency data accesses. HBase is a column-oriented store and runs on top of HDFS in a distributed fashion. HBase can provide fast, random read/write access to users and applications in near real-time.

Mahout [15] is the machine learning and data mining library for Hadoop. It implements machine learning and data mining algorithms, such as collaborative filtering, classification, clustering and dimensionality reduction using MapReduce.

Oozie [16], developed at Yahoo, is a workflow coordination service to coordinate, schedule and manage tasks executed on Hadoop. The tasks are represented as action nodes on a Directed Acyclic Graph (DAG), and the DAG sequence is used to control the subsequent actions. You can have several different action nodes within your Oozie workflows such as steps for chaining events, Pig and Hive tasks, MapReduce tasks or HDFS actions.

These Hadoop components are presented in Fig. 7 below.

Apart from those, there are various other Apache Projects built around the Hadoop framework and have become a part of the Hadoop Ecosystem. For a complete list, see [17].

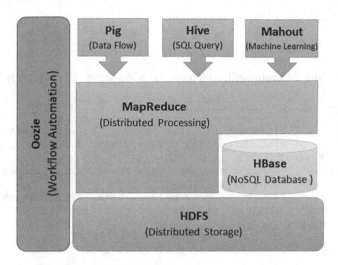

Fig. 7 Simplified Hadoop-MapReduce ecosystem

Additional Examples

We wrap up this chapter by providing additional MapReduce examples.

Example: Inverted Index

An inverted index consists of a list of all the unique words that appear in any document, and for each word, a list of the documents in which it appears. The inverted index is useful for fast retrieval of relevant information. Let's look at building an inverted index for a set of tweets based on their hashtags and how we can map the solution as a MapReduce.
Input Data:

```
"It's not too late to vote. #ElectionDay"
"Midtown polling office seeing a steady flow of voters!
  #PrimaryDay"
"Today's the day. Be a voter! #ElectionDay"
"Happy #PrimaryDay"
"Say NO to corruption & vote! #ElectionDay"
"About to go cast my vote...first time #ElectionDay"
```

MapReduce mapping:

$$map : tweet \rightarrow (hashtag, tweet)$$

$$reduce : (hashtag, \{list(tweet)\}) \rightarrow hashtag, \{list(tweet)\}$$

Map Output:

```
("ElectionDay", "It's not too late to vote. #ElectionDay")
("PrimaryDay", "Midtown polling office seeing a steady flow of
   voters! #PrimaryDay")
("ElectionDay", "Today's the day. Be a voter! #ElectionDay ")
("PrimaryDay", "Happy #PrimaryDay")
("ElectionDay", "Say NO to corruption & vote! #ElectionDay")
("ElectionDay", "About to go cast my vote...first time
   #ElectionDay")
```

Reduce Input:

Reducer 1:

```
("ElectionDay", "It's not too late to vote. #ElectionDay")
("ElectionDay", "Today's the day. Be a voter! #ElectionDay ")
("ElectionDay", "Say NO to corruption & vote! #ElectionDay")
("ElectionDay", "About to go cast my vote...first time
   #ElectionDay")
```

Reducer 2:

```
("PrimaryDay", "Midtown polling office seeing a steady flow of
   voters! #PrimaryDay")
("PrimaryDay", "Happy #PrimaryDay")
```

Reduce Output:

```
("ElectionDay", [ "It's not too late to vote.
                  #ElectionDay",
                  "Today's the day. Be a voter!
                  #ElectionDay ",
                  "Say NO to corruption & vote!
                  #ElectionDay",
                  "About to go cast my vote...first time
                  #ElectionDay"])

("PrimaryDay", [ "Midtown polling office seeing a
                 steady flow of voters! #PrimaryDay",
                 "Happy #PrimaryDay"])
```

Example: Relational Algebra (Table JOIN)

MapReduce can be used to join two database tables based on common criteria. Let's take an example. We have two tables, where the first contains an employee's personal information primary keyed on SSN and the second table includes the employee's income again keyed on SSN. We would like to compute average income in each city in 2016. This computation requires a JOIN operation on these two tables. We will map the problem to a two-phase MapReduce solution. The first phase effectively creates a JOIN on the two tables using two map functions (one for each table), and the second phase gathers the relevant data for calculating desired statistics.

Input Data:

        Table 1: (SSN, {Personal Information})
                111222:(Stephen King; Sacramento, CA)
                333444:(Edward Lee; San Diego, CA)
                555666:(Karen Taylor; San Diego, CA)
        Table 2: (SSN, {year, income})
                111222:(2016,$70000),(2015,$65000),(2014,$6000),...
                333444:(2016,$72000),(2015,$70000),(2014,$6000),...
                555666:(2016,$80000),(2015,$85000),(2014,$7500),...

MapReduce Mapping:

Stage 1 (table JOIN)

$$map_{table1} : record_{table1} \rightarrow (SSN, City)$$

$$map_{table2} : record_{table2} \rightarrow (SSN, Income_{2016})$$

$$reduce : (SSN, \{City, Income_{2016}\}) \rightarrow SSN, (City, Income_{2016})$$

Stage 2

$$map : SSN, (City, Income_{2016}) \rightarrow (City, Income_{2016})$$

$$reduce : City, \{list (Income_{2016})\} \rightarrow City, avg(Income_{2016})$$

Stage 1
Map Output:

Mapper 1a: (SSN, city)

    (111222, "Sacramento, CA")
    (333444, "San Diego, CA)
    (555666, "San Diego, CA)

Mapper 1b: (SSN, income 2016)

    (111222, $70000)
    (333444, $72000)
    (555666, $80000)

Reduce Input: (SSN, city), (SSN, income)

    (111222, "Sacramento, CA")
    (111222, $70000)
    (333444, "San Diego, CA")
    (333444, $72000)
    (555666, "San Diego, CA")
    (555666, $80000)

Reduce Output: (SSN, [city, income])

    (111222, ["Sacramento, CA", 70000])
    (333444, ["San Diego, CA", 72000])
    (555666, ["San Diego, CA", 80000])

Stage 2:

Map Input: (SSN, [city, income])

    (111222, ["Sacramento, CA", 70000])
    (333444, ["San Diego, CA", 72000])
    (555666, ["San Diego, CA", 80000])

Map Output: (city, income)

    ("Sacramento, CA", 70000)
    ("San Diego, CA", 72000)
    ("San Diego, CA", 80000)

Reduce Input: (city, income)

Reducer 2a:

    ("Sacramento, CA", 70000)

Reducer 2b:

    ("San Diego, CA", 72000)
    ("San Diego, CA", 80000)

Reduce Output: (city, average [income])

Reducer 2a:

    ("Sacramento, CA", 70000)

Reducer 2b:

    ("San Diego, CA", 76000)

The reader is encouraged to think how the solution will differ if the employee is allowed to have multiple addresses, i.e., there can be multiple addresses per SSN in Table 1.

Advanced Example: Graph Algorithm (Single Source Shortest Path)

This example assumes that the reader has familiarity with the graph algorithm terminology, such as vertices, edges, adjacency lists, etc. MapReduce can be used to calculate statistics iteratively where each iteration can use the previous iteration's

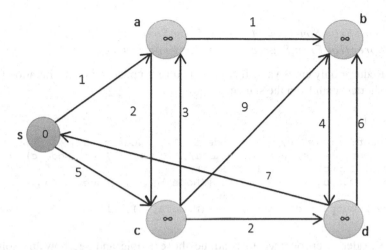

Fig. 8 Input graph for single source shortest path problem

output as its input. This kind of iterative MapReduce is useful for applications including graph problems. For example, given Fig. 8, we would like to calculate the single source shortest path from source vertex 's' to all other vertices in the graph. The shortest path is defined as a path between two vertices in a graph such that the sum of the weights of its constituent edges is minimized.

We will be using MapReduce iterative approach to solve this problem, where each iteration, starting from the origin, will be 'radiating' information 'one edge hop' distance at a time.

Input Data Format:

$$Node :< id, cost From Source, prev Hop From Source, Adjacency List >$$

$$Adjacency List : \{neighbor Node, cost To Neighbor Node\}$$

Initial Input Data:

```
Node s: <s, 0, - , {(Node a, 1), (Node c, 5)}>
Node a: <a, ∞, - , {(Node b, 1), (Node c, 2)}>
Node b: <b, ∞, - , {(Node d, 4)}>
Node c: <c, ∞, - , {(Node a, 3), (Node b, 9), (Node d, 2)}>
Node d: <d, ∞, - , {(Node s, 7), (Node b, 6)}>
```

MapReduce Iteration Mapping:

$map1 : Node.id : Node \rightarrow \{list(Node.neighbor Node.id : (Node.id, SUM$
$(Node.cost To Neighbor Node, Node.cost From Source)))\}$

$map2 : Node.id : Node \rightarrow \{list (Node.Neighbor Node.id, Node.Neighbor$
$Node)\}$

$reduce : Node.id : \{list (prev Hop From Source, cost From Source, Node)\} \rightarrow$
$Node.id : Node'$

where

$$Node'.costFromSource = MIN(costFromSource),$$
$$Node'.prevHopFromSource = prevHopFromSource_{min}$$

Note: Reducer only emits value if Node structure is updated, i.e., the iteration found a new shortest path from the source.

Iteration 1:

```
Map Input: s: <s, 0, - , {(Node a, 1), (Node c, 5)}>
Map Output: (a: s, 1), (a: Node a), (c: s, 5), (c, Node c)
Reduce 1 Input: a: (s, 1, Node a)
Reduce 1 Output: a: <a, 1, s, {(Node b, 1), (Node c, 2)}>
Reduce 2 Input: c: (s, 5, Node c)
Reduce 2 Output: c: <c, 5, s, {(Node a, 3), (Node b, 9), (Node
   d, 2)}>
```

The reader is encouraged to continue the example and see how the solution converges in four iterations.

References

1. [Online]. "Mind-blowing growth & power of big data - Business Insider" Available: http://www.businessinsider.com/mind-blowing-growth-and-power-of-big-data-2015-6
2. EMC Digital Universe with Research & Analysis by IDC. The Digital Universe of Opportunities: Rich Data and the Increasing Value of the Internet of Things. 2014
3. [Online]. "Volume, velocity, and variety: Understanding the three V's of big data," in DIY-IT Available: http://www.zdnet.com/article/volume-velocity-and-variety-understanding-the-three-vs-of-big-data/
4. G. Sanjay, G. Howard, and L. Shun-Tak, "The Google File system," in ACM SIGOPS Operating Systems Review - Volume 37 Issue 5, December 2003
5. D. Jeff and G. Sanjay, "MapReduce: Simplified Data Processing on Large Clusters," in Communications of the ACM – 50th Anniversary Issue, Vol. 51 No. 1, Pages 107–113, 2008.
6. [Online]. "Apache Hadoop" Available: http://hadoop.apache.org/
7. [Online]. "Apache Hadoop YARN" Available: http://hadoop.apache.org/docs/current/hadoop-yarn/hadoop-yarn-site/YARN.html
8. [Online]. "Hadoop Basics—Creating a MapReduce Program," DZone Available: https://dzone.com/articles/hadoop-basics-creating
9. "Data-Intensive Text Processing with MapReduce" by Jimmy Lin and Chris Dyer, University of Maryland, College Park, Manuscript prepared April 11, 2010
10. "MapReduce Patterns, Algorithms, and Use Cases" by Ilya Katsov in Highly Scalable Blog, 2012 (https://highlyscalable.wordpress.com/2012/02/01/MapReduce-patterns/)
11. [Online]. "Apache Spark" Available: http://spark.apache.org/
12. [Online]. "Apache Hive" Available: https://hive.apache.org/
13. [Online]. "Apache Pig" Available: https://pig.apache.org/
14. [Online]. "Apache HBase" Available: https://hbase.apache.org/
15. [Online]. "Apache Mahout" Available: http://mahout.apache.org/
16. [Online]. "Apache Oozie" Available: http://oozie.apache.org/
17. [Online]. "The Hadoop Ecosystem Table" Available: https://hadoopecosystemtable.github.io/

The Realm of Graphical Processing Unit (GPU) Computing

Vivek K. Pallipuram and Jinzhu Gao

Abstract The goal of the chapter is to introduce the upper-level Computer Engineering/Computer Science undergraduate (UG) students to general-purpose graphical processing unit (GPGPU) computing. The specific focus of the chapter is on GPGPU computing using the Compute Unified Device Architecture (CUDA) C framework due to the following three reasons: (1) Nvidia GPUs are ubiquitous in high-performance computing, (2) CUDA is relatively easy to understand versus OpenCL, especially for UG students with limited heterogeneous device programming experience, and (3) CUDA experience simplifies learning OpenCL and OpenACC. The chapter consists of nine pedagogical sections with several active-learning exercises to effectively engage students with the text. The chapter opens with an introduction to GPGPU computing. The chapter sections include: (1) Data parallelism; (2) CUDA program structure; (3) CUDA compilation flow; (4) CUDA thread organization; (5) Kernel: Execution configuration and kernel structure; (6) CUDA memory organization; (7) CUDA optimizations; (8) Case study: Image convolution on GPUs; and (9) GPU computing: The future. The authors believe that the chapter layout facilitates effective student-learning by starting from the basics of GPGPU computing and then leading up to the advanced concepts. With this chapter, the authors aim to equip students with the necessary skills to undertake graduate-level courses on GPU programming and make a strong start with undergraduate research.

Relevant core courses: Computer Systems Architecture and Advanced Computer Systems courses.
Relevant PDC topics: Table 1 lists the relevant PDC concepts covered along with their Bloom levels.

V. K. Pallipuram (✉) · J. Gao
University of the Pacific, Stockton, CA, USA
e-mail: vpallipuramkrishnamani@pacific.edu; jgao@pacific.edu

© Springer International Publishing AG, part of Springer Nature 2018   191
S. K. Prasad et al. (eds.), *Topics in Parallel and Distributed Computing*,
https://doi.org/10.1007/978-3-319-93109-8_8

Table 1 PDC concepts across chapter sections and their Bloom levels

	Chapter section								
PDC concept	8.1	8.2	8.3	8.4	8.5	8.6	8.7	8.8	8.9
Data parallelism	C								
GPGPU devices		C	A	A	A	A	A	A	
nvcc compiler			A						
Thread management				A	A				
Parallel patterns							A	A	
Performance evaluation							A		
Performance optimization							A	A	
CUDA		A		A	A	A	A	A	
Advancements in GPU computing									K

Learning outcomes: By the end of this chapter, students will be able to:

- Explain CUDA concepts including thread management, memory management, and device management.
- Identify performance bottlenecks in CUDA programs and calculate performance achieved in floating-point operations per second (FLOPS).
- Develop CUDA programs and apply optimizations pertaining to memory hierarchy, instructions, and execution configuration.

Context for use: The book chapter is envisioned for upper-level Computer Science/Computer Engineering undergraduate courses/electives on systems, advanced computer systems architecture, and high-performance computing with GPUs. The book chapter is also intended as a "quick start" guide to general-purpose GPU programming in undergraduate research.

The general-purpose graphical processing units (commonly referred to as GPG-PUs) are throughput-oriented devices. Unlike the conventional central processing units (CPUs) that employ a significant portion of the silicon wafer for caches, GPGPU devices devote a large chunk of the chip real-estate to computing logic. Consequently, GPGPU devices today feature several hundreds of cores dedicated to performing accelerated computing. To unleash the tremendous power in these computing cores, programmers must create several hundreds of thousands of threads. This task warrants programmers devise creative techniques for task decomposition, data partitioning, and synchronization. The GPGPU computing includes an additional challenge of CPU-GPGPU device communication, which stems from the fact that CPU and GPGPU device memories are typically disjoint. In most GPGPU programs, the CPU (host) prepares and transfers the data to the GPGPU device via the Peripheral Interconnect Express (PCIe) bus for computations. Once the GPGPU device finishes all of the computations, it sends the processed data back to the CPU host via the PCIe bus. A few recent architectures from AMD feature accelerated processing units (APUs) that integrate CPU and GPU in a single chip. AMD calls this approach as heterogeneous unified memory access (hUMA) where the CPU and GPU memory spaces are unified, thereby avoiding any explicit data transfers between the CPU host and GPU device. However, such integration leads to CPU-GPU competition for the on-chip resources, leading to limited performance benefits. This chapter builds the GPGPU programming concepts using the disjoint CPU-GPGPU memory model.

To enable programmers to perform general purpose computing with GPUs, NVIDIA introduced the Compute Unified Device Architecture (CUDA) [1] in 2006, ultimately replacing the notion of "express-it-as-graphics" approach to GPU computing. CUDA is appropriately classified as a parallel computing platform and programming model – it helps programmers to develop GPGPU programs written in common languages such as C, C++, and Fortran by providing an elegant set of keywords and language extensions. Additionally, CUDA provides useful libraries such as the CUDA Basic Linear Algebra Subroutines (cuBLAS) library [2] for GPGPU accelerated linear algebra routines and the CUDA Deep Neural Network (cuDNN) library [3] for GPGPU accelerated primitives for deep neural networks. At the time of this writing, CUDA is in its current avatar CUDA 9 and is freely available for Linux, Windows, and Mac OSX.

GPGPU programming has continued to evolve ever since the introduction of CUDA. Open Computing Language (OpenCL [4]) was released in 2009 as a royalty-free standard for parallel programming on diverse architectures such as GPGPUs, multi-core CPUs, field programmable gate arrays (FPGAs), and digital signal processors (DSPs). Using a set of platform specific modifications, OpenCL allows programmers to adapt their codes for execution across a variety of heterogeneous platforms. Both CUDA and OpenCL tend to be verbose, for instance CUDA traditionally requires programmers to perform explicit data transfers between the CPU host and GPGPU device. The CPU host explicitly calls the GPGPU device functions (called kernels) to execute the parallel tasks. OpenCL, with its cross-platform requirements, further requires programmers to explicitly create device-related task queues. To reduce such burden on programmers, OpenACC [5] standard was officially released in 2013 as a paradigm to simplify CPU-GPGPU programming. OpenACC offers compiler directives (for example, `#pragma acc`) that are strategically placed across the source code to automatically map computations to the GPGPU device. The software advancements are not only limited to GPGPU programming models – software libraries such as Thrust [6] accelerate GPGPU code development by providing helper classes and functions for common algorithms such as sort, scan, reduction, and transformations, enabling programmers to focus on the high-level details of the application.

This chapter introduces students to the basics of GPGPU computing using the CUDA programming model. Section "Data Parallelism" introduces the concept of data parallelism, which is critical for GPGPU computing. Section "CUDA Program Structure" explains the typical structure of a CUDA program. Section "CUDA Compilation Flow" describes the compilation flow of CUDA programs. Sections "CUDA Thread Organization" and "Kernel: Execution Configuration and Kernel Structure" describe the CUDA thread organization and CUDA kernel configuration, respectively. Section "CUDA Memory Organization" details the GPGPU memory organization as viewed by a CUDA program. Section "CUDA Optimizations" expounds several CUDA optimization techniques employed by programmers to maximize the application performance. The chapter concludes in section "Case Study: Image Convolution on GPUs" on convolution with GPGPU devices as a case study. By the end of this chapter, students will be able to explain computation mapping to CUDA threads, write GPGPU device kernels, and employ optimization strategies to achieve high application performance using GPGPU devices.

Data Parallelism

Several scientific applications today process large amounts of data. Some example applications include molecular dynamics, deep neural networks, and image processing, among others. A sequential scan of the data for such applications on a conventional CPU may incur significant application runtime. Fortunately, several scientific applications offer data parallelism, meaning the data can be partitioned into several chunks that can be executed independent of each other. The data parallelism is the primary source of scalability for many programs. In a molecular dynamics simulation, the electrostatic potential of atoms at grid points is evaluated in parallel. In a neural network simulation, the voltages of firing neurons at a given neuron layer are independent, and therefore can be evaluated in parallel. Several image-processing applications offer data parallelism via independent evaluation of pixel attributes. Life teaches us several lessons – including data parallelism! When the professor allows students to collaborate for an assignment, students divide work (equally) with each other and complete the assignment in a short time. Similarly, teaching assistants divide the grading work among themselves to reduce the grading time.

Active Learning Exercise 1 – Identify five common activities in day-to-day life that express significant data parallelism.

GPGPU devices work extremely well with applications that offer significant data parallelism. In fact, the very primitive job of a GPU device, i.e. graphics rendering, is extremely data parallel. Consider a simple example of vector-vector addition to illustrate the concept of data parallelism. Figure 1 shows the addition of two vectors A and B; the result is stored in vector C. The corresponding elements of vectors A and B are added using processing elements, PEs. Clearly, each processing element, PE_i, works independently of any other processing element. Therefore, each data element of vector C can be evaluated in parallel, thereby utilizing data parallelism inherent in this operation.

Matrix-matrix multiplication is another frequently used mathematical operation that offers significant data parallelism. Consider the multiplication of two matrices, $A_{m \times n}$ and $B_{n \times p}$; the result is stored in the matrix $C_{m \times p}$. Each element c_{ij} of $C_{m \times p}$ is evaluated by computing the scalar product (also called the inner product in the context of Euclidean space) between the ith row of $A_{m \times n}$ and jth column of matrix $B_{n \times p}$. Equation 1 summarizes c_{ij} computation.

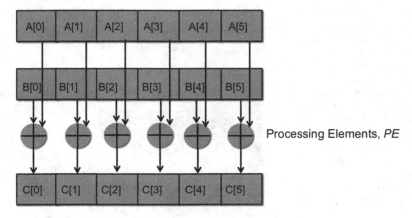

Fig. 1 Addition of two vectors A and B to elucidate data parallelism. The processing elements, PEs, work independently to evaluate elements in C

$$c_{ij} = \sum_{k=1}^{n} a_{ik} \times b_{kj} \qquad (1)$$

A careful inspection of the above equation reveals that computation of any c_{ij} is independent of the others; therefore, c_{ij} can be computed in parallel. Matrix-matrix multiplication is an interesting operation because it can be parallelized in a variety of ways. For example, one can create $m \times p$ processing elements (PEs) where each PE_{ij} computes a specific matrix element, c_{ij}. Consider another example, where the PEs perform partial product computations and then add the partial product results from other pertinent PEs to obtain the final result, c_{ij}. Ahead in this chapter, we study the parallelization of matrix-matrix multiplication on the GPGPU device.

There are several computationally-intensive mathematical operations used in engineering and science that offer data-parallelism. Some examples include reduction, prefix sum, scan, sorting, among many others. Not surprisingly, many scientific applications are composed of these computationally-intensive operations. By parallelizing these operations, programmers can achieve significantly high performance for their scientific codes and simulations.

Active Learning Exercise 2 – Perform a research on the following operations and explain how they offer data-parallelism: reduction, prefix sum, scan, and sorting.

CUDA Program Structure

In this section, we explore the structure of common CUDA programs. First, we explore a simple real-world example and then transfer the intuition to CUDA programs. Consider the example of multiple graders sharing the grading workload

Fig. 2 Program flow of a typical CUDA program interleaved with host portion (executed by a single CPU thread) and device portion (executed by several thousands of GPU threads). The host-to-device (H2D) and device-to-host (D2H) communications occur between the interleaved portions denoting data transfers from CPU-to-GPGPU and GPGPU-to-CPU, respectively

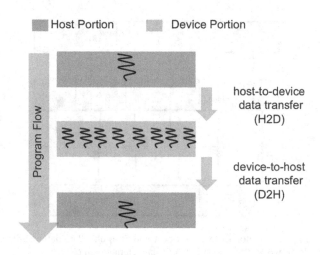

for a large class. Let us assume that the instructor collects the student assignments and distributes them equally to all of the graders. There are multiple scenarios that can arise in this case. In Scenario A, the graders complete the grading job easily without any doubts and/or clarifications with respect to grading. In this scenario, the instructor gets the graded assignments expeditiously. In Scenario B, the graders may have questions on grading and they visit the instructor's office for clarification. Due to this instructor-grader communication, the grading is slower than Scenario A. In another Scenario C, the graders may choose to communicate with each other and avoid long trips to the instructor's office, thereby finishing the job faster than Scenario B. The structure of typical CUDA programs is no different than the structure of grading scenarios – in what follows, we describe the layout of a typical CUDA program.

Figure 2 illustrates the structure of a typical CUDA program, which has two primary interleaved sections namely, the host portion and the device portion. Depending on the application, these interleaved sections may be repeated several times. A single CPU thread executes the host portion, while the GPGPU device executes the device portion of the CUDA program. At the start of the program, the CPU host portion prepares the data to be executed on the GPGPU device. After the data preparation, the CPU host transfers the data to the GPGPU device memory via host-to-device (H2D) transfer operation, which is performed over the PCIe bus. After the GPGPU device portion finishes operating on the data, the processed data is transferred back to the CPU host memory via device-to-host (D2H) operation. A CUDA program may contain several interleaved host and device portions (similar to the multiple graders case discussed above); however a prudent programmer must be wary of communication costs incurred due to frequent H2D and D2H transfers. To maximize the performance of CUDA programs, it is recommended to minimize the host-device communications.

Let us investigate the structure of our first CUDA program. Listing 1 provides the complete CUDA program for vector-vector addition. Line 1 includes the cuda.h header file that provides GPGPU device-related functions. Inside the main() function, note the host and device pointer variables declaration in lines 14 and 15, respectively. It is recommended to provide the h_ prefix for the host pointer variables and d_ prefix for the device pointer variables. These prefixes enable programmers to avoid the accidental de-referencing of device pointers by the host and vice-versa, which cause the programs to break with error messages. The host portion of the CUDA program prepares the data for the GPGPU device execution (see Lines 18 through 29) and allocates the host and device memories for computations (Lines 19 though 25). For seamless programming, CUDA provides the cudaMalloc() function (similar to host's malloc() function) for allocating device pointers in the GPGPU device memory.

Once the host portion of the code finishes the necessary preprocessing steps, it initiates a host-to-device data transfer via the cudaMemcpy() function call (Lines 31 and 32). The cudaMemcpy() function inputs the destination pointer, source pointer, number of bytes to be transferred, and the data transfer direction (host-to-device, device-to-host, etc.). Readers are encouraged to pay special attention to cudaMemcpy() function and its parameters. Incorrect function parameters can also lead to incorrect referencing of pointers. In Listing 1 on lines 31 and 32, the destination pointer arguments are the device pointers d_a and d_b, respectively, the source pointer arguments are the host pointers h_a and h_b, respectively, and the data transfer direction is specified by the cudaMemcpyHostToDevice flag, denoting host-to-device data transfer.

Listing 1 An example CUDA program illustrating vector-vector addition. Note the interleaved CPU-host and GPGPU device portions. Host-device communications occur between the interleaved host and device portions of the CUDA program.

```
1.  #include <cuda.h>
2.  #include <stdio.h>
3.  //Device kernel
4.  __global__
5.  void gpu_kernel(int *d_a, int *d_b, int *d_c, int vec_size){
6.      int tid = threadIdx.x + blockIdx.x*blockDim.x;
7.      if(tid <vec_size) {
8.          d_c[tid] = d_a[tid] + d_b[tid];
9.      }
10. } //end device kernel
11. int main(int argc, char **argv){
12.     //declare variables
13.     int i,vec_size;
14.     int *h_a,*h_b,*h_c; //data pointers for host-section
15.     int *d_a,*d_b,*d_c; //data pointers for device-section
16.     //Host-Section prepares the data
17.     vec_size=1000;
18.     //Host-portion prepares the host data and allocates device pointers
19.     h_a=(int *)malloc(sizeof(int)*vec_size);
20.     h_b=(int *)malloc(sizeof(int)*vec_size);
21.     h_c=(int *)malloc(sizeof(int)*vec_size);
22.     //Allocate GPGPU device pointers
23.     cudaMalloc((void **)&d_a, sizeof(int)*vec_size);
24.     cudaMalloc((void **)&d_b, sizeof(int)*vec_size);
25.     cudaMalloc((void **)&d_c, sizeof(int)*vec_size);
26.     //Host-portion prepares the data
27.     for (i=0; i<vec_size; i++)   {
28.         h_a[i]=i; h_b[i]=i;
29.     }
```

```
30.    //CPU host transfers the data to GPGPU device memory
31.    cudaMemcpy(d_a,h_a,sizeof(int)*vec_size,cudaMemcpyHostToDevice);
32.    cudaMemcpy(d_b,h_b,sizeof(int)*vec_size,cudaMemcpyHostToDevice);
33.    //CPU host invokes the GPGPU device portion
34.    gpu_kernel<<<1, 1000>>>(d_a,d_b,d_c,vec_size);
35.    //GPGPU device transfers the processed data to CPU host
36.    cudaMemcpy(h_c,d_c,sizeof(int)*vec_size,cudaMemcpyDeviceToHost);
37.    //CPU host-portion resumes operation
38.    for(i=0; i<vec_size; i++){
39.        printf(''\n C[%d]=%d'',i,h_c[i]);
40.    }
41.    free(h_a);
42.    free(h_b);
43.    free(h_c);
44.    cudaFree(d_a);
45.    cudaFree(d_b);
46.    cudaFree(d_c);
47.    return 0;
48. }
```

After transferring the data to the GPGPU device, the host portion invokes the
GPGPU device kernel in Line 34. A device kernel is a GPGPU device function
that is callable from the host and executed by the GPGPU device. A device kernel
invocation is also referred to as a *kernel launch*. The calling name (gpu_kernel)
specifies the name of the device kernel. The angular brackets (<<< >>>) specify
the GPGPU device execution configuration, which mainly consists of the number
of thread blocks and the number of threads per block to operate on the input
data. We discuss threads and thread blocks in detail in section "CUDA Thread
Organization". In this example when the gpu_kernel is launched, one thread
block containing 1000 CUDA threads are created that execute the kernel function
concurrently. More details on GPGPU device kernels and execution configuration
appear in section "Kernel: Execution Configuration and Kernel Structure". The lines
4–9 are executed as the device portion of the code on the GPGPU device. The
__global__ keyword specifies that the following function (gpu_kernel in
our case) is a device kernel. The in-built variables threadIdx, blockIdx, and
blockDim in Line 6 enable programmers to access the threads' global indices. In
this program, a thread with index tid operates on tid-th element of the vectors
A, B, and C. It should be noted that GPGPU device kernel calls are asynchronous,
meaning that after the kernel launch, the control immediately returns to the host
portion. In this program after the kernel launch, the host portion invokes the
cudaMemcpy() function (Line 36), which waits for all of the GPGPU threads to
finish the execution. After the GPGPU device finishes execution, the control returns
to line 36 where the device transfers the processed data (vector C) to the host.
Note that in this device-to-host transfer, the host pointer (h_c) is the destination,
the device pointer (d_c) is the source, and the direction of the data transfer is
device-to-host denoted by the cudaMemcpyDeviceToHost flag. The lines 44
through 46 release the device memory variables via the cudaFree() function
call.

CUDA Compilation Flow

Now that we understand the structure of CUDA programs, let us study how CUDA programs are compiled and a single executable is generated in a Linux-based environment. NVIDIA's nvcc compiler facilitates the splitting, compilation, preprocessing, and merging of CUDA source files to create an application's executable. Although the nvcc compiler enables transparent code compilation, an understanding of the compilation flow can enable further performance improvement. The nvcc compiler in the Linux environment recognizes a selected set of input files given in Table 2. In what follows, we study a high-level compilation flow of CUDA source programs.

Figure 3 provides a high-level abstraction of the CUDA compilation process. The nvcc compiler, in conjunction with a compatible host code compiler such as gcc/g++, splits the compilation of CUDA source programs into two trajectories namely, the host trajectory and the device trajectory. These trajectories are not completely disjoint and often interact with each other via intermediate 'stub' functions. The host trajectory extracts the host code, host stub functions (functions that set up the kernel launch when the device kernel is invoked by the host), and compiles the host code to produce the .o object file. The device trajectory includes multiple steps such as device code extraction, host stub extraction, and device code optimization. The nvopenacc command inputs the intermediate compilation files (.cpp3.i) to produce the virtual architecture assembly file (.ptx) that contains a generic device instruction set. Next, the ptxas assembly command generates the .cubin file: the real architecture binary for a specific GPGPU device. The fatbinary stage combines multiple .cubin files (each targeting a different GPGPU device) into a .fatbin binary file. This binary file is ultimately linked with the host .o object file to create the final executable file, a.out. When a.out is executed, an appropriate .cubin file is selected from .fatbin for the target GPGPU device.

The CUDA toolkit documentation [7] provides a highly detailed explanation of the compilation process. The nvcc compiler also offers programmers with

Table 2 List of input files recognized by the nvcc compiler in Linux-based environment

Input file type	Description
.cu	CUDA source file containing host and device portions
.c	C source file
.cpp, .cc, .cxx	C++ source file
.gpu	Intermediate device-code only file
.o	Object file
.a	Library file
.so	Shared object files (not included in executable)
.res	Resource file

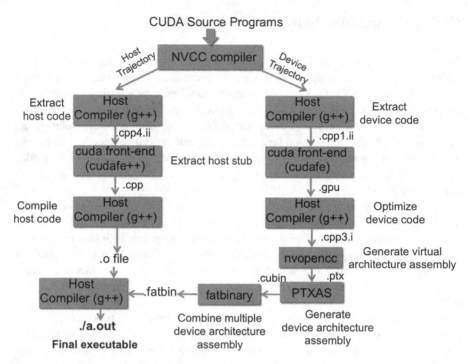

Fig. 3 A high-level abstraction of nvcc compilation process. The nvcc compiler breaks the compilation process into two trajectories: host trajectory and device trajectory

several compiler switches to control the code generation. Here, we only discuss two important switches: --gpu-architecture and --gpu-code. These switches allow for the GPGPU device architecture evolution. Before describing the role of these compiler switches, let us define the term *Compute Capability*. The Compute Capability of a device is represented by a version number that identifies the supported hardware features of the GPGPU device. The Compute Capability is used during the compilation process to determine the set of instructions for the target GPGPU device. The purpose of the above-mentioned nvcc compiler switches is as follows.

--gpu-architecture (short: -arch): This switch enables the selection of a virtual architecture, thereby controlling the output of the nvopencc command. A virtual architecture is a *generic* set of instructions for the virtual GPGPU device with the desired compute capabilities. By itself, the virtual architecture does not represent a specific GPGPU device. Some example values of --gpu-architecture switch are: compute_20 (Fermi support); compute_30 (Kepler support); compute_35 (recursion via dynamic parallelism); compute_50 (Maxwell support).

--gpu-code (short: -code): The switch enables the selection of a specific GPGPU device (the actual GPU architecture). Some examples values include:

Table 3 Examples of code generation using `--gpu-architecture` and `--gpu-code` switches

Example	Description

nvcc vector.cu
`--gpu-architecture=compute_30`
`--gpu-code=sm_30,sm_35`
	The `fatbinary` includes two cubins;
	one cubin corresponding to each architecture.

nvcc vector.cu
`--gpu-architecture=compute_30`
`--gpu-code=compute_30,`
`sm_30,sm_35`
	The same as the above with the inclusion
	of PTX assembly in the `fatbinary`.

nvcc vector.cu
`--gpu-architecture=compute_30`
`--gpu-code= sm_20,sm_30`
	Fails because `sm_20` is lower than the
	virtual architecture `compute_30`

`sm_20` (Fermi support); `sm_30` (Kepler support); `sm_35` (recursion via dynamic parallelism); `sm_50` (Maxwell support).

In what follows, we outline the general guidelines used to set values of the above mentioned compiler switches for different types of code generation. The `--gpu-architecture` switch takes a specific value, whereas the `--gpu-code` switch can be set to multiple architectures. In such a case, `.cubin` files are generated for each architecture and included in the `fatbinary`. The `--gpu-code` switch can include a single virtual architecture, which causes the corresponding PTX code to be added to the `fatbinary`. The NVIDIA documentation suggests keeping the value of `--gpu-architecture` switch as low as possible to maximize the number of actual GPGPU devices. The `--gpu-code` switch should preferably be higher than the selected virtual architecture. Table 3 provides several compilation examples for code generation. We encourage readers to peruse the Nvidia software development kit (SDK) for sample Makefiles and adapt them for their respective applications and GPGPU devices.

Active Learning Exercise 3 – Write a compilation command for generating a `fatbinary` with PTX included for Fermi and Kepler architectures.

CUDA Thread Organization

A CUDA program follows Single Program, Multiple Data (SPMD) methodology where several thousands of threads work concurrently to execute the same kernel function on different data elements. However, different groups of threads may be executing different sections of the same CUDA kernel. To enable CUDA threads to access the relevant data elements upon which to operate, it is imperative to fully understand the CUDA thread organization. The CUDA threads are organized in a two-level hierarchy of *grids* and *blocks*. A grid is a three-dimensional collection

Fig. 4 Two examples of
CUDA grids and thread
blocks. When `Kernel1` is
called, it launches a 2 × 2
grid of 2 × 2 thread blocks.
When `Kernel2` is called, it
launches a 1D grid with two
1D thread blocks with each
block containing 5 threads

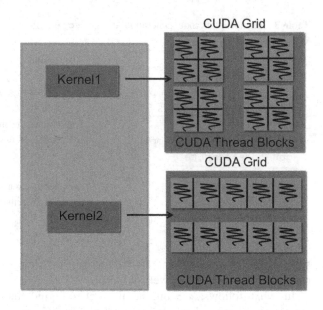

of one or more blocks and a block is a three-dimensional collection of several
threads. When a kernel function is called, a grid containing multiple thread blocks
is launched on the GPGPU device (Fig. 4). As shown in the same figure, when
the kernel function `Kernel1` is called, a two-dimensional grid of thread blocks
(2 blocks each in x and y dimensions) is launched on the GPGPU device. In this
example, each thread block is a two-dimensional arrangement of threads with two
threads in both the x and y dimensions. The `Kernel2` function call launches a
CUDA grid with two thread blocks, where each thread block is a one-dimensional
arrangement of five threads. For illustration purposes, the above examples work with
only four or five threads per block. Readers should note that GPGPU devices require
a minimum number of threads per block depending on the Compute Capability.

First, let us investigate CUDA grids. As mentioned earlier, each grid is a three-
dimensional arrangement of thread blocks. When the kernel function is launched,
the first parameter in execution configuration, `<<<`**`dimGrid`**`, ..>>>`, specifies
the dimensions of the CUDA grid. The size of grid dimensions depends on the
Compute Capability of the GPGPU device. In CUDA programs, the dimensions of
the grids can be set using the C structure, `dim3`, which consists of three fields:
x, y, and z for x, y, and z dimensions, respectively. By setting the dimensions
of CUDA grids in the execution configuration, we automatically set the values of
x, y, and z fields of the predefined variable, `gridDim`. This variable is used in
the kernel function to access the number of blocks in a given grid dimension. The
blocks in each dimension are then accessed via the predefined variable, `blockIdx`,
which also contains three fields: x, y, and z. The variable `blockIdx.x` takes on
values ranging from 0 to `gridDim.x-1`; `blockIdx.y` takes on values ranging
from 0 to `gridDim.y-1`; and `blockIdx.z` takes on values ranging from 0 to
`gridDim.z-1`. Table 4 provides examples of CUDA grid initialization using the

Table 4 Examples of CUDA grid initialization using dim3 structure. The corresponding values (range of values) of gridDim and blockIdx variables are shown

Example	Description	gridDim variable			blockIdx variable		
		x	y	z	x	y	z
dim3 dimGrid1(32,1,1)	1D grid with 32 thread-blocks	32	1	1	0–31	0	0
dim3 dimGrid2(16,16,1)	2D grid with 16 blocks in x and y dimensions	16	16	1	0–15	0–15	0
dim3 dimGrid3(16,16,2)	3D grid with 16 blocks in x and y dimensions and 2 blocks in z dimension	16	16	2	0–15	0–15	0–1

dim3 structure and illustrates the values of gridDim and blockIdx variables. Note that the unused dimensions in the dim3 structure are set to one.

The dimensions of a CUDA grid can also be set at runtime. For instance, if a programmer requires 256 threads per block to work on n elements, the dimensions of the grid can be set as:

<<<dimGrid(round_up(n,256)),..>>>. Note that round_up() function is required to launch enough thread blocks to operate on all of the n elements.

Active Learning Exercise 4 – Initialize a three-dimensional CUDA grid with two blocks in each dimension. Give the values of pertinent predefined variables.

Next, we turn our attention to CUDA thread blocks. As mentioned before, the CUDA thread blocks are three-dimensional arrangements of threads. The second parameter in the execution configuration, <<<dimGrid, **dimBlock**,..>>>, specifies the dimensions of a single thread block. Similar to grids, the thread block dimensions can also be set using the dim3 structure. It should be noted that the total number of threads in a block should not exceed 1024. Once the block dimensions are set, the x, y, and z fields of the in-built variable, blockDim are initialized. Each field of blockDim variable denotes the number of threads in x, y, and z dimensions. Each thread in a given thread block is then accessed using the predefined variable, threadIdx. Akin to the blockIdx variable, the threadIdx variable has three fields namely, threadIdx.x varying from 0 to blockDim.x-1, threadIdx.y varying from 0 to blockDim.y-1, and threadIdx.z varying from 0 to blockDim.z-1. Table 5 provides examples of block dimension initialization and the corresponding values of blockDim and threadIdx variables.

Active Learning Exercise 5 – Initialize a 2D CUDA block with 16 threads in each dimension. Give the values of pertinent predefined variables.

As discussed before, the maximum grid and block dimensions depend on the Compute Capability of the GPGPU device. It is always a good idea to verify these values for newer architectures. Table 6 provides the maximum device specific values for Compute Capability 3.x, 5.x, and 6.x devices.

Active Learning Exercise 6 – Investigate the device specific values of earlier compute capabilities, i.e. 1.x and 2.x. Also provide one GPGPU device from these compute capabilities. What are the significant changes in device specific values for Compute Capability 2.x onwards? Make a note on how these changes influence the GPGPU programming.

Kernel: Execution Configuration and Kernel Structure

As readers may recall, several thousands of threads created by the programmer in a CUDA program concurrently execute a special device function, the kernel. The host portion of the CUDA program asynchronously calls the CUDA kernel, meaning that the control immediately returns to the host portion after the kernel launch. During the kernel execution, the host portion may perform some computations (thereby overlapping computations) or may choose to wait for the GPGPU device to finish operating on the data. An example of kernel launch is as follows:

```
gpu_kernel <<<dimGrid,dimBlock>>> (arg1, arg2,..,argN);
```

In the above statement, the GPGPU device kernel named `gpu_kernel` is executed by all of the threads created in the CUDA program. The number of threads created is a function of the kernel execution configuration specified by the `dimGrid` and `dimBlock` (`dim3` type) variables configured by the programmer (see Tables 4 and 5 for examples). As discussed in the foregoing section, the `dimGrid` variable specifies the number of CUDA blocks arranged in x, y, and z dimensions of a CUDA grid, whereas the `dimBlock` variable specifies the number of CUDA threads arranged in x, y, and z dimensions in a CUDA block. A general procedure for setting an execution configuration is follows.

1. Set the thread block dimensions and the number of threads in each dimension such that the total number of threads in a block does not exceed 1024. Pay attention to GPGPU device specific limits (see Table 6).
2. Calculate the number of thread blocks required in each grid dimension.

Table 5 Examples of CUDA block initialization using `dim3` structure. The corresponding values (range of values) of `blockDim` and `threadIdx` variables are shown

Example	Description	blockDim variable			threadIdx variable		
		x	y	z	x	y	z
`dim3 dimblock1(32,1,1)`	1D block with 32 threads	32	1	1	0–31	0	0
`dim3 dimblock2(32,32,1)`	2D block with 32 threads in x and y dimensions	32	32	1	0–31	0–31	0
`dim3 dimblock3(32,32,2)`	Incorrect. The number of threads in the block exceeds 1024.	–	–	–	–	–	–

Table 6 Limitations on device specific parameters for Compute Capability 3.x, 5.x, and 6.x devices

Device parameter	Maximum number
Maximum number of grid dimensions	3
Grid maximum in x dimension	$2^{31} - 1$
Grid maximum in y and z dimensions	$2^{16} - 1$
Maximum number of block dimensions	3
Block maximum in x and y dimensions	1024
Block maximum in z dimension	64
Maximum threads per block	1024
Example GPGPU device (3.x)	Kepler GK110
Example GPGPU device (5.x)	Maxwell GM200
Example GPGPU device (6.x)	Pascal GP102

Once the execution configuration is set and the kernel is launched, it is customary for each thread to 'know' its local and global thread identification numbers (IDs). It is via these thread IDs that different threads access their respective portions of the data. As discussed in section "CUDA Thread Organization", threads can access their IDs inside the device kernel function using in-built variables: `gridDim`, `blockDim`, `blockIdx`, and `threadIdx`. These variables are set when the execution configuration is passed to the kernel during the kernel launch. The methodology of setting the execution configuration usually depends on the type of parallel patterns in an application. Simple parallel patterns such as vector-vector addition, prefix sum, etc. may only require one-dimensional execution configuration. Whereas more complex patterns such as matrix-matrix multiplication, two-dimensional image convolution, etc. intuitively lend themselves to two-dimensional execution configuration. More complex applications that operate on three-dimensional data are parallelized using a three-dimensional execution configuration. In what follows, we use two example parallel patterns illustrating one-dimensional and two-dimensional execution configurations, namely vector-vector addition and matrix-matrix multiplication. We study how the execution configuration is set and the threads are accessed inside the device kernel function for these two parallel patterns. These examples help us build our intuition for one- and two-dimensional grids and blocks, which can be easily extended to three-dimensional execution configuration.

Consider addition of two vectors A and B, each containing n elements. The result of addition is stored in vector C as illustrated by Fig. 1. We use 1D blocks and grids for this case, given that our working arrays A, B, and C are one-dimensional arrays. An example execution configuration with 256 threads per block appears in Listing 2.

Listing 2 The example illustrates an execution configuration with 256 threads per block for vector-vector addition. The example also shows how a thread accesses its global index/identifier (ID) in the CUDA grid.

```
// Auxiliary C function for rounding up
int round_up(int numerator, int denominator) {
```

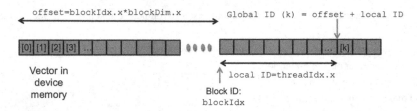

Fig. 5 The illustration shows how a thread accesses its global ID and the corresponding data element in the vector

```
return (numerator+denominator −1)/denominator;
}

//Inside main

// Step 1: Set the block configuration
1. dim3 dimBlock(256, 1, 1);
//Step 2: Set the grid configuration
2. dim3 dimGrid (round_up(n,256), 1, 1);
//GPU kernel call
3. gpu_kernel <<<dimGrid, dimBlock>>>(A, B, C);
:
:
//Inside gpu_kernel function (device portion)
:
//The local thread ID in a given block
A. local_tid = threadIdx.x;
//The global thread ID in the entire grid
B. global_tid = local_tid + blockIdx.x*blockDim.x;
:
//Array access
AA. C[global_tid] = A[global_tid] + B[global_tid];
```

In Listing 2, Line 1 sets the x dimension of the thread block to 256 and the remaining unused fields (y and z) are set to one. In Line 2, the x dimension of the grid is set to `round_up(n,256)`, whereas the unused y and z dimensions are set to 1. The rounding up operation (using `round_up()`) is performed to create enough number of thread blocks to execute all of the n data elements. Inside the `gpu_kernel` function, Line A performs the access of the local thread ID, i.e. the thread's ID in its block. Line B shows how a thread accesses its global thread ID. In general, the global thread ID in any dimension follows the formula: `global_tid = local_tid + offset`. In this case, the offset equals `blockIdx.x*blockDim.x` and local ID equals `threadIdx.x`. Each thread then accesses a unique element of vectors A, B, and C using the global thread ID (`global_tid`) in Line AA. Figure 5 illustrates the global thread ID access discussed above.

Next, we consider the example of matrix-matrix multiplication to illustrate two-dimensional execution configuration. For simplicity, assume multiplication of two 2D matrices $A_{n \times n}$ and $B_{n \times n}$ of dimensions $n \times n$ each. The result of this multiplication is stored in another 2D matrix of the same dimensions, $C_{n \times n}$. For the purpose of illustration, assume 16×16 as the thread block dimensions. Readers should recall that the number of threads per block should not exceed 1024. The `dim3` type variables, `dimGrid` and `dimBlock`, are configured as shown in Listing 3.

Listing 3 Configuration of dimGrid and dimBlock in the host portion; and access of local and global thread IDs in the device portion.

```
// Preparing the execution configuration inside host portion of the code
// Step 1: Set the block configuration
1. dim3 dimBlock(16, 16, 1);
// Step 2: Set the grid configuration
2. dim3 dimGrid (round_up(n,16), round_up(n,16), 1);
//GPU kernel call
3. gpu_kernel <<<dimGrid, dimBlock>>>(A, B, C);
:
:
// Inside gpu_kernel function (device portion)
:
//The local thread ID in x-dimension in a given block
A. local_tidx = threadIdx.x;
//The local thread ID in y-dimension in a given block
B. local_tidy=threadIdx.y;
//The global thread ID in x-dimension in the entire grid
C. global_tidx = local_tidx + blockIdx.x*blockDim.x;
//The global thread ID in y-dimension in the entire grid
D. global_tidy = local_tidy + blockIdx.y*blockDim.y;
:
// Array access
AA. a=A[global_tidx][global_tidy];b=B[global_tidx][global_tidy];
```

In the example shown in Listing 3, a dim3 structure (dimBlock) for 2D CUDA block is declared with 16 threads in x and y dimensions, respectively; the unused z dimension is set to 1. Because the matrices are square with n elements in x and y dimensions, the CUDA grid consists of round_up(n, 16) number of CUDA blocks in x and y dimensions; the unused z dimension is set to 1 (Line 2). Inside the gpu_kernel, the local and global thread IDs in x and y dimensions are accessed as shown in lines A through D. The global element access using the global thread IDs is elucidated in Line AA. Figure 6 illustrates the above discussed concept for two-dimensional thread ID access.

In the foregoing examples and parallel patterns similar to them, readers should ensure that the threads cover all of the data elements and the number of idle threads is minimized. For instance, consider an example of addition of two vectors with 1000 elements each. A choice of 256 threads per block results in four thread blocks, thereby creating 1024 threads for the entire application. Because the threads with global IDs 0 through 999 operate on the corresponding data elements 0 through 999, the threads with IDs 1000 through 1023 remain idle. Similarly, a choice of 200 threads per block results in 5 thread blocks with no idle threads. However, there is more to execution configuration than simply creating sufficient number of threads. The number of threads per block and thread blocks affect the number of concurrent thread groups (a group of 32 concurrent threads is called a warp) active on a streaming multiprocessor. This concept is discussed in detail in section "CUDA Memory Organization".

Active Learning Exercise 7 – Create a 2D grid with 2D blocks for operation on an image of size 480 × 512. Elucidate, how each thread accesses its ID and its corresponding pixel element (x, y). How can you extend this process for a color image 480 × 512 × 3 where the third dimension corresponds to the red, green, and blue (RGB) color channels?

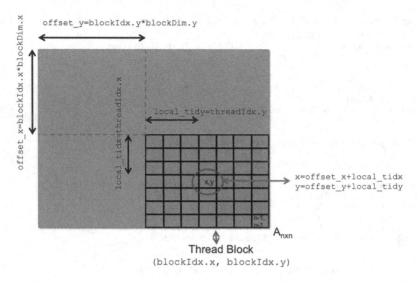

Fig. 6 The illustration shows how a thread accesses its global 2D ID (x, y) and the corresponding data element (x, y) in a two-dimensional matrix, $A_{n \times n}$

CUDA Memory Organization

The GPGPU devices are throughput-oriented architectures, favoring compute-logic units over memory units. The GPGPU device's main memory (also called the *device memory*) is usually separate from the GPGPU device. Consequently, most of the CUDA programs observe a performance bottleneck due to frequent device memory accesses. Therefore, programmers pursuing high-performance on GPGPU devices must have a deep understanding of the device memory hierarchy. A sound understanding of the CUDA memory hierarchy enables programmers to perform optimizations effectively. In what follows, we discuss the device memory hierarchy with respect to the CUDA programming model.

Figure 7 shows an abstract representation of a CUDA GPGPU device with its streaming multiprocessors interacting with the device memory. Henceforth, we refer to this memory hierarchy as the CUDA memory hierarchy.

As shown in Fig. 7, a GPGPU device contains multiple streaming processors (SMs), each containing multiple CUDA cores. In a typical CUDA program, the thread blocks are launched on the SMs while the CUDA cores execute the threads in a thread block. The CUDA memory hierarchy follows a pyramid fashion from the fastest but smallest memory units to the slowest but largest memory units as under:

- **On-chip Registers ($\approx$32 K registers per SM)** – In a SM, each CUDA core has exclusive access to its own set of registers. The register accesses are blazingly fast, each access taking only one clock cycle. The lifetime of registers is the

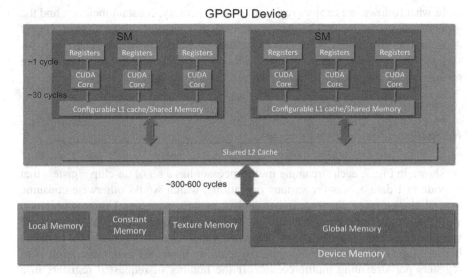

Fig. 7 The CUDA memory hierarchy: at the lowest level, CUDA cores inside SMs have access to fast registers. All of the CUDA cores in a given SM have shared access to L1 cache/shared memory (fast but slower than registers). All the SMs share the L2 cache (if present). The farthest memory unit from the GPGPU device is the device memory, which consists of special memory units including local memory, cached constant memory, texture memory, and global memory

lifetime of a thread. The automatic variables in the CUDA kernel are allotted registers depending on the device's Compute Capability. The leftover registers spill into the device's local memory, which resides in the off-chip device memory.

- **On-chip Shared memory (≈64 KB per SM)** – Further away from the registers is the shared memory shared by all of the CUDA cores in a SM. The accesses to shared memory are also fast; an access typically takes ≈30 clock cycles. The shared memory persists for the lifetime of a thread block.
- **Off-chip Device Memory (typically several GB)** – The largest and perhaps the most important memory unit of all is the GPGPU device memory, which resides in the off-chip random access memory (RAM). The device memory further consists of sub-units including:

  - **Local memory** for storing 'spilled' register variables.
  - **Cached constant memory** for storing constant values.
  - **Texture memory** with specialized hardware for filtering operations.
  - **Global memory** accessible to the entire GPGPU device via CUDA memory transfer functions.

  Accesses to the device memory typically take 300–600 clock cycles. However, a CUDA program can obtain significant performance boost due to L1/L2 caches in recent GPGPU architectures. The device memory persists for the lifetime of the entire program.

In what follows, we explore registers, shared memory, constant memory, and the global memory in detail. The texture memory is operated via the Texture Object APIs and its usefulness is limited in general-purpose computing. Therefore, we skip the discussion on texture memory, although readers are encouraged to explore texture memory discussed in the CUDA programming guide [7].

Registers

As shown in Fig. 7, each streaming multiprocessor has a set of on-chip registers that provide fast data access for various operations, which would otherwise consume several clock cycles due to frequent device memory accesses. Upon compilation with the nvcc compiler, the automatic variables declared in a CUDA kernel are stored in registers. However, not all automatic variables reap the benefits of registers because the GPGPU device's Compute Capability limits the maximum number of registers per streaming multiprocessor. If the number of requested registers in a CUDA kernel exceeds the device's capability, the leftover variables spill into the local memory (in off-chip device memory). Thereafter, any subsequent accesses to these variables may consume several hundreds of clock cycles. With recent advancements in the GPGPU device architecture and inclusion of caches, this performance artifact can be alleviated, however it is application-specific.

The number of registers used by threads in a CUDA kernel in conjunction with the number of threads per block also has a major performance implication – to what extent are the SMs occupied? The GPGPU devices realize parallelism via *warps*, a group of 32 concurrent threads. All of the threads in a warp execute the same instruction. Although, different warps may be executing different instructions of the same kernel. A streaming multiprocessor can have several active warps that can execute concurrently – when a set of warps executes memory instructions, the other set of warps performs useful computations. This level of concurrency amortizes the global memory latency. The *multiprocessor occupancy* is defined as the ratio of the number of active warps on SM to the maximum number of warps that can reside on a SM. Consequently, this ratio can at most be equal to 1 and a high value of multiprocessor occupancy is desirable to ensure high concurrency.

With the above background, let us study how the number of registers per thread and the number of threads per block affect the multiprocessor occupancy. Consider the Kepler K20Xm GPGPU device architecture, which belongs to Compute Capability 3.5. For this device, the maximum number of registers per SM is equal to 65536 and the maximum number of warps per SM is equal to 64. Using the nvcc compiler's Xptxas switch, we can determine the number of registers used and the amount of spill into the local memory. An illustration appears in Listing 4 where we compile a CUDA program, convolve.cu. As shown in the listing, the total number of registers per thread is 23 and there is no spill into the device's local memory.

Listing 4 An illustration of nvcc compiler's Xptxas option to determine the number of registers used and the amount of register spill into the local memory.

```
bash -4.2# nvcc -Xptxas -v -arch=sm_35 convolve.cu
ptxas info    : 0\,bytes gmem
ptxas info    : Compiling entry function '_Z8convolvePiiiPfiS_' for 'sm_35'
ptxas info    : Function properties for _Z8convolvePiiiPfiS_
0\,bytes stack frame, 0\,bytes spill stores, 0\,bytes spill loads
ptxas info    : Used 23 registers, 360\,bytes cmem[0]
```

The multiprocessor occupancy for a given kernel is obtained via Eqs. 2 through 5.

$$registers\_per\_block = registers\_per\_thread \times threads\_per\_block \quad (2)$$

$$total\_blocks = \frac{(max\_registers\_per\_SM)}{(registers\_per\_block)} \quad (3)$$

$$resident\_warps = min\left(maximum\_warps, \frac{total\_blocks \times threads\_per\_block}{32}\right) \quad (4)$$

$$occupancy = \frac{resident\_warps}{maximum\_warps} \quad (5)$$

For the example in Listing 4, let us assume that the CUDA kernel is launched with 256 threads per block. The total number of registers per block is: $23 \times 256 = 588$ registers. For this example, a SM in theory can execute a total of 11 blocks. The total number of resident warps is $min(64, \frac{11 \times 256}{32}) = 64$, thereby yielding multiprocessor occupancy equal to 1. Equations 6 through 11 show the calculations for multiprocessor occupancy if the threads in the above example were to use 100 registers.

$$registers\_per\_thread = 100; threads\_per\_block = 256 \quad (6)$$

$$registers\_per\_SM = 65536; maximum\_warps = 64 \quad (7)$$

$$registers\_per\_block = 100 \times 256 = 25600 \quad (8)$$

$$total\_blocks = \left\lfloor \frac{65536}{25600} \right\rfloor = 2 \quad (9)$$

$$resident\_warps = min(64, \frac{2 \times 256}{32}) = 16 \quad (10)$$

$$occupancy = \frac{16}{64} = 25\% \quad (11)$$

NVIDIA's CUDA occupancy calculator facilitates the occupancy calculations and elucidates the impact of varying thread block size and register count per thread on the multiprocessor occupancy. We discuss the occupancy calculator in detail in section "CUDA Optimizations".

Active Learning Exercise 8 – For a Compute Capability device 3.0, the `nvcc` compiler reports a usage of 50 registers per thread. If the thread block size is 512, what is the multiprocessor occupancy? Make sure to use the NVIDIA GPU data for the device related constants (maximum registers per SM, warp size, maximum number of warps per SM, etc.). Will the occupancy be any better if the kernel were to use 128 threads per block?

Shared Memory

NVIDIA GPGPU devices offer 64 KB on-chip shared memory that is used to cache frequently accessed data. The shared memory is slower than registers ($\approx$30 cycles per access versus 1 cycle per access for registers). However unlike registers, shared memory is accessible to all the threads in a thread block. The shared memory space is commonly used for thread collaboration and synchronization. These accesses, if performed via global memory, would typically consume several hundreds of clock cycles, thereby reducing the performance.

The kernel functions should be 'aware of' whether the variables are located in the device memory or in the shared memory. Programmers can statically allocate shared memory inside the kernel using the `__shared__` qualifier. Some examples

Table 7 Examples of CUDA shared memory declaration

Example	Syntax	Description
1	`__shared__ float a;`	The variable a is allocated in shared memory and is accessible to all threads inside a thread block
2	`__shared__ float A[BLOCKSIZE][BLOCKSIZE]` `//All threads` `load a value` `tidx=threadIdx.x;` `tidy=threadIdx.y;` `global_tidx=` `tidx+blockIdx.x` `*blockDim.x;` `global_tidy=` `tidy+blockIdx.y` `*blockDim.y;` `A[tidx][tidy]=` `global_A[global_tidx]` `[global_tidy];`	A two-dimensional array A is declared in the shared memory. The dimensions are BLOCKSIZE x BLOCKSIZE where BLOCKSIZE is the number of threads per block. All of the threads inside the thread block can access this array. This type of allocation is usually performed when each thread inside a thread block loads a value from the device global memory to shared memory, thereby optimizing the global memory bandwidth
3	`__shared__ float *A;` `A=(float *)malloc` `(sizeof(float)` `*BLOCKSIZE);`	Incorrect because array A is not static. See text for dynamic shared memory allocation

of static shared memory allocation appear in Table 7. In the first example, a simple shared memory variable, a is declared. Example 2 shows how a 2D shared memory variable is declared inside a kernel function. All of the threads in a thread block have access to this 2D shared memory variable. Example 2 also shows how local threads in a thread block load the corresponding global data element into this shared variable. The last example shows an incorrect way of dynamically allocating a shared memory variable.

It is also possible to dynamically allocate variables in the shared memory. The third parameter of execution configuration (the first two parameters are for specifying the dimensions of grid and thread blocks, respectively) specifies the size of the shared memory to be dynamically allocated inside the kernel function. Additionally, the dynamic shared memory variable inside the kernel function is declared with the extern qualifier. For example, consider that the BLOCKSIZE parameter is determined at runtime – in this case, example 3 in Table 7 for allocating array A will not work. Programmers can specify the size of the shared memory in the execution configuration during the kernel call as shown in Listing 5.

Note that it is also possible to perform multiple dynamic shared memory allocations by specifying the combined size of required arrays in the execution configuration. Inside the kernel function, a single shared memory array is used with appropriate offsets (using array sizes) to access the individual shared memory arrays.

Listing 5 An illustration of dynamic shared memory allocation by specifying the amount of memory to be allocated in the execution configuration. The corresponding shared memory variable declaration has extern qualifier.

```
__global__ void kernel(kernel-args) {
:
extern __shared__ float A[];
:
}
int main () {
:       kernel<<<dimGrid,dimBlock,sizeof(float)*BLOCKSIZE>>>(kernel-args);
:
}
```

Next, we study how threads within a thread block synchronize their accesses to the shared memory for thread collaboration. The threads in a thread block can synchronize via the __syncthreads() function, which provides a barrier for all of the threads in a thread block. Unless all the threads in a thread block finish executing the code preceding the __syncthreads(), the execution does not proceed ahead. This concept is illustrated by Fig. 8. More on __syncthreads() function appears in section "CUDA Optimizations" where we discuss shared memory optimization for algorithms that re-use the data (matrix-matrix multiplication for instance).

Active Learning Exercise 9 – Declare a BLOCKSIZE sized shared memory variable called mask inside of a CUDA kernel. Outline the methodology for allocating shared memory space for the shared variable, mask.

Active Learning Exercise 10 – In the foregoing section, we mentioned a method of allocating multiple shared memory variables inside a CUDA kernel. The

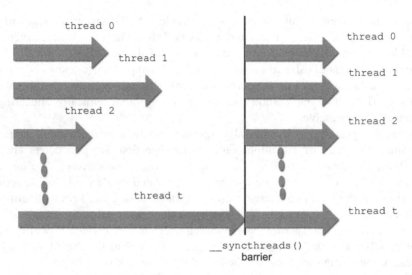

Fig. 8 Threads 0 to t inside a thread block synchronizing via the __syncthreads() function. All of the preceding statements before the __syncthreads() statement must be executed by all the threads in a thread block

methodology is as follows: (a) Specify the overall shared memory size in bytes in the execution configuration. This step is similar to the dynamic shared memory allocation method. (b) Declare a single extern __shared__ variable in the CUDA kernel. (c) Using the individual shared variable sizes as offsets, access the appropriate base addresses using the shared variable declared in Step b. Employ the outlined methodology to reserve a shared memory space for three variables: float A (*k* elements), float B (*l* elements), and float C (*m* elements).

In addition to shared memory, there are other mechanisms that enable threads to communicate with each other. The preceding discussion examines how threads within a block synchronize using the shared memory and __synchthreads() function. The threads within a warp can also synchronize and/or communicate via warp vote functions and warp shuffle functions. As readers may recall, a warp is a group of 32 concurrent threads.

The vote functions allow *active* threads within a warp to perform reduce-and-broadcast operation. The active threads within a warp are all threads that are in the intended path of warp execution. The threads that are not in this path are disabled (inactive). The vote functions allow active threads to compare an input integer from each participating thread to zero. The result of comparison is then broadcast to all of the participating threads in the warp. The warp voting functions are as follows.

- __all(int input): All participating threads compare input with zero. The function returns a non-zero value if and only if all active threads evaluate the input as non-zero.

- `__any(int input)`: The function is similar to `__any(input)`, however the function returns a non-zero value if and only if any one of the active threads evaluates the `input` as non-zero.
- `__ballot(int input)`: The function compares the `input` to zero on all active threads and returns an integer whose N^{th} bit is set when the Nth thread of the warp evaluates the `input` as non-zero.

The shuffle functions (`__shfl()`) allow all active threads within a warp to exchange data while avoiding shared memory all together. At a time, threads exchange 4 bytes of data; exchanges of 8 byte data is performed by calling shuffle functions multiple times. The exchanges are performed with respect to a thread's *lane ID*, which is an integer number from 0 to $warpSize - 1$. Some of the shuffle functions are as follows:

- `__shfl(int var, int srcLane, int width=warpSize)`: This function allows an active thread to look up the value of variable `var` in the source thread whose ID is given by `srcLane`. If the `width` is less than `warpSize` then each subsection of the warp acts as a separate entity with starting lane ID of 0. If `srcLane` is outside the $[0 : width - 1]$, then the function calculates the source as $srcLane\%width$.
- `__shfl_up(int var, unsigned int delta, int width=warpSize)`: The function calculates the lane ID of the source thread by subtracting `delta` from the current thread's lane ID and returns the value `var` held by the source thread. If the `width` is less than `warpSize` then each subsection of the warp acts as a separate entity with starting lane ID of 0. The source index does not wrap around the value of `width`, therefore lower `delta` lanes are unchanged.
- `__shfl_down(int var, unsigned int delta, int width=warpSize)`: This function is similar to `__shfl_up()` function, except that `__shfl_up()` computes the source lane ID by adding `delta` to the current thread's lane ID. Similar to `__shfl_up()`, the function does not wrap around for upper values of `delta`.
- `__shfl_xor(int var, int laneMask, int width=warpSize)`: This function calculates the source's lane ID by performing bitwise-XOR of the caller's lane ID and `laneMask`. The value held by the resulting source is returned into `var`. If `width` is less than `warpSize`, then each group of `width` threads is able to access elements from earlier groups of threads. However, if a group attempts to access later groups' elements, then the function returns their own value of the variable, `var`.

The warp vote and shuffle functions typically find their application when programmers wish to perform reduction or scan operations. Note that our discussion thus far comprised intra-block and intra-warp synchronizations. The synchronization between two blocks can only be accomplished via global memory accesses, which consumes significant amount of time. Programmers must pay attention to the type of applications they are porting to the GPGPU devices – applications that involve significant memory accesses and frequent global memory synchronization may perform better on the CPU host instead on the GPGPU device.

Constant Memory

The constant memory resides in the device memory and is cached. This memory space is used for storing any constant values frequently accessed by the kernel function, which would otherwise consume several clock-cycles if done via the device global memory. The constant memory is also useful for passing immutable arguments to the kernel function. The current GPGPU architectures provide L1 and L2 caches for global memory, making the constant memory less lucrative. However, constant memory can provide performance boost for earlier GPGPU architectures. To declare constant memory variables inside a .cu file, programmers must declare global variables with __constant__ prefix. For example,

```
__constant__ float pi=3.14159;
```

The host portion (CPU) is capable of changing a constant memory variable since a constant variable is constant only with respect to the GPGPU device. The host performs any changes to the constant memory via cudaMemcpyToSymbol() function:

```
template <class T> cudaError_t cudaMemcpyToSymbol (
const T & symbol,  // Destination address
const void &src,  // source address
size_t count,  // the number of bytes to copy
size_t offset,  // Offset from the start of symbol
enum cudaMemcpyKind kind ); // kind is cudaMemcpyHostToDevice
```

Active Learning Exercise 11 – Consider a host variable h_Cosine, a one-dimensional vector of constant size, Bins, initialized with cosine function values at Bins number of angles between 0 and 2π. Declare a constant memory variable d_Cosine of a fixed size equal to Bins. Perform a host-to-device copy from h_Cosine to d_Cosine.

Global Memory

In section "CUDA Program Structure", we explored how to manage the device global memory using cudaMalloc and cudaMemcpy functions. In this section, we study these functions in more depth. The device global memory is easily the most important unit with respect to the CUDA architecture. It is the largest memory unit where all (or at least, most) of the data for GPGPU processing is stored. Because this memory unit is located in the off-chip RAM, frequent accesses to the device global memory constitutes one of the major performance limiting factors in GPGPU computing. As discussed before, the CPU host and GPGPU device memories are usually disjoint. The host portion of a CUDA program explicitly allocates the device global memory for device variables. Throughout the program, the host portion communicates with the GPGPU device by copying data to-and-from the device global memory. In what follows, we discuss CUDA functions that enable programmers to allocate the device memory variables and perform host-device communications.

C programmers are already aware of the procedure for allocating and deallocating memory regions using the malloc() and free() functions, respectively. The CUDA programming model provides simple C extensions to facilitate device global memory management using the cudaMalloc() and cudaFree() functions. The syntaxes appear under.

```
// cudaMalloc: host portion allocates device global memory for device variables
cudaError_t cudaMalloc(
void **devPtr, // Host pointer address that will store the
// allocated device memory s address
size_t size) // size number of bytes to be allocated in device memory
```

```
// cudaFree: host portion 'frees' the device global memory
cudaError_t cudaFree( void *devPtr);
// The host pointer address storing the allocate device memory's
// address to be freed
```

The data transfer between the host portion of the code and device portion of the code is performed via the cudaMemcpy() function as follows:

```
// cudaMemcpy: Data transfer between the host and GPGPU device
cudaMemcpy(
void *dst_ptr, // destination address
const void *src, // source address
size_t count, // number of bytes to be transferred
cudaMemcpyKind kind) // enum type kind where kind can be
// cudaMemcpyHostToHost (0), cudaMemcpyHostToDevice (1),
// cudaMemcpyDeviceToHost (2), cudaMemcpyDeviceToDevice (3)
```

Readers are encouraged to exercise caution with de-referencing the device pointers inside the host portion, which can prove fatal for the CUDA program. Seasoned CUDA programmers avoid such mistakes by adding h_ prefix for the host pointers and d_ prefix for the device pointers. Additionally, readers are strongly encouraged to free the allocated device global memory pointers because the GPGPU device does not have a smart operating system for garbage collection. A complete reboot may be the only way to recover the lost device global memory.

CUDA Optimizations

The CUDA programming model is not known for straight-forward GPGPU application development. A naïve and sloppy CUDA program may provide little to no performance benefits at all! To develop an efficient CUDA application, programmers must be highly intimate with the device architecture to reap its complete benefits. Fortunately, researchers have meticulously studied different applications on GPGPU architectures to provide a generic set of strategies to perform GPGPU

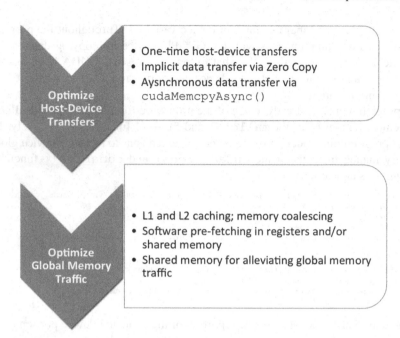

Fig. 9 A list of commonly used memory-level optimization strategies to alleviate host-device and global memory traffic

program optimization. Although, the strategies may vary from one application to another. In general, CUDA provides three primary optimization strategies namely, Memory-level optimization, Execution Configuration-level optimization, and Instruction-level optimization. In addition, CUDA also offers program structure optimization via unified memory. In what follows, we discuss each of these optimization strategies.

Memory-Level Optimization

While CUDA programming model provides several memory-level optimizations, we discuss memory optimization strategies to alleviate common performance bottlenecks arising due to host-device transfers and global memory traffic. These memory-level optimization strategies are listed in Fig. 9.

Memory-level optimization: Host-device transfers – One memory optimization strategy is to reduce the frequent transfers between the host and the device since the host-to-device bandwidth is usually an order of magnitude lower than the device-to-device bandwidth. It is highly beneficial to transfer all of the relevant data to the device memory for processing (even if it requires multiple kernel calls) and later transfer the data back to the host memory once all of the operations are finished.

Overlapping the kernel execution with data transfers using Zero-Copy can further optimize the host-device bandwidth. In this technique, the data transfers are performed implicitly as needed by the device kernel code. To enable Zero-Copy, the GPGPU device should support the host-mapped memory. The CUDA programming model provides `cudaHostAlloc()` and `cudaFreeHost()` functions to allocate and free the paged host memory. The mapping of the paged host memory into the address space of the device memory is performed by passing `cudaHostAllocMapped` parameter to `cudaHostAlloc()` function. The GPGPU device kernel implicitly accesses this mapped memory via the device pointer returned by the `cudaHostGetDevicePointer()` function. The functions for managing the page mapped memory are as follows.

Listing 6 Functions for zero-copy between host and device.
```
cudaError_t cudaHostAlloc(
void **ptr , // Host pointer to be paged
size_t size , // Size of the paged memory in bytes
unsigned int flags ); // cudaHostAllocMapped for zero copy.


cudaHostGetDevicePointer(
void ** devptr , // Device pointer for GPGPU to access the paged memory
void *hostptr , // The host pointer of the paged memory
unsigned int flags ); // flags is meant for any extensions , zero for now
```

Listing 7 illustrates Zero-Copy between the CPU host and the GPGPU device. The readers are also encouraged to read about `cudaMemcpyAsync()` [7] function for asynchronous host-device data transfers.

Listing 7 Illustration of Zero-Copy between the CPU host and GPGPU device. The memory copy is performed implicitly whenever the device accesses the host mapped memory via the device pointer (d_nfire) returned by `cudaHostGetDevicePointer()` function.
```
int main(){
:
// host vector to be page mapped
char *h_nfire;
// device pointer for the mapped memory
char *d_nfire;
cudaAllocHost((void **)&h_nfire , sizeof(char)*num_neurons , cudaHostAllocMapped ));
cudaHostGetDevicePointer((void **)&d_nfire ,(void *)h_nfire ,0));
:
kernel <<dimGrid , dimBlock >>>(d_nfire , num_neurons );
:
}
```

Memory-level optimization: Caching in L1 and L2 caches; and coalescing – The more recent GPGPU devices (Compute Capability 2 and higher) offer caches for global memory namely the L1 and L2 caches. For Compute Capability devices 2.x, by using the `nvcc` compiler flag `dlcm`, programmers can enable either both L1 and L2 caches by default (`-Xptxas dlcm=ca`) or L2 cache alone (`-Xptxas dlcm=cg`). A cache line is 128 bytes and is aligned with a 128-byte segment in the device memory. If both L1 and L2 caches are enabled, then the memory accesses are serviced via 128-byte transactions. If only L2 cache is enabled, then the memory accesses are serviced via 32-byte transactions. If the request size is 8 bytes, then the 128-byte transaction is broken into two requests, one for each half-warp. If the request size is 16 bytes, then the 128-byte transaction is broken into four requests, one for each quarter warp (8 threads). Each memory request is broken into cache line requests, which are serviced independently. The cache behavior for GPGPU devices is similar to general-purpose processors. If there is a cache hit, the request is served at the throughput of L1 or L2 cache. A cache miss results in a request that is serviced at the device global memory throughput.

Compute Capability 3.x, 5.x, and 6.x devices usually allow global memory caching in L2 cache alone. However some 3.5 and 3.7 devices allow programmers to opt for the L1 cache as well. The L2 caching for Compute Capability devices 3.x and above is similar to Compute Capability 2.x devices.

A progam optimally utilizes the global memory when the accesses lead to as many cache hits as possible. In such a case, threads within a warp complete memory accesses in fewer transactions. This optimized global memory access pattern is generally called *coalesced access*. The term *global memory coalescing* had significant importance to Compute Capability 1.x devices, where coalesced access rules were highly stringent. However, with the introduction of caches in recent GPGPU architectures, the term coalescing has become obscure. To achieve global memory 'coalescing' in recent GPGPU architectures, programmers must strive to write cache-friendly codes that perform aligned accesses. Similar to CPU architectures, good programming practices lead to optimal GPGPU codes.

Active Learning Exercise 12 – Perform research on Compute Capability 1.x devices; and write down the rules for coalesced global memory accesses.

Memory-level optimization: Software-prefetching using registers and shared memory – The device global memory is an order of magnitude slower than registers and shared memory. Programmers can use the register and shared memory space for caching frequently used data from the device global memory. This technique is referred to as *software prefetching*; avid assembly language programmers among readers may already be aware of this technique.

Memory-level optimization: Shared memory to alleviate global memory traffic – The judicious use of shared memory space to reduce the global memory traffic is a highly important technique especially for algorithms that exploit data locality, matrix-matrix multiplication and several image processing applications for instance. Here, we discuss how the shared memory space can be used to enhance the global memory throughput using matrix-matrix multiplication as a case study. Readers should recall the concept of matrix-matrix multiplication: any two matrices $A_{m \times n}$ and $B_{n \times p}$ are multiplied to yield a matrix, $C_{m \times p}$. Any element c_{ij} in matrix $C_{m \times p}$ is obtained by computing the scalar product between the ith row of matrix A and jth column of matrix B. Let us first consider a naïve matrix-matrix multiplication and find out why it sub-optimally utilizes the global memory bandwidth. Listing 8 shows the naïve implementation that multiplies two matrices of equal dimensions (width x width each).

Listing 8 A naïve implementation of matrix-matrix multiplication kernel.
```
1.  __global__ void
2.  matrixmul_kernel(float *d_A, float *d_B, float *d_C,int width) {
3.  int row, col, k=0;
```

```
4. float temp=0;
//thread accesses global row
5. row = threadIdx.x + blockIdx.x*blockDim.x;
//thread accesses global col
6. col = threadIdx.y + blockIdx.y*blockDim.y;
7. if(row <width && col < width){   //out of bound threads must not work
8.   temp=0;
9.   for (k=0;k<width;k++){
10.      temp+=d_A[row*width + k]*d_B[k*width + col];
11.  }
12. d_C[row*width+col]=temp;
13. }
14. }
```

A careful inspection of the kernel function in Listing 8 reveals that the performance bottleneck is in lines 9 and 10. Note that in each iteration of the for loop in Line 9, a thread performs two global memory loads (loads elements d_A[row*width +k] and d_B[k*width + col], respectively) and performs two floating-point operations (multiplies the two loaded elements and adds the product with the temp variable). Let us define the term computation-to-global memory access (CGMA) ratio, which is the ratio of the total number of computations to the total number of global memory accesses. The CGMA ratio is often used to characterize a GPGPU kernel as a computation-bound kernel or a communication-bound kernel. In our example of naïve matrix-matrix multiplication, the CGMA ratio is (2 floating-point operations per 2 floating-point accesses) equal to 1. This ratio is too small to reap the maximum benefits of a throughput-oriented architecture. For instance, if the GPGPU device memory has a bandwidth of 200 GB/s, then the kernel in Listing 8 performs computations at the rate of 50 giga-floating point operations per second (GFLOPS). This computation throughput does not do justice to modern day GPGPU devices with peak performance as high as 10 TFLOPS for single-precision.

It is clear from the above example that the CGMA ratio for matrix-matrix multiplication needs to improve, possibly by boosting the global memory bandwidth. In what follows, we discuss 'tiled' matrix-matrix multiplication using shared memory, which enables us to improve the global memory bandwidth for this operation. Prior to delving into the GPGPU implementation, let us investigate the concept of 'tiling'. To perform matrix-matrix multiplication, the matrices can be broken into smaller tiles that are multiplied together to yield partial results. The partial results from pertinent tile-multiplication are then added to obtain the final result.

For example, consider multiplication of two matrices, $M_{4 \times 4}$ and $N_{4 \times 4}$; the result is stored in the matrix, $P_{4 \times 4}$ (see Fig. 10). The matrix P can be broken into four tiles where tile-1 comprises elements $P_{0,0}$, $P_{0,1}$, $P_{1,0}$, and $P_{1,1}$; tile-2 comprises elements $P_{0,2}$, $P_{0,3}$, $P_{1,2}$, and $P_{1,3}$, and so on. Consider the evaluation of tile-1 elements; Fig. 10 shows the tile-1 elements of matrix P enclosed in the square box. The tile-1 elements are evaluated in two steps: In the first step, the curved tiles over matrices M and N (see Fig. 10) are multiplied together to yield the partial result for tile-1 elements $P_{0,0}$ through $P_{1,1}$ (first two terms in the right hand side of the equations in Fig. 10). In the second step, the tile over matrix M moves to the right (see Fig. 11) and the tile over matrix N moves down (see Fig. 11) to compute the next set of

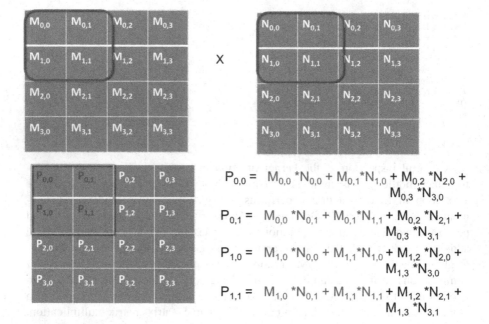

Fig. 10 The tiles in matrices M and N (curved tiles) multiply to yield the partial results for the tile in matrix P (highlighted in square box)

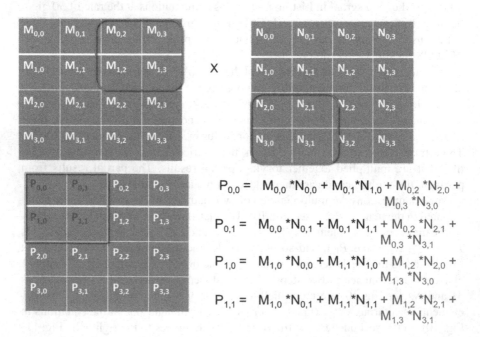

Fig. 11 The tiles in matrices M and N (curved tiles) multiply to yield the partial results for the tile in matrix P (square box)

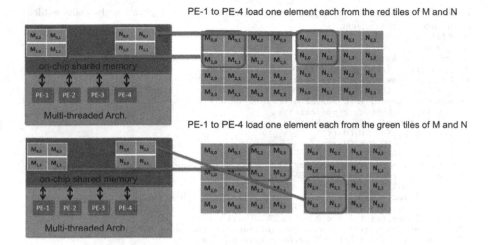

Fig. 12 A general depiction of matrix-matrix multiplication on a multi-threaded architecture with shared memory

partial results (last two terms in the right hand side of the equations in Fig. 11). The partial results from the above two steps are added to produce the complete result for tile-1 elements. This tile movement is in agreement with the concept of matrix-matrix multiplication where we compute the scalar product between the rows of the first matrix (M in this case) and the columns of the second matrix (N in this case). The evaluation of the other tiles is similar to this tile-1 example. Readers are encouraged to compute the results for the remaining tiles for practice.

In general, how does tiling help with parallelization of matrix-matrix multiplication? To obtain an answer to this question, consider a multi-threaded computing architecture (see Fig. 12) that stores the operand matrices in the off-chip memory, which resides far away from the computing architecture. Consequently, accesses to this off-chip memory is slow. Let us assume that this architecture is also equipped with on-chip shared memory that provides faster access versus the off-chip memory. The architecture contains four processing elements (PEs) that share the on-chip memory. For the foregoing example of multiplying matrices $M_{4 \times 4}$ and $N_{4 \times 4}$, envision the following scenario. Each one of the four PEs loads a curved tile element from matrices M and N into the shared memory as depicted in Fig. 12 (top). PE_1 loads $M_{0,0}$ and $N_{0,0}$; PE_2 loads $M_{0,1}$ and $N_{0,1}$; and so on. After this collaborative loading, the shared memory now contains the curved tiles from M and N for the computation of the first set of partial result. Each PE computes its partial result via shared memory look-up: PE_1 computes the partial result for $P_{0,0}$, PE_2 computes the partial result for $P_{0,1}$ and so on. Similarly, the PEs cooperatively load the next set of curved tile elements (see Fig. 12 bottom) to evaluate the second set of partial result. This collaborative loading has clearly reduced the number of trips to the farther, off-chip memory, thereby providing tremendous benefits. Do we have an architecture that facilitates this tiling operation? GPGPU devices are great fit!

Listing 9 The shared memory implementation of matrix-matrix multiplication, also called as tiled matrix-matrix multiplication.

```
0. #define TILEWIDTH 16
1. __global__ void
2. matrixmul_kernel(float *d_A, float *d_B, float *d_C,int width) {
3. __shared__ float Ashared[TILEWIDTH][TILEWIDTH];
//shared memory to load shared tile from matrix A
4. __shared__ float Bshared[TILEWIDTH][TILEWIDTH];
//shared memory to load shared tile from matrix B
5. int bx=blockIdx.x, by=blockIdx.y;
6. int tx=threadIdx.x, ty=threadIdx.y;
7. int row=bx*TILEWIDTH+tx;
8. int col=by*TILEWIDTH+ty;
9. float temp=0;
10.//Loop over the tiles Ashared and Bshared to compute an element in d_C
11. for (int i=0;i < width/TILEWIDTH; i++){
//threads collaboratively load Ashared
12.     Ashared[tx][ty] = d_A[row*width + i*TILEWIDTH + ty];
//threads collaboratively load Bshared
13.     Bshared[tx][ty] = d_B[(i*TILEWIDTH+tx)*width + col];
14.     __syncthreads(); //wait for threads int the block to finish
15.//Loop over the tiles and perform computations
16.     for(int k=0;k<TILEWIDTH;k++){
17.         temp+=Ashared[tx][k]*Bshared[k][ty];
18.     }
19.     __syncthreads(); //wait for threads in the block to finish
20. }
21. d_C[row*width + col] = temp;
22. }
```

Listing 9 provides the kernel for the shared memory implementation. In Listing 9, Line 0 sets the width of the tile via #define TILEWIDTH 16. For simplicity, we assume that the program creates thread blocks of dimensions, TILEWIDTH*TILEWIDTH. Lines 3 and 4 statically declare two shared variables, Ashared and Bshared. Because these variables reside in the shared memory space, all the threads in a thread block will have access to these variables. Lines 5 and 6 store the thread block IDs (in x and y dimensions) and thread IDs (in x and y dimensions) in variables bx, by, tx, and ty, respectively. In Lines 7 and 8, each thread calculates the global row (row) and global column (col) indices of the target element in d_C.

Figure 13 shows the conceptual representation of calculating the matrix indices for tiled matrix multiplication. A for loop over counter, i in Line 11 performs tile traversal over the matrices. Because the matrix d_A is traversed horizontally, the tile traversal requires an offset of i*TILEWIDTH in the horizontal direction for each iteration of the counter, i. A single thread with local ID (tx, ty) then accesses the element (row, i*TILEWIDTH + ty) in matrix d_A and loads it in the shared array Ashared[tx][ty] (Line 12). Similarly, the matrix d_B is traversed in vertical direction; therefore the tile traversal requires an offset of i*TILEWIDTH in vertical direction for each iteration of counter, i. Correspondingly, a thread with local ID (tx, ty) accesses the element (i*TILEWIDTH + tx, col) in matrix d_B and loads it in shared array, Bshared[tx][ty] (Line 13). Note that the threads in a thread block must wait for all the other participant threads to load their respective elements. This synchronization is provided by __syncthreads() in Line 14. After loading the shared arrays with relevant matrix elements, each

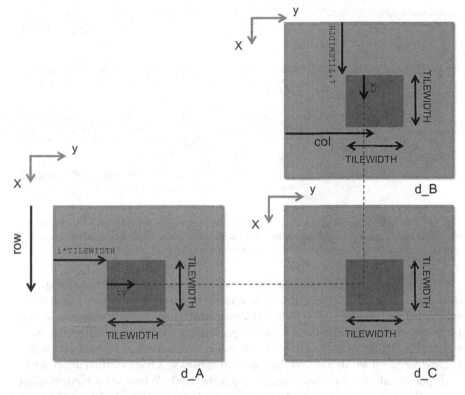

Fig. 13 Conceptual representation of index calculation for tiled matrix multiplication

thread evaluates its partial result in Lines 16 through 18. Line 19 provides the synchronization such that all the threads in a thread block finish their respective computations. At the end of the for loop (Line 20), each thread loads the complete result of its respective element (row,col) in matrix, d_C (Line 21).

Readers should carefully observe that each thread performs exactly two global loads, one for each matrix in lines 12 and 13. After these global loads, each thread performs TILEWIDTH multiplications and TILEWIDTH additions (i.e., TILEWIDTHx2 floating-point operations) in lines 16–18. Therefore, this kernel performs TILEWIDTH floating-point computations for every floating-point global memory access, thereby providing TILEWIDTH times boost to the CGMA ratio (recall that CGMA ratio for the naïve implementation is 1). On a GPGPU device with 200 GB/s global memory bandwidth, the kernel provides a performance of

$$\frac{200 \ GB/s}{4B \ per \ floating-point} \times (TILEWIDTH = 16) = 800 \ GFLOPS!$$

Active Learning Exercise 13 – A kernel performs 100 floating-point operations for every 10 floating-point global memory accesses. What is the CGMA ratio for this kernel? Assuming that the GPGPU device has a global memory bandwidth equal to 150 GB/s, what is the kernel performance in GFLOPS?

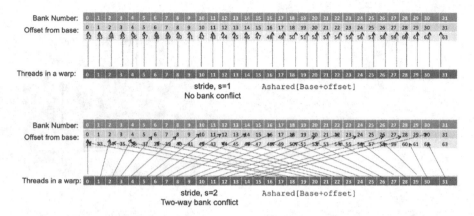

Fig. 14 This illustration shows examples of two stride accesses, $s = 1$ (top) and $s = 2$ (bottom). For the stride access, $s = 1$, each thread in a warp accesses a unique bank. For stride access, $s = 2$, there is two-way bank conflict between the first half-warp and the second half-warp

With our previous discussion on the use of shared memory to alleviate the global memory congestion, it is clear that judicious use of the shared memory can provide substantial performance boost for CUDA programs. However, programmers should be aware of a potential shared memory performance bottleneck called the *bank conflict*. In GPGPU devices, the shared memory is divided into 32 banks such that successive 32-bit words are stored in successive banks. A bank conflict arises when multiple threads within a warp access the same bank. Whenever a bank conflict arises, the accesses to the shared memory bank are serialized. An n-way bank conflict arises when n threads in a warp access the same bank – such accesses are completed in n serial steps. If two threads access the addresses within the same 32-bit word, then the word is broadcast to the threads, thereby avoiding a bank conflict. Similarly, a bank conflict is avoided when all the threads in a warp or a half-warp access the same word. In such a case, the word is broadcast to the entire warp or the half-warp. Bank conflicts usually arise when threads access the shared memory with some stride, s. For example:

```
extern __shared__ float Ashared[];
data=Ashared[Base+s*thread_id];
```

Figure 14 shows examples of two stride accesses, $s = 1$ and $s = 2$. As shown in the same figure, the shared memory is divided into 32 banks with successive words stored in successive banks. The bank-0 stores words 0 and 32, bank-1 stores, 1 and 33, and so on. For stride $s = 1$, each thread (0 through 31) in a warp accesses a unique bank (0 through 31), therefore there is no bank conflict in this case. However, for stride $s = 2$, the threads in the first half-warp (0–15) have a two-way bank conflict with the threads in the second half-warp (16–31). For example, the thread with ID equal to 0 (belonging to the first half-warp) accesses a word at offset 0 from the base address (stored in bank-0) and the thread with ID equal to 16 (belonging to the second half-warp) accesses a word at offset 32 from the base address (also stored in bank-0), leading to a two-way bank conflict.

Active Learning Exercise 14 – This activity summarizes our understanding of the CGMA ratio and shared memory bank conflict. Assume that the kernel given in Listing 10 is executed on a GPGPU device with global memory bandwidth equal to 200 GB/s. Calculate the CGMA ratio and the performance achieved by this kernel in GFLOPS. Notice the use of two shared memory variables, fire and fired. Is there a potential for bank conflict(s)? Why or why not?

Listing 10 Kernel code for active learning Exercise 14.

```
1.  __global__ void kernel(float *level1_I,float *level1_v,
float *level1_u,mytype *L1_firings, mytype2 *myfire,int Ne){
2.      extern __shared__ bool fire[];
3.      __shared__ bool fired;
4.      int k = threadIdx.x + blockIdx.x*blockDim.x;
5.      int j= threadIdx.x;
6.      auto float level1v, level1u;
7.      if(j==0)
8.                    fired =1;
9.      __syncthreads();
10.     if(k<Ne){
11.             level1v = level1_v[k];
12.             level1u = level1_u[k];
13.             if (level1v>30) {
14.                     L1_firings[k]=0;
15.                     level1v=-55;
16.                     level1u=level1u+4;
17.             }
18.             level1v=level1v+0.5*(level1v*(0.04*level1v+5)
                          +140-level1u+level1_I[k]);
19.             level1u=level1u+0.02*(0.2*(level1v)-level1u);
20.             level1_v[k] = level1v; level1_u[k] = level1u;
21.             fire[j] = L1_firings[k];
22.             fired&=fire[j];
23.             __syncthreads();
24.     }
25. }
```

Execution Configuration-Level Optimization

This level of optimization targets the parameters appearing in the kernel execution configuration (<<< >>>) and serves two primary performance objectives: (1) maximize the multiprocessor occupancy and (2) enable concurrent execution via streams. In what follows, we discuss these two performance objectives.

Maximizing multiprocessor occupancy – As discussed in section "CUDA Memory Organization", on-chip, fast memories such as registers and shared memory can provide tremendous performance boost. However, the catch lies in their limited quantity, which is dependent on the device's Compute Capability. The limited number of registers and shared memory limits the number of thread blocks (and therefore, the number of warps) that can reside on a streaming multiprocessor (SM), affecting the multiprocessor occupancy. Readers should recall that the multiprocessor occupancy is the ratio of the total number of warps residing on an SM to the maximum number of warps that can reside on an SM. While a high multiprocessor occupancy does not always imply high performance, nonetheless it is a good measure of concurrency. Therefore, CUDA programmers must strive to create grids and thread blocks for kernels such that the multiprocessor occupancy is generally high. Although this process may involve some experimentation with multiple execution configurations.

How can I achieve high multiprocessor occupancy, whilst not spending time performing meticulous calculations as shown in Eqs. 6, 7, 8, 9, 10, and 11? NVIDIA has a wonderful and simple tool called the CUDA occupancy calculator [7] to perform all of this mathematical work! The CUDA occupancy calculator allows users to select the Compute Capability and shared memory configuration for their GPGPU devices. Once these device configurations are selected, the CUDA occupancy calculator automatically fills the device related constants such as active threads per SM, active warps per SM, etc. The programmer then provides kernel information including the number of registers per thread (identified using the Xptxas nvcc switch discussed in section "CUDA Memory Organization"), the amount of shared memory per block, and the number of threads per block information to the occupancy calculator. After receiving the above pertinent kernel information, the occupancy calculator provides the multiprocessor occupancy value (in percentage) and graphically displays the impact of varying block size, shared memory usage per block, and register count per thread on the multiprocessor occupancy.

For the CUDA kernel in Listing 10, let us assume that the target architecture belongs to Compute Capability 3.5 and the shared memory configuration is 16 KB (48 KB for L1 cache). The nvcc compilation with Xptxas option for this kernel yields 20 registers per thread. If we assume a thread block size equal to 192 and shared memory per block equal to 192 bytes, then CUDA occupancy calculator provides us with multiprocessor occupancy value equal to 94%. Figure 15 shows the impact of varying block size, shared memory usage, and register count on occupancy, as given by the occupancy calculator. These figures suggest that for the thread block size equal to 256, we can expect the occupancy to reach 100%.

Readers are also encouraged to explore CUDA APIs such as cudaOccupancyMaxActiveBlocksPerMultiprocessor [7] for calculating the multiprocessor occupancy for CUDA kernels.

Active Learning Exercise 15 – Analyze the multiprocessor occupancy for the tiled matrix-matrix multiplication example. Assuming Compute Capability devices 3 and 5, use the CUDA occupancy calculator to obtain the multiprocessor occupancy values for thread block sizes: 128, 256, 512, and 1024.

Concurrent execution using streams – Readers should recall that frequent host-device transfers are significant bottlenecks that appear in CUDA programs. The CUDA streams provide a way to hide the data transfer latency by overlapping the memory transfers with kernel invocations. A *stream* consists of a sequence of instructions that execute in-order; these sequences include host-device transfers, memory allocations, and kernel invocations. For devices with Compute Capability 2.0 and above, streams enable programmers to perform device-level concurrency – while all of the instruction sequences within a stream execute in-order, multiple streams may have instruction sequences executing out-of-order. Therefore, instruc-

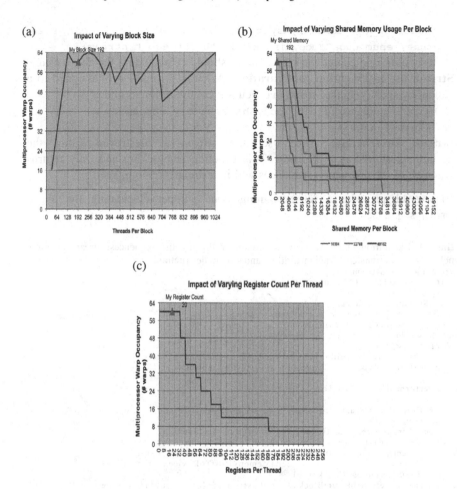

Fig. 15 Impact of thread block size, shared memory per block usage, and register count per thread on multiprocessor occupancy. (**a**) Impact of block size on occupancy. (**b**) Impact of shared memory on occupancy. (**c**) Impact of register count on occupancy

tion sequences from different streams can be issued concurrently. For instance, when a single stream performs kernel invocation, the other stream completes any data transfer operation. It should be noted that relative execution order of instruction sequences across streams is unknown.

CUDA streams are of type `cudaStream_t` type and generally follow the coding sequence given under:

- **Stream creation**: : `cudaStreamCreate()` function call creates a stream:

  ```
  cudaError_t cudaStreamCreate(cudaStream_t *stream);
  ```
- **Stream use in asynchronous data transfer**: A stream can also perform asynchronous data transfers using `cudaMemcpyAsync()` function as follows:

```
cudaError_t cudaMemcpyAsync(void *dst, const void *src, size_
t count, enum cudaMemcpyKind kind, cudaStream_t stream);
```
It should be noted that the host memory should be pinned for the above usage.
- **Stream use in execution configuration**: A kernel invocation is assigned to a stream by specifying the stream in execution configuration as under:
```
kernel <<<dimGrid,dimBlock,SharedMemory,stream>>>
(<kernel-args>);
```
- **Stream Destruction**: After use, the stream is destroyed using the cudaStreamDestroy() function. This function is blocking and only returns when all of the instruction sequences within a stream are completed.

Listing 11 provides a self-explaining code snippet elucidating the above described sequence.

Listing 11 Illustration of two concurrent streams following the sequences: stream creation, asynchronous data transfer, kernel invocation, and stream destruction.

```
//Creating two streams
1.  int size=1024; //1024 data items per stream
2.  :
3.  cudaStream_t stream[2];
4.  //Allocate host and device memory
5.  float *h_data[2],*d_data[2];
//one host-device pair for each stream
6.  for(i=0;i<2;i++) {
7.      cudaMallocHost((void**)&h_data[i],sizeof(float)*size);
8.      cudaMalloc((void**)&d_data[i],sizeof(float)*size);
9.  }
10. //Perform initialization
11. :
12. //Follow the stream sequences except for destruction
13. for(i=0;i<2;i++) {
14.     cudaStreamCreate(&stream[i]); //create stream i
// ith stream initializes async. host-to-device transfer
15.     cudaMemcpyAsync(d_data[i],h_data[i],sizeof(float)*size,
                            cudaMemcpyHostToDevice,stream[i]);
// ith stream invokes the kernel
16.     kernel<<<dimGrid,dimBlock,shared,stream[i]>>>(d_data[i],size);
17. cudaMemcpyAsync(h_data[i],d_data[i],sizeof(float)*size,
                            cudaMemcpyDeviceToHost,stream[i]);
//ith stream initializes async. device-to-host transfer
18. }
19. //Streams synchronize. Blocks until streams finish
20. cudaStreamDestroy(stream[0]);
21. cudaStreamDestroy(stream[1]);
22. //free pointers
23. }
```

Active Learning Exercise 16 – Write a CUDA program that creates n streams to perform vector-vector addition. Hint: The ith stream operates on the data starting from &d_A[i*data_per_stream] and &d_B[i*data_per_stream].

Instruction-Level Optimization

This level of optimization targets the optimization of arithmetic instructions and branching statements in a CUDA kernel. The arithmetic operations can be easily

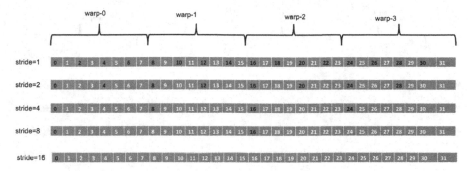

Fig. 16 An illustration of participating threads (highlighted in black) within hypothetical warps of size equal to 8 threads. In each iteration of the `for` loop, there is at least one divergent warp

optimized using `fast math` [1] functions. The branching statement optimization, however, requires meticulous handling of statements to avoid an artifact known as *divergent warps*. Readers should recall that all of the threads within a warp execute the same instruction. A warp is divergent if the threads inside a warp follow different execution paths (for example, first half-warp satisfies the `if` statement while the second half-warp satisfies the `else` statement). In such a case, divergent paths are serialized, which results in reduced performance. To illustrate this concept, we discuss an important parallel pattern called reduction, which derives a single value by applying an operation (addition, for instance) to all of the elements in an array. Listing 12 provides the code snippet of a reduction kernel (Variant 1), which is prone to producing divergent warps. Readers are encouraged to verify that the code will produce the correct reduction result.

Listing 12 Reduction kernel snippet (Variant 1) that produces divergent warps.
```
1. __shared__ float partialSum[BLOCKSIZE];
2. :
3. int t = threadIdx.x;
4. for(int stride = 1; stride < blockDim.x; stride*=2){
5.     __syncthreads();
6.     if(t%(2*stride)==0)
7.         partialSum[t]+=partialSum[t+stride];
8.}
```

To analyze this example, let us assume that our hypothetical GPGPU device supports 8 threads per warp. Further assume that reduction is performed using blocks of size 32 threads. Figure 16 illustrates the participating threads (highlighted in black) within a warp in each iteration of the `for` loop (stride varies from 1 to 16). As seen in the same figure, there is at least one divergent warp in each iteration of the `for` loop. Specifically, strides 1, 2, and 4 include four divergent warps each; whereas strides 8 and 16 include two and one divergent warps, respectively. The entire execution of the `for` loop leads to $4+4+4+2+1 = 15$ divergent warps. As discussed before, divergent warps are serialized, thereby reducing the performance.

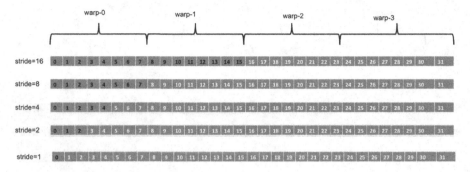

Fig. 17 Illustration of participating threads (highlighted in black) within hypothetical warps of size equal to 8 threads. In first two iterations, none of the warps are divergent. Divergent warps (one each) occur in last three iterations

Listing 13 provides the code snippet of a reduction kernel (Variant 2 that reduces the number of divergent warps). Figure 17 illustrates the participating threads within a warp in each iteration of the `for` loop (`stride` varies from 16 to 1).

Listing 13 Reduction kernel snippet (Variant 2) that reduces the number of divergent warps.
```
1.  __shared__ float partialSum[BLOCKSIZE];
2.  int t = threadIdx.x;
3.  for(int stride = blockDim.x/2; stride >= 1; stride /=2){
4.    __syncthreads();
5.    if(t< stride)
6.        partialSum[t]+=partialSum[t+stride];
7.  }
```

As seen in Fig. 17, none of the 8-thread warps are divergent in the first two iterations of the `for` loop. The divergent warps (one each) occur only in the last three iterations, thereby leading to a total of three divergent warps (versus 15 divergent warps in Variant 1). Therefore, Variant 2 provides higher performance versus Variant 1.

Active Learning Exercise 17 – Assume that there are 256 threads per block; calculate the total number of divergent warps for Variant 1 and Variant 2 of the reduction kernel. Is the scenario any better for 512 threads per block?

Program Structure Optimization: Unified Memory

In our programs so far, we performed explicit (with the exception of Zero-Copy) data transfers between the CPU host and GPGPU device via `cudaMemcpy` function. Needless to say, this process may be very lengthy and highly error-prone for large programs. *Unified memory* is a nice feature introduced in CUDA 6.0 that enables programmers to perform implicit data transfers between the host and the device. Unified memory introduces the concept of *managed memory* wherein the memory is allocated on both the host and the device under the supervision of the

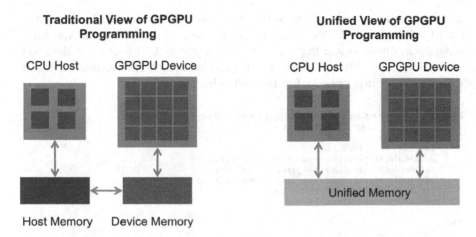

Fig. 18 Difference in 'developer's view' between explicit data transfers and unified memory data transfers

device driver. The device driver ensures that these two sets of data remain coherent throughout the program execution. In essence, the user just maintains a single pointer for both the CPU host and GPGPU device. A data transfer is implicitly triggered before the kernel launch and another one immediately after the kernel termination. Readers should note that the unified memory operation is similar to explicit host-device transfers, with the exception that the device driver automatically manages data transfers in unified memory. Unified memory alleviates programmers with the burden of meticulous host-device transfer management, allowing them to write shorter codes and focus more on the program logic. Unified memory should not be confused with Zero-Copy where the data transfer is triggered whenever the device kernel accesses the data. Figure 18 summarizes the difference in 'developer's view' between an explicit data transfer (shown on the left) and unified memory data transfer (shown on the right).

Programmers can allocate managed memory in two ways:

1. Dynamically via the cudaMallocManaged() function call.
2. Statically by declaring global variable with the prefix: __managed__.

The syntax for cudaMallocManaged() is as follows.

```
template <class T> cudaMallocManaged(
T **dev_ptr, // address of the memory pointer
 size_t bytes, // size in bytes of the required memory
 unsigned flags) // Either cudaMemAttachGlobal for all kernels to access or
// cudaMemAttachHost to make the variable local to declaring host
// and kernels invoked by the declaring host.
```

Listing 14 illustrates unified memory using vector-vector addition as example. While the kernel construction is the same as Listing 1, notice the changes in the main() function. Using cudaMallocManaged(), lines 4–6 allocate the space for variables a, b, and c on both the CPU host and GPGPU device. The host input is initialized in lines 8–10 and the device output is evaluated by the kernel call in Line 12. Prior to accessing the modified values of the variables, programmers must ensure that the kernel has terminated. This check is done via cudaDeviceSynchronize() function call in Line 13. The variables a, b, and c are freed via cudaFree() function call in lines 15–17.

Listing 14 Vector-vector addition code snippet illustrating unified memory.

```
1.  int main(int argc, char **argv) {
2.    int *a,*b,*c;
3.    int vec_size=1000, i;
4.    cudaMallocManaged(&a,vec_size*sizeof(int));
5.    cudaMallocManaged(&b,vec_size*sizeof(int));
6.    cudaMallocManaged(&c,vec_size*sizeof(int));
7.    //Host-portion prepares the data
8.    for (i=0; i<vec_size; i++) {
9.      a[i]=i; b[i]=i;
10.   }
11.   //Run the GPU Kernel
12.   gpu_kernel <<<1,vec_size >>>(a,b,c,vec_size);
13.   cudaDeviceSynchronize(); //Wait for the GPU to finish execution.
14.   //Free pointers
15.   cudaFree(a);
16.   cudaFree(b);
17.   cudaFree(c);
18.   return 0;
19. }
```

The example in Listing 14 shows substantial simplification of the vector-vector addition code structure using the unified memory concept. Although, programmers must note that unified memory is not a performance optimization. Proficient CUDA programmers with a command on explicit host-device transfers and Zero-Copy optimization technique can achieve high-performance for their applications.

In this section, we discussed several optimization strategies that CUDA programmers can employ to achieve significant application performance. It is worth noting that the choice of optimization varies across applications. While the techniques covered here are quite comprehensive, we have not fully exhausted the list of possible strategies. For instance, dynamic parallelism allows a CUDA kernel to create child kernels, thereby avoiding kernel synchronization in the host portion and any host-device transfers. The high-performance computing (HPC) community continually augments the optimization strategy list via exhaustive research efforts. Readers are encouraged to stay abreast with scientific publications. Several applications share 'parallelization logic' that helps programmers avoid re-inventing the wheel.

Case Study: Image Convolution on GPUs

In this section, we study a parallel pattern that commonly arises in various scientific applications namely, the convolution. The convolution algorithm frequently occurs in signal processing contexts such as audio processing, video processing, and image filtering, among others. For example, images are convolved with convolution kernels (henceforth referred to as convolution masks to avoid ambiguity with the CUDA kernel) to detect sharp edges. The output of a linear time invariant (LTI) design is obtained via convolution of the input signal with the impulse response of the LTI design. The convolution operation has two interesting aspects that make it highly lucrative for the GPGPU device. First, the convolution operation is highly data parallel – different elements of the input data can be evaluated independent of the other elements. Second, the convolution operation on a large input (a large image or an audio signal for instance) leads to significantly large number of operations. In what follows, we first provide a brief mathematical background on this highly important mathematical operation. Then, we explore how the convolution operation can be effectively deployed on GPGPU devices.

Convolution is a mathematical array operation (denoted with asterisk, *) where each output element (P[j]) is a weighted sum of neighboring elements of the target input element (N[j]). The weights are defined by an input array called, the convolution mask. The weighted sum, P[j], is calculated by aligning the center of the convolution mask over the target element, N[j]. The input mask usually consists of odd number of elements so that equal numbers of neighboring elements surround the target element in all directions.

Let us consolidate our understanding of the convolution operation via an example. For simplicity, let us assume that we need to convolve an array of eight elements, N, with a convolution mask of five elements, M. Figure 19 illustrates the convolution procedure. Notice the evaluation of element, P[2] (top) – the center of the mask (M[2]) is aligned with the target input element N[2] (dark gray); next the overlapping elements of P and M are multiplied and the products are added to obtain the weighted sum:

$$P[2] = N[0] \times M[0] + N[1] \times M[1] + N[2] \times M[2] + N[3] \times M[3] + N[4] \times M[4]$$

Notice the evaluation procedure of the element, P[1] (Fig. 19 bottom). Similar to evaluation of P[2], the center of the mask, M[2] is aligned with the target input element N[1] (highlighted in dark gray). However, the mask element, M[0] flows out of array, N. In such a case, the overflowing elements of the mask are multiplied with 'ghost elements', g_i, which are customarily set to zero. The element, P[1] in this case is evaluated as:

$$g1 = 0$$
$$P[1] = g1 \times M[0] + N[0] \times M[1] + N[1] \times M[2] + N[2] \times M[3] + N[3] \times M[4]$$

This process is performed on all of the array elements to obtain the convolution output, P.

The convolution operation can also be extended to higher dimensions. Figure 20 shows the convolution of a two-dimensional matrix, $N_{5 \times 5}$ with a two-dimensional convolution mask, $M_{5 \times 5}$. Consider the evaluation of element, P[1][1]. As shown in Fig. 20, the center of the convolution mask, M[1][1], aligns with the target element, N[1][1]. The overlapping elements of matrices M and N are then multiplied and the products are added to obtain the weighted sum as:

$$P[1][1] = M[0][0] \times N[0][0] + M[0][1] \times N[0][1] + M[0][2] \times N[0][2] +$$

convolution operator: *

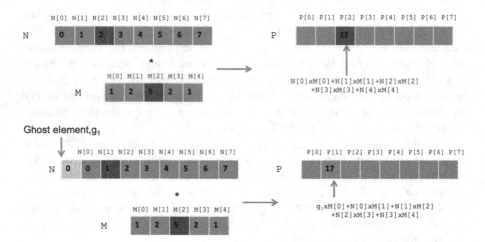

Fig. 19 An illustration of one-dimensional convolution operation. The evaluation of element P[2] (see top) involves internal elements N[0] through N[4]. However, the evaluation of element P[1] (see bottom) requires a ghost element, g_1, which is customarily set to zero. The aligned elements are highlighted in gray and the center and target elements of M and N are highlighted in dark gray

$$M[1][0] \times N[1][0] + M[1][1] \times N[1][1] + M[1][2] \times N[1][2]+$$
$$M[2][0] \times N[2][0] + M[2][1] \times N[2][1] + M[2][2] \times N[2][2]$$

Notice the evaluation of element P[0][1] as shown in the same figure with mask element, M[1][1] aligned with the target element, N[0][1]. The mask elements M[0][0], M[0][1], and M[0][2] flow beyond the bounds of matrix, N. Therefore, the overflowing mask elements are multiplied with ghost elements, g_1, g_2, g_3, which are all set to zero. The element P[0][1] is evaluated as:

$$g1 = g2 = g3 = 0$$
$$P[0][1] = M[0][0] \times g1 + M[0][1] \times g2 + M[0][2] \times g3+$$
$$M[1][0] \times N[0][0] + M[1][1] \times N[0][1] + M[1][2] \times N[0][2]+$$
$$M[2][0] \times N[1][0] + M[1][1] \times N[2][1] + M[2][2] \times N[1][2]$$

The above process is performed on all of the array elements to obtain the convolution output, P. As illustrated through examples in Figs. 19 and 20, it is clear that: (a) convolution operation is highly data parallel; (b) convolution operation can be computationally intensive for large input sizes; and (c) programmers must pay special attention to boundary conditions, i.e. when the convolution mask elements flow beyond the bounds of the input data.

Active Learning Exercise 18 – Perform the convolution of the two vectors, A and B given as: $A = [-1, 0, 1]$ $B = [-3, -2, -1, 0, 1, 2, 3]$.

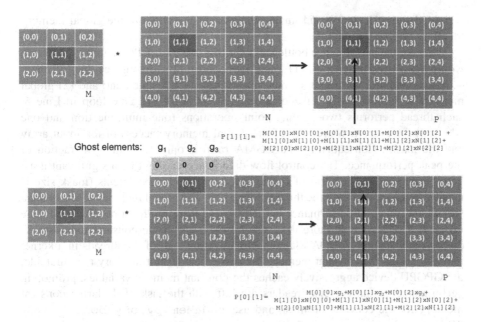

Ghost elements: g_1 g_2 g_3

$$P[1][1]= \begin{array}{l} M[0][0] \times N[0][0]+M[0][1] \times N[0][1]+M[0][2] \times N[0][2] + \\ M[1][0] \times N[1][0]+M[1][1] \times N[1][1]+M[1][2] \times N[1][2]+ \\ M[2][0] \times N[2][0]+M[2][1] \times N[2][1]+M[2][2] \times N[2][2] \end{array}$$

$$P[0][1]= \begin{array}{l} M[0][0] \times g_1+M[0][1] \times g_2+M[0][2] \times g_3+ \\ M[1][0] \times N[0][0]+M[1][1] \times N[0][1]+M[1][2] \times N[0][2]+ \\ M[2][0] \times N[1][0]+M[2][1] \times N[1][1]+M[2][2] \times N[1][2] \end{array}$$

Fig. 20 Illustration of two dimensional convolution operation. The evaluation of element P[1][1] (see top) involves internal elements of N highlighted in gray (target element is highlighted in dark gray). However, the evaluation of element P[0][1] requires a ghost elements, g_1, g_2, g_3, which are customarily set to zero

Now that we are mathematically equipped to perform the convolution operation, let us study how it can be performed on the GPGPU devices. For simplicity, let us perform one-dimensional convolution. The arguments for a CUDA convolution kernel include the following arrays: N (input), M (mask), and output, P. In addition, the kernel requires the width of array N, let this variable be width; and width of the convolution mask, let this variable be mask_width. A naïve implementation of the one-dimensional convolution kernel appears in Listing 15.

Listing 15 A naïve implementation of one-dimensional convolution kernel.

```
1.  __global__ void kernel(float *N, float *M, float *P,
                            int width, int mask_width) {
2.    int tid = threadIdx.x + blockIdx.x*blockDim.x;
3.    int start_point = tid -mask_width/2;    //place the mask center on N[tid]
4.    float temp=0;
5.    for( int i=0; i<mask_width; i++) {       //loop over the mask
6.      if(start_point + i >=0 && start_point + i < width) //check boundary
7.        temp+=N[start_point +i]*M[i];
8.    }
9.    P[tid]=temp;
10.}
```

As seen in Listing 15, each thread obtains its global thread ID, tid in Line 2. Because the center of the mask is placed on the target element N[tid], the starting element of the mask, M[0] is aligned with N[tid - mask_width/2]. Line 3 sets the starting point to tid - mask_width/2. Lines 5 through 8 perform the weighted sum calculation and finally, the answer is written to the global memory location, P[tid] (Line 9).

What are the performance bottlenecks for this naïve kernel? A careful inspection would yield two bottlenecks: (1) There is a control flow divergence due to Line 6 – threads within a warp may or may not satisfy the if statement; and (2) global memory is sub-optimally utilized. In each iteration of the for loop in Line 5, each thread performs two floating-point operations (one multiplication and one addition) for every two accesses of the global memory (access of the input array and the mask). Consequently, the CGMA ratio is only 1, yielding a fraction of the peak performance. The control flow divergence may not be a significant issue here because only a handful of threads process the ghost elements (mask size is usually much smaller than the thread block size). The global memory accesses are a significant source of performance bottleneck and therefore must be alleviated. One immediate remedy is to store the convolution mask in the constant memory. As discussed in section "CUDA Memory Organization", all of the threads in a kernel globally access the constant memory. Because the constant memory is immutable, the GPGPU device aggressively caches the constant memory variables, promoting performance. As an exercise, readers are left with the task of declaring constant memory for the convolution mask and use cudaMemcpyToSymbol() to copy the host mask pointer, h_M to the device constant memory, M.

A careful inspection of the naïve convolution kernel in Listing 15 also suggests that threads within a block tend to share the access to array elements. For instance in Fig. 19, elements required to evaluate P[2] are N[0] through N[4]. Similarly, elements needed to evaluate P[3] are N[1] through N[5]. Therefore, consecutive threads in a warp evaluating elements P[2] and P[3] require common access to elements N[2] through N[4]. The threads in a block can access the shared computational elements via shared memory. Specifically, the threads in a block load their respective elements into the shared memory, reducing the number of trips to the global memory unlike the naïve convolution. Despite of this cooperative loading, some of the threads may need access to the elements loaded by the adjacent thread blocks. Additionally, some threads within a block may require access to ghost elements. This issue is illustrated in Fig. 21. In the same figure, consider the thread blocks of size four threads, array N of size equal to 15, and a convolution mask of size equal to 5. The convolution operation requires four blocks: block-0 operates on elements 0 through 3; block-1 operates on elements 4 through 7, and so on. Consider block-0 for example – the evaluation of elements 2 and 3 clearly require elements 4 and 5, which are loaded into the shared memory by threads in block-1. We refer to these elements as halo elements (highlighted in gray). In addition to halo elements, threads 0 and 1 need access to ghost elements (highlighted in vertical bars).

With the introduction of L2 caches in modern GPGPU devices, the access to the halo elements is greatly simplified; whereas the ghost elements can be tackled using the code logic. When the threads in block-1 load elements N[4] through N[7], it is a reasonable assumption that these values will also be stored in the L2 cache. Consequently with high probability, block-0 can find its halo elements (N[4] and N[5]) in the L2 cache, thereby optimizing global memory accesses. Similarly,

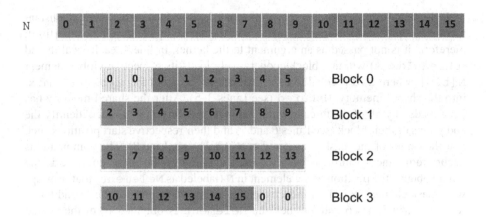

Fig. 21 Illustration of thread blocks of size 4 requiring access to halo and ghost elements

block-1 can also find the halo elements, 8 and 9 when block-2 threads load their respective elements ($N[8]$ through $N[11]$) into the shared memory.

To summarize, an optimized CUDA kernel can alleviate the global memory traffic using three strategies: (1) by storing the convolution mask in the constant memory, which is aggressively cached, (2) by requiring threads in a block to load their respective elements into the shared memory; these elements will also be cached in L2, and (3) access the halo elements via L2 cache. The optimized CUDA kernel for convolution operation appears in Listing 16.

Listing 16 Optimized convolution kernel that makes use of: constant memory to cache the convolution mask, L2 cache to enable threads access the elements loaded by neighboring thread blocks, and shared memory for collaborative load of elements by threads in a block.

```
1.  __global__ void convolution_kernel(float *N, float *P,
                                 int width, int mask_width) {
2.   int tid = threadIdx.x + blockIdx.x*blockDim.x;
3.   __shared__ float Nshared[BLOCKSIZE];
4.   Nshared[threadIdx.x]=N[tid]; //each thread loads its respective element in
//shared memory
5.   __syncthreads(); //make sure all threads finish loading before proceeding
6.   int myblock_start = blockIdx.x*blockDim.x;
7.   int nextblock_start=(blockIdx.x+1)*blockDim.x;
8.   int start = tid - mask_width/2; //places the center of mask on N[tid]
9.   float temp=0;
10.  for (int i=0;i<mask_width;i++){ //loop over the mask
11.     int Nelement=start + i; //element overlapping with i
12.     if(Nelement >=0 && Nelement < width) { //boundary check
13.        if(Nelement >=myblock_start &&Nelement < nextblock_start){
//Nelement present in shared memory
14.            temp+=Nshared[threadIdx.x+i-mask_width/2]*M[i]; }
15.        else {
                  //not in shared memory. Access using L2 cache
16.            temp+=N[Nelement]*M[i];
17.        }
18.     }
19.  }
20.  P[tid]=temp; //write the answer to global memory
21. }
```

In Listing 16, note that the convolution mask, M resides in the device constant memory (copied into the constant memory of the device by the host in host portion); therefore, it is not passed as an argument to the kernel. In line 4, each local thread (threadIdx.x) within a block cooperatively loads its respective global element N[tid], where tid is equal to threadIdx + blockIdx.x*blockDim.x, into the shared memory, Nshared (see Lines 3–5). After the shared memory has been loaded by all of the threads within a block, the threads in a block identify the end points of their block (see Lines 6 and 7) and their respective start positions such that the center of the mask is centered at N[tid] (see Line 8). The computations occur from Line 10 through 19 – for each iteration of the mask counter, i, the thread obtains the position of the element in N (labeled as Nelement) that overlaps with mask element, M[i]. If this element is within the bounds of the thread block (calculated in Lines 6 and 7), then the Nelement is obtained from the shared memory variable, Nshared (see Lines 13 and 14). However, if Nelement lies outside of the block boundaries, then the corresponding element in N is obtained via a global memory access (see Lines 15 through 17). With high probability, this global memory location is cached in L2, therefore served with L2 cache throughput. The final computation result is written back to the global memory in Line 20.

Active Learning Exercise 19 – Extend the optimized 1D convolution kernel to perform 2D convolution. Assume modern GPGPU devices that allows for general L2 caching.

In section "Case Study: Image Convolution on GPUs", we discussed an interesting parallel pattern, the convolution, which appears frequently in several scientific applications and simulations. Due to its inherent data parallelism and computation-intensiveness, the convolution operation is a great fit for GPGPU devices. Readers are also encouraged to investigate other parallel patterns including prefix sums and sparse matrix multiplication for a comprehensive understanding of GPGPU device optimizations.

We conclude our discussion on the CUDA programming model. In this chapter, we discussed the CUDA thread model and CUDA memory hierarchy, which are critical to writing effective CUDA programs. We studied different optimization strategies to attain a significant fraction of the device's peak performance. We completed our discussion on CUDA with convolution as a case study, which highlights the importance of optimizations such as constant memory, shared memory, and general L2 caching. The exploration of CUDA optimizations is figuratively endless – several applications continue to emerge that are re-organized or re-written for GPGPU computing, thereby making it a truly disruptive technology.

GPU Computing: The Future

In summary, this chapter covers major topics in GPGPU computing using the CUDA framework for upper-level Computer Engineering/Computer Science undergraduate (UG) students. Starting with the concept of data parallelism, we explained in detail the CUDA program structure, compilation flow, thread organization, memory organization, and common CUDA optimizations. All of these concepts were put together in section "Case Study: Image Convolution on GPUs" where we discussed convolution on GPGPUs as a case study. We organized the previous eight sections in a way that promotes active learning, encourages students to apply their knowledge and skills immediately after learning, and prepares them for more advanced topics in HPC. We hope that, after studying this chapter and finishing all active learning exercises, the students will have a good understanding of GPGPU computing and will be able to program GPGPUs using the CUDA paradigm.

Over the years, with a humble start as graphics-rendering devices, GPUs have evolved into powerful devices that support tasks that are more general, more sophisticated, and more computationally intensive. After decades of competition in the GPU world, NVIDIA and AMD are the two major players left. Their GPUs have been used to build the world's fastest and greenest supercomputers. In April 2016, NVIDIA unveiled the world's first deep-learning supercomputer in a box. Supported by a group of AI industry leaders, the company's new products and technologies are focusing on deep learning, virtual reality and self-driving cars. Equipped with the NVIDIA Tesla P100 GPU, the servers can now deliver the performance of hundreds of CPU server nodes. Taking advantage of the new Pascal architecture, the updated NVIDIA SDK provides extensive supports in deep learning, accelerated computing, self-driving cars, design visualization, autonomous machines, gaming, and virtual reality. Supporting these key areas will definitely attract more researchers and developers to this exciting field and enable them to create efficient solutions for problems that were considered unsolvable before. In the coming years, the evolution of GPUs will follow this increasing trend in terms of GPU processing power, software capabilities, as well as the diversity of GPU-accelerated applications.

References

1. CUDA zone.
   https://developer.nvidia.com/cuda-zone. Last Accessed 11 Feb. 2018
2. cuBLAS.
   https://developer.nvidia.com/cublas. Last Accessed 11 Feb. 2018
3. Nvidia cuDDN. GPU Accelerated Deep Learning.
   https://developer.nvidia.com/cudnn. Last Accessed 11 Feb. 2018
4. OpenCL overview.
   https://www.khronos.org/opencl/. Last Accessed 11 Feb. 2018
5. OpenACC. More sciene, less programming.
   https://www.openacc.org/. Last Accessed 11 Feb. 2018
6. Thrust. https://developer.nvidia.com/thrust.
   https://developer.nvidia.com/cudnn. Last Accessed 11 Feb. 2018
7. Nvidia.
   www.nvidia.com. Last Accessed 11 Feb. 2018

Managing Concurrency in Mobile User Interfaces with Examples in Android

Konstantin Läufer and George K. Thiruvathukal

Abstract In this chapter, we explore various parallel and distributed computing topics from a user-centric software engineering perspective. Specifically, in the context of mobile application development, we study the basic building blocks of interactive applications in the form of events, timers, and asynchronous activities, along with related software modeling, architecture, and design topics.

Relevant software engineering topics: software requirements: functional requirements (C), nonfunctional requirements (C) software design: user interface patterns (A), concurrency patterns (A), testing patterns (A), architectural patterns (C), dependency injection (C), design complexity (C); software testing: unit testing (A), managing dependencies in testing (A); cross-cutting topics: web services (C), pervasive and mobile computing (A)

Relevant parallel and distributed computing topics: algorithmic problems: asynchrony (C); architecture classes: simultaneous multithreading (K), SMP (K); parallel programming paradigms and notations: task/thread spawning (A); semantics and correctness issues: tasks and threads (C), synchronization (A); concurrency defects: deadlocks (C), thread safety/race conditions (A); cross-cutting topics: why and what is parallel/distributed computing (C), concurrency (A), nondeterminism (C)

Learning outcomes: The student will be able to model and design mobile applications involving events, timers, and asynchronous activities. The student will be able to implement these types of applications on the Android platform. The student will develop an understanding of nonfunctional requirements.

Context for use: A semester-long intermediate to advanced undergraduate course on object-oriented development. Assumes prerequisite CS2 and background in an object-oriented language such as Java, C++, or C#.

K. Läufer (✉) · G. K. Thiruvathukal
Department of Computer Science, Loyola University Chicago, Chicago, IL, USA
e-mail: laufer@cs.luc.edu; gkt@cs.luc.edu

243

Background and Motivation

In this chapter, we will explore various parallel and distributed computing topics from a user-centric software engineering perspective. Specifically, in the context of mobile application development, we will study the basic building blocks of interactive applications in the form of events, timers, and asynchronous activities, along with related software modeling, architecture, and design topics.

Based on the authors' ongoing research and teaching in this area, this material is suitable for a five-week module on concurrency topics within a semester-long intermediate to advanced undergraduate course on object-oriented development. It is possible to extend coverage by going into more depth on the online examples [17] and studying techniques for offloading tasks to the cloud [19]. The chapter is intended to be useful to instructors and students alike.

Given the central importance of the human-computer interface for enabling humans to use computers effectively, this area has received considerable attention since around 1960 [26]. Graphical user interfaces (GUIs) emerged in the early 1970s and have become a prominent technical domain addressed by numerous widget toolkits (application frameworks for GUI development). Common to most of these is the need to balance ease of programming, correctness, performance, and consistency of look-and-feel. Concurrency always plays at least an implicit role and usually becomes an explicit programmer concern when the application involves processor-bound, potentially long-running activities controlled by the GUI. Here, long-running means anything longer than the user wants to wait for before being able to continue interacting with the application. This chapter is about the concepts and techniques required to achieve this balance between correctness and performance in the context of GUI development.

During the last few years, mobile devices such as smartphones and tablets have displaced the desktop PC as the predominant front-end interface to information and computing resources. In terms of global internet consumption (minutes per day), mobile devices overtook desktop computers in mid-2014 [5], and "more websites are now loaded on smartphones and tablets than on desktop computers" [14] as of October 2016. Google also announced [3] that it will be displaying mobile-friendly web sites higher in the search results, which speaks to the new world order. These mobile devices participate in a massive global distributed system where mobile applications offload substantial resource needs (computation and storage) to the cloud.

In response to this important trend, this chapter focuses on concurrency in the context of mobile application development, especially Android, which shares many aspects with previous-generation (and desktop-centric) GUI application frameworks such as Java AWT and Swing yet. (And it almost goes without saying that students are more excited about learning programming principles via technologies like Android and iOS, which they are using more often than their desktop computers.)

While the focus of this chapter is largely on concurrency within the mobile device itself, the online source code for one of our examples [19] goes beyond the on-device experience by providing versions that connect to RESTful web services (optionally hosted in the cloud) [6]. We've deliberately focused this chapter around the on-device experience, consistent with "mobile first" thinking, which more generally is the way the "Internet of Things" also works [1]. This thinking results in proper separation of concerns when it comes to the user experience, local computation, and remote interactions (mediated using web services).

It is worth taking a few moments to ponder why mobile platforms are interesting from the standpoint of parallel and distributed computing, even if at first glance it is obvious. From an architectural point of view, the landscape of mobile devices has followed a similar trajectory to that of traditional multiprocessing systems. The early mobile device offerings, even when it came to smartphones, were single core. At the time of writing, the typical smartphone or tablet is equipped with four CPU cores and a graphics processing unit (GPU), with the trend of increasing cores (to at least 8) expected to continue in mobile CPUs. In this vein, today's–and tomorrow's–devices need to be considered serious parallel systems in their own right. (In fact, in the embedded space, there has been a corresponding emergence of parallel boards, similar to the Raspberry Pi.)

The state of parallel computing today largely requires the mastery of two styles, often appearing in a hybrid form: *task parallelism* and *data parallelism*. The emerging mobile devices are following desktop and server architecture by supporting both of these. In the case of task parallelism, to get good performance, especially when it comes to the user experience, concurrency must be disciplined. An additional constraint placed on mobile devices, compared to parallel computing, is that unbounded concurrency (threading) makes the device unusable/unresponsive, even to a greater extent than on desktops and servers (where there is better I/O performance in general). We posit that learning to program concurrency in a resource-constrained environment (e.g. Android smartphones) can be greatly helpful for writing better concurrent, parallel, and distributed code in general. More importantly, today's students really want to learn about emerging platforms, so this is a great way to develop new talent in languages and systems that are likely to be used in future parallel/distributed programming environments.

Roadmap

In the remainder of this chapter, we first summarize the fundamentals of thread safety in terms of concurrent access to shared mutable state.

We then discuss the technical domain of applications with graphical user interfaces (GUIs), GUI application frameworks that target this domain, and the runtime environment these frameworks typically provide.

Next, we examine a simple interactive behavior and explore how to implement this using the Android mobile application development framework. To make our presentation relevant to problem solvers, our running example is a bounded click counter application (more interactive and exciting than the examples commonly found in concurrency textbooks, e.g., atomic counters and bounded buffers) that can be used to keep track of the capacity of, say, a movie theater.

We then explore more interesting scenarios by introducing timers and internal events. For example, a countdown timer can be used for notification of elapsed time, a concept that has almost uniquely emerged in the mobile space but has applications in embedded and parallel computing in general, where asynchronous paradigms have been present for some time, dating to job scheduling, especially for longer-running jobs.

We close by exploring applications requiring longer-running, processor-bound activities. In mobile app development, a crucial design goal is to ensure UI responsiveness and appropriate progress reporting. We demonstrate techniques for making sure that computation proceeds but can be interrupted by the user. These techniques can be generalized to offload processor-bound activities to cloud-hosted web services.[1]

Fundamentals of Thread Safety

Before we discuss concurrency issues in GUI applications, it is helpful to understand the underlying fundamentals of thread safety in situations where two or more concurrent threads (or other types of activities) access shared mutable state.

Thread safety is best understood in terms of *correctness*: An implementation is *correct* if and only if it conforms to its specification. The implementation is *thread-safe* if and only if it continues to behave correctly in the presence of multiple threads [28].

Example: Incrementing a Shared Variable

Let's illustrate these concepts with perhaps the simplest possible example: incrementing an integer number. The specification for this behavior follows from the definition of increment: *After performing the increment, the number should be one greater than before.*

Here is a first attempt to implement this specification in the form of an instance variable in a Java class and a `Runnable` instance that wraps around our increment code and performs it on demand when we invoke its run method (see below).

```
1 int shared = 0;
2
3 final Runnable incrementUnsafe = new Runnable() {
4   @Override public void run() {
```

---

[1] This topic goes beyond the scope of this chapter but is included in the corresponding example [19].

```
5     final int local = shared;
6     tinyDelay();
7     shared = local + 1;
8   }
9 };
```

To test whether our implementation satisfies the specification, we can write a simple test case:

```
1 final int oldValue = shared;
2 incrementUnsafe.run();
3 assertEquals(oldValue + 1, shared);
```

In this test, we perform the increment operation in the only thread we have, that is, the main thread. Our implementation passes the test with flying colors. Does this mean it is thread-safe, though?

To find out, we will now test two or more concurrent increment operations, where the instance variable shared becomes shared state. Generalizing from our specification, the value of the variable should go up by one for each increment we perform. We can write this test for two concurrent increments

```
1 final int threadCount = 2;
2 final int oldValue = shared;
3 runConcurrently(incrementUnsafe, threadCount);
4 assertEquals(oldValue + threadCount, shared);
```

where runConcurrently runs the given code concurrently in the desired number of threads:

```
1 public void runConcurrently(
2     final Runnable inc, final int threadCount) {
3   final Thread[] threads = new Thread[threadCount];
4   for (int i = 0; i < threadCount; i += 1) {
5     threads[i] = new Thread(inc);
6   }
7   for (final Thread t : threads) {
8     t.start();
9   }
10  for (final Thread t : threads) {
11    try {
12      t.join();
13    } catch (final InterruptedException e) {
14      throw new RuntimeException("interrupted during join");
15    }
16  }
17 }
```

But this test does not always pass! When it does not, one of the two increments appears to be lost. Even if its failure rate were one in a million, the specification is violated, meaning that *our implementation of increment is not thread-safe.*

Interleaved Versus Serialized Execution

Let's try to understand exactly what is going on here. We are essentially running two concurrent instances of this code:

```
1 /*f1*/ int local1 = shared;          /*f2*/ int local2 = shared;
2 /*s1*/ shared = local1 + 1;          /*s2*/ shared = local2 + 1;
```

(For clarity, we omit the invocation of tinyDelay present in the code above; this invokes Thread.sleep(0) and is there just so we can observe and discuss this phenomenon in conjunction with the Java thread scheduler.)

The instructions are labeled f_n and s_n for fetch and set, respectively. Within each thread, execution proceeds sequentially, so we are guaranteed that f_1 always comes before s_1 and f_2 always comes before s_2. But we do not have any guarantees about the relative order across the two threads, so all of the following interleavings are possible:

- $f_1\ s_1\ f_2\ s_2$: increments shared by 2
- $f_1\ f_2\ s_1\ s_2$: increments shared by 1
- $f_1\ f_2\ s_2\ s_1$: increments shared by 1
- $f_2\ f_1\ s_1\ s_2$: increments shared by 1
- $f_2\ f_1\ s_2\ s_1$: increments shared by 1
- $f_2\ s_2\ f_1\ s_1$: increments shared by 2

This kind of situation, where the behavior is nondeterministic in the presence of two or more threads is also called a *race condition.*[2]

Based on our specification, the only correct result for incrementing twice is to see the effect of the two increments, meaning the value of shared goes up by two. Upon inspection of the possible interleavings and their results, the only correct ones are those where both steps of one increment happen before both steps of the other increment.

Therefore, to make our implementation thread-safe, we need to make sure that the two increments do not overlap. Each has to take place *atomically.* This requires one to go first and the other to go second; their execution has to be *serialized* or *sequentialized* (see also [10] for details on the *happens-before* relation among operations on shared memory).

---

[2]When analyzing race conditions, we might be tempted to enumerate the different possible interleavings. While it seems reasonable for our example, this quickly becomes impractical because of the combinatorial explosion for larger number of threads with more steps.

Using Locks to Guarantee Serialization

In thread-based concurrent programming, the primary means to ensure atomicity is mutual exclusion by *locking*. Most thread implementations, including *p-threads (POSIX threads)*, provide some type of locking mechanism.

Because Java supports threads in the language, each object carries its own lock, and there is a `synchronized` construct for allowing a thread to execute a block of code only with the lock held. While one thread holds the lock, other threads wanting to acquire the lock on the same object will join the *wait set* for that object. As soon as the lock becomes available—when the thread currently holding the lock finishes the synchronized block—, another thread from the wait set receives the lock and proceeds. (In particular, there is no first-come-first-serve or other fairness guarantee for this wait set.)

We can use locking to make our implementation of increment atomic and thereby thread-safe [20]:

```
1  final Object lock = new Object();
2
3  final Runnable incrementSafe = new Runnable() {
4    @Override public void run() {
5      synchronized (lock) {
6        final int local = shared;
7        tinyDelay();
8        shared = local + 1;
9      }
10   }
11 };
```

Now it is guaranteed to pass the test every time.

```
1  final int threadCount = 2;
2  final int oldValue = shared;
3  runConcurrently(incrementUnsafe, threadCount);
4  assertEquals(oldValue + threadCount, shared);
```

We should note that thread safety comes at a price: There is a small but not insignificant overhead in handling locks and managing their wait sets.

The GUI Programming Model and Runtime Environment

As we mentioned above, common to most GUI application framework is the need to balance ease of programming, correctness, performance, and consistency of look-and-feel. In this section, we will discuss the programming model and runtime environment of a typical GUI framework.

In a GUI application, the user communicates with the application through input events, such as button presses, menu item selections, etc. The application responds to user events by invoking some piece of code called an *event handler* or *event listener*. To send output back to the user, the event handler typically performs some action that the user can observe, e.g., displaying some text on the screen or playing a sound.

The GUI Runtime Environment

Real-world GUI applications can be quite complex in terms of the number of components and their logical containment hierarchy. The GUI framework is responsible for translating *low-level events* such as mouse clicks and key presses to *semantic events* such as button presses and menu item selections targeting the correct component instances. To manage this complexity, typical GUI frameworks use a producer-consumer architecture, in which an internal, high-priority system thread places low-level events on an event queue, while an application-facing *UI thread*[3] takes successive events from this queue and delivers each event to its correct target component, which then forward it to any attached listener(s). The UML sequence diagram in Fig. 1 illustrates this architecture.

Because the event queue is designed to be thread-safe, it can be shared safely between producer and consumer. It coalesces and filters groups of events as appropriate, maintaining the following discipline:

- *Sequential (single-threaded)* processing: At most one event from this queue is dispatched simultaneously.
- *Preservation of ordering:* If an event A is enqueued to the event queue before event B, then event B will not be dispatched before event A.

Concretely, the UI thread continually takes events from the event queue and processes them. Here is the pseudo-code for a typical UI thread.

```
1 run() {
2   while (true) {
3     final Event event = eq.getNextEvent();
4     final Object src = event.getSource();
5     ((Component) src).processEvent(event);
6   }
7 }
```

---

[3]In some frameworks, including Java AWT/Swing, the UI thread is known as *event dispatch thread (EDT)*.

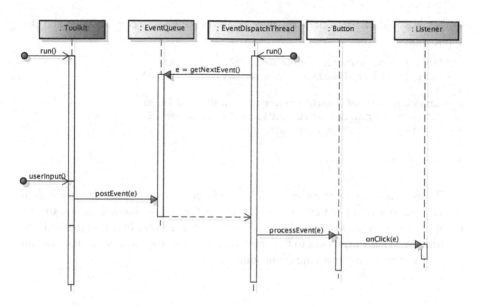

Fig. 1 UML sequence diagram showing the producer-consumer architecture of a GUI. Stick arrowheads represent asynchronous invocation, while solid arrowheads represent (synchronous) method invocation

The target component, e.g., `Button`, forwards events to its listener(s).

```
1 processEvent(e) {
2   if (e instanceof OnClickEvent) {
3     listener.onClick(e);
4   }
5   ...
6 }
```

While this presentation is mostly based on Java's AWT for simplicity, Android follows a similar approach with `MessageQueue` at the core and some responsibilities split between `Handler` and `Looper` instances [27].

This general approach, where requests (the events) come in concurrently, get placed on a request queue, and are dispatched sequentially to handlers, is an instance of the *Reactor design pattern* [30].

The Application Programmer's Perspective

Within the GUI programming model, the application programmer focuses on creating components and attaching event listeners to them. The following is a very simple example of the round-trip flow of information between the user and the application.

```
1 final Button button = new Button("press me");
2 final TextView display = new TextView("hello");
3
4 increment.setOnClickListener(new OnClickListener() {
5   @Override public void onClick(final View view) {
6     display.setText("world");
7   }
8 });
```

The event listener mechanism at work here is an instance of the *Observer design pattern* [8]: Whenever the event source, such as the button, has something to say, it notifies its observer(s) by invoking the corresponding event handling method and passing itself as the argument to this method. If desired, the listener can then obtain additional information from the event source.

Thread Safety in GUI Applications: The Single-Threaded Rule

Generally, the programmer is oblivious to the concurrency between the internal event producer thread and the UI thread. The question is whether there is or should be any concurrency on the application side. For example, if two button presses occur in very short succession, can the two resulting invocations of `display.setText` overlap in time and give rise to thread safety concerns? In that case, should we not make the GUI thread-safe by using locking?

The answer is that typical GUI frameworks are already designed to address this concern. Because a typical event listener accesses and/or modifies the data structure constituting the visible GUI, if there were concurrency among event listener invocations, we would have to achieve thread safety by serializing access to the GUI using a lock (and paying the price for this). It would be the application programmer's responsibility to use locking whenever an event listener accesses the GUI. So we would have greatly complicated the whole model without achieving significantly greater concurrency in our system.

We recall our underlying producer-consumer architecture, in which the UI thread processes one event at a time in its main loop. This means that event listener invocations are already serialized. Therefore, we can achieve thread safety directly and without placing an additional burden on the programmer by adopting this simple rule:

The application must always access GUI components from the UI thread.

This rule, known as the *single-threaded rule*, is common among most GUI frameworks, including Java Swing and Android. In practice, such access must happen either during initialization (before the application becomes visible), or within event listener code. Because it sometimes becomes necessary to create additional threads (usually for performance reasons), there are ways for those threads to schedule code for execution on the UI thread.

Android actually *enforces* the single-threaded GUI component access rule by raising an exception if this rule is violated at runtime. Android also enforces the "opposite" rule: It prohibits any code on the UI thread that will block the thread, such as network access or database queries [11].

Using Java Functional Programming Features for Higher Conciseness

To ensure compatibility with the latest and earlier versions of the Android platform, the examples in this chapter are based on Java 6 language features and API. As of October 2017, Android Studio 3.0 supports several recently introduced Java language features, including *lambda expressions* and *method references*; for details, please see [13].

These features can substantially improve both the conciseness and clarity of callback code, such as runnable tasks and Android event listeners. For example, given the equivalence between a single-method interface and a lambda expression with the same signature as the method, we can rewrite `incrementSafe` from section "Using Locks to Guarantee Serialization" and `setOnClickListener` from section "The Application Programmer's Perspective" more concisely:

```
1 final Runnable incrementSafe = () ->
2   synchronized (lock) {
3     final int local = shared;
4     tinyDelay();
5     shared = local + 1;
6   };
```

```
1 increment.setOnClickListener(
2   (final View view) -> display.setText("world")
3 );
```

Single-Threaded Event-Based Applications

In this section, we will study a large class of applications that will not need any explicit concurrency at all. As long as each response to an input event is short, we can keep these applications simple and responsive by staying within the Reactor pattern.

We will start with a simple interactive behavior and explore how to implement this using the Android mobile application development framework [9]. Our running example will be a bounded click counter application that can be used to keep track of the capacity of, say, a movie theater. The complete code for this example is available online [18].

The Bounded Counter Abstraction

A *bounded counter* [16], the concept underlying this application, is an integer counter that is guaranteed to stay between a preconfigured minimum and maximum value. This is called the *data invariant* of the bounded counter.

$$min \leq counter \leq max$$

We can represent this abstraction as a simple, passive object with, say, the following interface:

```
1 public interface BoundedCounter {
2    void increment();
3    void decrement();
4    int get();
5    boolean isFull();
6    boolean isEmpty();
7 }
```

In following a *test-driven* mindset [2], we test implementations of this interface using methods such as this one, which ensures that incrementing the counter works properly:

```
1 @Test
2 public void testIncrement() {
3    decrementIfFull();
4    assertFalse(counter.isFull());
5    final int v = counter.get();
6    counter.increment();
7    assertEquals(v + 1, counter.get());
8 }
```

In the remainder of this section, we'll put this abstraction to good use by building an interactive application on top of it.

The Functional Requirements for a Click Counter Device

Next, let's imagine a device that realizes this bounded counter concept. For example, a greeter positioned at the door of a movie theater to prevent overcrowding would require a device with the following behavior:

- The device is preconfigured to the capacity of the venue.
- The device always displays the current counter value, initially zero.
- Whenever a person enters the movie theater, the greeter presses the *increment* button; if there is still capacity, the counter value goes up by one.
- Whenever a person leaves the theater, the greeter presses the *decrement* button; the counter value goes down by one (but not below zero).
- If the maximum has been reached, the *increment* button either becomes unavailable (or, as an alternative design choice, attempts to press it cause an error). This behavior continues until the counter value falls below the maximum again.
- There is a *reset* button for resetting the counter value directly to zero.

A Simple Graphical User Interface (GUI) for a Click Counter

We now provide greater detail on the user interface of this click counter device. In the case of a dedicated hardware device, the interface could have tactile inputs and visual outputs, along with, say, audio and haptic outputs.

As a minimum, we require these interface elements:

- Three buttons, for incrementing and decrementing the counter value and for resetting it to zero.
- A numeric display of the current counter value.

Optionally, we would benefit from different types of feedback:

- Beep and/or vibrate when reaching the maximum counter value.
- Show the percentage of capacity as a numeric percentage or color thermometer.

Instead of a hardware device, we'll now implement this behavior as a mobile software app, so let's focus first on the minimum interface elements. In addition, we'll make the design choice to disable operations that would violate the counter's data invariant.

These decisions lead to the three *view states* for the bounded click counter Android app (see Fig. 2: In the initial (minimum) view state, the decrement button is disabled. In the counting view state of the, all buttons are enabled. Finally, in the maximum view state, the increment button is disabled; we assume a maximum value of 10). In our design, the reset button is always enabled.

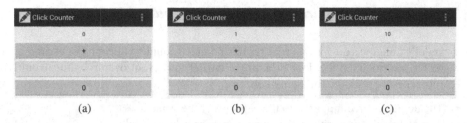

        (a)   (b)   (c)

Fig. 2 View states for the click counter. (**a**) Minimum state. (**b**) Counting state. (**c**) Maximum state

Understanding User Interaction as Events

It was fairly easy to express the familiar bounded counter abstraction and to envision a possible user interface for putting this abstraction to practical use. The remaining challenge is to tie the two together in a meaningful way, such that the interface uses the abstraction to provide the required behavior. In this section, we'll work on bridging this gap.

Modeling the Interactive Behavior

As a first step, let's abstract away the concrete aspects of the user interface:

- Instead of touch buttons, we'll have *input events*.
- Instead of setting a visual display, we'll *modify a counter value*.

After we take this step, we can use a UML state machine diagram [29] to model the dynamic behavior we described at the beginning of this section more formally.[4] Note how the touch buttons correspond to events (triggers of *transitions*, i.e., arrows) with the matching names.

The behavior starts with the *initial pseudostate* represented by the black circle. From there, the counter value gets its initial value, and we start in the minimum state. Assuming that the minimum and maximum values are at least two apart, we can increment unconditionally and reach the counting state. As we keep incrementing, we stay here as long as we are at least two away from the maximum state. As soon as we are exactly one away from the maximum state, the next increment takes us to that state, and now we can no longer increment, just decrement. The system mirrors this behavior in response to the decrement event. There is a surrounding global state to support a single reset transition back to the minimum state. Figure 3 shows the complete diagram.

---

[4]A full introduction to the Unified Modeling Language (UML) [29] would go far beyond the scope of this chapter. Therefore, we aim to introduce the key elements of UML needed here in an informal and pragmatic manner. Various UML resources, including the official specification, are available at http://www.uml.org/. Third-party tutorials are available online and in book form.

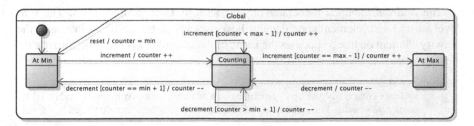

Fig. 3 UML state machine diagram modeling the dynamic behavior of the bounded counter application

As you can see, the three model states map directly to the view states from the previous subsection, and the transitions enabled in each model state map to the buttons enabled in each view state. This is not always the case, though, and we will see examples in a later section of an application with multiple model states but only a single view state.

GUI Components as Event Sources

Our next step is to bring the app to life by connecting the visual interface to the interactive behavior. For example, when pressing the increment button in a non-full counter state, we expect the displayed value to go up by one. In general, the user can trigger certain events by interacting with view components and other event sources. For example, one can press a button, swipe one's finger across the screen, rotate the device, etc.

Event Listeners and the Observer Pattern

We now discuss what an event is and what happens after it gets triggered. We will continue focusing on our running example of pressing the increment button.

The visual representation of an Android GUI is usually auto-generated from an XML source during the build process.[5] For example, the source element for our increment button looks like this; it declaratively maps the onClick attribute to the onIncrement method in the associated activity instance.

```
1  <Button
2    android:id="@+id/button_increment"
3    android:layout_width="fill_parent"
4    android:layout_height="wrap_content"
5    android:onClick="onIncrement"
6    android:text="@string/label_increment" />
```

---

[5]It is also possible—though less practical—to build an Android GUI programmatically.

The *Android manifest* associates an app with its main activity class. The top-level `manifest` element specifies the Java package of the activity class, and the activity element on line 5 specifies the name of the activity class, `ClickCounter-Activity`.

```
1  <manifest
2      xmlns:android="http://schemas.android.com/apk/res/android"
3      package="edu.luc.etl.cs313.android.clickcounter" ...>
4      ...
5      <application ...>
6          <activity android:name=".ClickCounterActivity" ...>
7              <intent-filter>
8                  <action android:name="android.intent.action.MAIN" />
9                  <category
10                     android:name="android.intent.category.LAUNCHER" />
11             </intent-filter>
12         </activity>
13     </application>
14 </manifest>
```

An *event* is just an invocation of an *event listener* method, possibly with an argument describing the event. We first need to establish the association between an event source and one (or possibly several) event listener(s) by *subscribing* the listener to the source. Once we do that, every time this source emits an event, normally triggered by the user, the appropriate event listener method gets called on each subscribed listener.

Unlike ordinary method invocations, where the caller knows the identity of the callee, the (observable) event source provides a general mechanism for subscribing a listener to a source. This technique is widely known as the *Observer design pattern* [8].

Many GUI frameworks follow this approach. In Android, for example, the general component superclass is View, and there are various types of listener interfaces, including `OnClickListener`. In following the *Dependency Inversion Principle (DIP)* [24], the `View` class owns the interfaces its listeners must implement.

```
1  public class View {
2      ...
3      public static interface OnClickListener {
4          void onClick(View source);
5      }
6      public void setOnClickListener(OnClickListener listener) {
7          ...
8      }
9      ...
10 }
```

Android follows an event source/listener naming idiom loosely based on the JavaBeans specification [15]. Listeners of, say, the onX event implement the OnXListener interface with the onX(Source source) method. Sources of this kind of event implement the setOnXListener method.[6] An actual event instance corresponds to an invocation of the onX method with the source component passed as the source argument.

Processing Events Triggered by the User

The Android activity is responsible for mediating between the view components and the POJO (plain old Java object) bounded counter model we saw above. The full cycle of each event-based interaction goes like this.

- By pressing the increment button, the user triggers the onClick event on that button, and the onIncrement method gets called.
- The onIncrement method interacts with the model instance by invoking the increment method and then requests a view update of the activity itself.
- The corresponding updateView method also interacts with the model instance by retrieving the current counter value using the get method, displays this value in the corresponding GUI element with unique ID textview_value, and finally updates the view states as necessary.

Figure 4 illustrates this interaction step-by-step.

```
1 public void onIncrement(final View view) {
2    model.increment();
3    updateView();
4 }
5 protected void updateView() {
6    final TextView valueView =
7      (TextView) findViewById(R.id.textview_value);
8    valueView.setText(Integer.toString(model.get()));
9    // afford controls according to model state
10   ((Button) findViewById(R.id.button_increment))
11     .setEnabled(!model.isFull());
12   ((Button) findViewById(R.id.button_decrement))
13     .setEnabled(!model.isEmpty());
14 }
```

What happens if the user presses two buttons at the same time? As discussed above, the GUI framework responds to at most one button press or other event trigger at any given time. While the GUI framework is processing an event, it

---

[6] Readers who have worked with GUI framework that supports multiple listeners, such as Swing, might initially find it restrictive of Android to allow only one. We'll leave it as an exercise to figure out which well-known software design pattern can be used to work around this restriction.

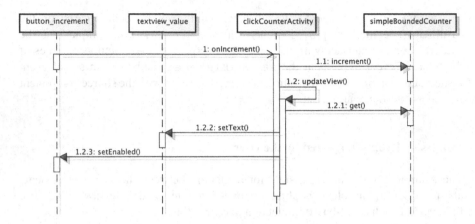

Fig. 4 Sequence diagram showing the full event-based interaction cycle in response to a press of the increment button. Stick arrowheads represent events, while solid arrowheads represent (synchronous) method invocation

places additional incoming event triggers on a queue and fully processes each one in turn. Specifically, only after the event listener method handling the current event returns will the framework process the next event. (Accordingly, activation boxes of different event listener method invocations in the UML sequence diagram must not overlap.) As discussed in section "Thread Safety in GUI Applications: The Single-Threaded Rule", this *single-threaded event handling* approach keeps the programming model simple and avoids problems, such as race conditions or deadlocks, that can arise in multithreaded approaches.

Application Architecture

This overall application architecture, where a component mediates between view components and model components, is known as *Model-View-Adapter (MVA)* [4], where the adapter component mediates all interactions between the view and the model. (By contrast, the *Model-View-Controller (MVC)* architecture has a triangular shape and allows the model to update the view(s) directly via update events.)

Figure 5 illustrates the MVA architecture. The solid arrows represent ordinary method invocations, and the dashed arrow represents event-based interaction. View and adapter play the roles of observable and observer, respectively, in the Observer pattern that describes the top half of this architecture.

Fig. 5 UML class diagram showing the Model-View-Adapter (MVA) architecture of the bounded click counter Android app. Solid arrows represent method invocation, and dashed arrows represent event flow

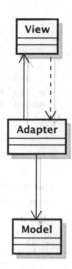

System-Testing GUI Applications

Automated system testing of entire GUI applications is a broad and important topic that goes beyond the scope of this chapter. Here, we complete our running example by focusing on a few key concepts and techniques.

In system testing, we distinguish between our application code, usually referred to as the *system under test (SUT)*, and the *test code*. At the beginning of this section, we already saw an example of a simple component-level unit test method for the POJO bounded counter model. Because Android view components support triggering events programmatically, we can also write system-level test methods that mimic the way a human user would interact with the application.

System-Testing the Click Counter

The following test handles a simple scenario of pressing the reset button, verifying that we are in the minimum view state, then pressing the increment button, verifying that the value has gone up and we are in the counting state, pressing the reset button again, and finally verifying that we are back in the minimum state.

```
1  @Test
2  public void testActivityScenarioIncReset() {
3      assertTrue(getResetButton().performClick());
4      assertEquals(0, getDisplayedValue());
5      assertTrue(getIncButton().isEnabled());
6      assertFalse(getDecButton().isEnabled());
7      assertTrue(getResetButton().isEnabled());
8      assertTrue(getIncButton().performClick());
9      assertEquals(1, getDisplayedValue());
10     assertTrue(getIncButton().isEnabled());
```

```
11  assertTrue(getDecButton().isEnabled());
12  assertTrue(getResetButton().isEnabled());
13  assertTrue(getResetButton().performClick());
14  assertEquals(0, getDisplayedValue());
15  assertTrue(getIncButton().isEnabled());
16  assertFalse(getDecButton().isEnabled());
17  assertTrue(getResetButton().isEnabled());
18  assertTrue(getResetButton().performClick());
19  }
```

The next test ensures that the visible application state is preserved under device rotation. This is an important and effective test because an Android application goes through its entire lifecycle under rotation.

```
1  @Test
2  public void testActivityScenarioRotation() {
3    assertTrue(getResetButton().performClick());
4    assertEquals(0, getDisplayedValue());
5    assertTrue(getIncButton().performClick());
6    assertTrue(getIncButton().performClick());
7    assertTrue(getIncButton().performClick());
8    assertEquals(3, getDisplayedValue());
9    getActivity().setRequestedOrientation(
10     ActivityInfo.SCREEN_ORIENTATION_LANDSCAPE);
11   assertEquals(3, getDisplayedValue());
12   getActivity().setRequestedOrientation(
13     ActivityInfo.SCREEN_ORIENTATION_PORTRAIT);
14   assertEquals(3, getDisplayedValue());
15   assertTrue(getResetButton().performClick());
16  }
```

System Testing In and Out of Container

We have two main choices for system-testing our app:

- *In-container/instrumentation testing* in the presence of the target execution environment, such as an actual Android phone or tablet emulator (or physical device). This requires deploying both the SUT and the test code to the emulator and tends to be quite slow. So far, Android's build tools officially support only this mode.
- *Out-of-container testing* on the development workstation using a test framework such as *Robolectric* that simulates an Android runtime environment tends to be considerably faster. This and other non-instrumentation types of testing can be integrated in the Android build process with a bit of extra effort.

Although the Android build process does not officially support this or other types of non-instrumentation testing, they can be integrated in the Android build process with a bit of extra effort.

Structuring Test Code for Flexibility and Reuse

Typically, we'll want to run the exact same test logic in both cases, starting with the simulated environment and occasionally targeting the emulator or device. An effective way to structure our test code for this purpose is the xUnit design pattern *Testcase Superclass* [25]. As the pattern name suggests, we pull up the common test code into an abstract superclass, and each of the two concrete test classes inherits the common code and runs it in the desired environment.

```
1 @RunWith(RobolectricTestRunner.class)
2 public class ClickCounterActivityRobolectric
3 extends AbstractClickCounterActivityTest {
4     // some minimal Robolectric-specific code
5 }
```

The official Android test support, however, requires inheriting from a specific superclass called `ActivityInstrumentationTestCase2`. This class now takes up the only superclass slot, so we cannot use the Testcase Superclass pattern literally. Instead, we need to approximate inheriting from our `AbstractClickCounterActivityTest` using delegation to a subobject. This gets the job done but can get quite tedious when a lot of test methods are involved.

```
1 public class ClickCounterActivityTest
2 extends ActivityInstrumentationTestCase2<ClickCounterActivity>
     {
3     ...
4     // test subclass instance to delegate to
5     private AbstractClickCounterActivityTest actualTest;
6
7     @UiThreadTest
8     public void testActivityScenarioIncReset() {
9         actualTest.testActivityScenarioIncReset();
10     }
11     ...
12 }
```

Having a modular architecture, such as model-view-adapter, enables us to test most of the application components in isolation. For example, our simple unit tests for the POJO bounded counter model still work in the context of the overall Android app.

Test Coverage

Test coverage describes the extent to which our test code exercises the system under test, and there are several ways to measure test coverage [31]. We generally want test coverage to be as close to 100% as possible and can measure this using suitable tools, such as JaCoCo along with the corresponding Gradle plugin.[7]

Interactive Behaviors and Implicit Concurrency with Internal Timers

In this section, we'll study applications that have richer, timer-based behaviors compared to the previous section. Our example will be a countdown timer for cooking and similar scenarios where we want to be notified when a set amount of time has elapsed. The complete code for a very similar example is available online [21].

The Functional Requirements for a Countdown Timer

Let's start with the functional requirements for the countdown timer, amounting to a fairly abstract description of its controls and behavior.

The timer exposes the following controls:

- One two-digit display of the form 88.
- One multi-function button.

The timer behaves as follows:

- The timer always displays the remaining time in seconds.
- Initially, the timer is stopped and the (remaining) time is zero.
- If the button is pressed when the timer is stopped, the time is incremented by one up to a preset maximum of 99. (The button acts as an increment button.)
- If the time is greater than zero and three seconds elapse from the most recent time the button was pressed, then the timer beeps once and starts running.
- While running, the timer subtracts one from the time for every second that elapses.
- If the timer is running and the button is pressed, the timer stops and the time is reset to zero. (The button acts as a cancel button.)

---

[7]More information on JaCoCo the JaCoCo Gradle plugin is available at http://www.eclemma.org/jacoco/ and https://github.com/arturdm/jacoco-android-gradle-plugin, respectively.

Fig. 6 View states for the countdown timer. (a) Initial stopped state with zero time. (b) Initial stopped state after adding some time. (c) Running (counting down) state. (d) Alarm ringing state

- If the timer is running and the time reaches zero by itself (without the button being pressed), then the timer stops counting down, and the alarm starts beeping continually and indefinitely.
- If the alarm is sounding and the button is pressed, the alarm stops sounding; the timer is now stopped and the (remaining) time is zero. (The button acts as a stop button.)

A Graphical User Interface (GUI) for a Countdown Timer

Our next step is to flesh out the GUI for our timer. For usability, we'll label the multifunction button with its current function. We'll also indicate which state the timer is currently in.

The screenshots in Fig. 6 show the default scenario where we start up the timer, add a few seconds, wait for it to start counting down, and ultimately reach the alarm state.

Modeling the Interactive Behavior

Let's again try to describe the abstract behavior of the countdown timer using a UML state machine diagram. As usual, there are various ways to do this, and our guiding principle is to keep things simple and close to the informal description of the behavior.

It is easy to see that we need to represent the current counter value. Once we accept this, we really don't need to distinguish between the stopped state (with counter value zero) and the counting state (with counter value greater than zero). The other states that arise naturally are the running state and the alarm state. Figure 7 shows the resulting UML state machine diagram.

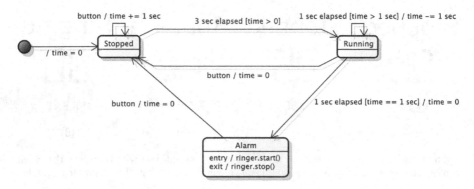

Fig. 7 UML state machine diagram modeling the dynamic behavior of the countdown timer application

As in the click counter example, these model states map directly to the view states shown above. Again, the differences among the view states are very minor and are aimed mostly at usability: A properly labeled button is a much more effective affordance than an unlabeled or generically labeled one.

Note that there are two types of (internal) timers at work here:

- *one-shot timers*, such as the three-second timer in the stopped state that gets restarted every time we press the multifunction button to add time
- *recurring timers*, such as the one-second timer in the running state that fires continually for every second that goes by

The following is the control method that starts a recurring timer that ticks approximately every second.

```
1  // called on the UI thread
2  public void startTick(final int periodInSec) {
3    if (recurring != null) throw new IllegalStateException();
4
5    recurring = new Timer();
6
7    // The clock model runs onTick every 1000 milliseconds
8    // by specifying initial and periodic delays
9    recurring.schedule(new TimerTask() {
10     @Override public void run() {
11       // fire event on the timer's internal thread
12       listener.onTick();
13     }
14   }, periodInSec * 1000, periodInSec * 1000);
15 }
```

Thread-Safety in the Model

Within the application model, each timer has its own internal thread on which it schedules the run method of its `TimerTask` instances. Therefore, other model components, such as the state machine, that receive events from either the UI and one or more timers, or more than one timer, will have to be kept thread-safe. The easiest way to achieve this is to use locking by making all relevant methods in the state machine object synchronized; this design pattern is known as *Fully Synchronized Object* [22] or *Monitor Object* [7, 28, 30].

```
1  @Override public synchronized void onButtonPress() {
2      state.onButtonPress();
3  }
4  @Override public synchronized void onTick() {
5      state.onTick();
6  }
7  @Override public synchronized void onTimeout() {
8      state.onTimeout();
9  }
```

Furthermore, update events coming back into the adapter component of the UI may happen on one of the timer threads. Therefore, to comply with the single-threaded rule, the adapter has to explicitly reschedule such events on the UI thread, using the `runOnUiThread` method it inherits from `android.app.Activity`.

```
1  @Override public void updateTime(final int time) {
2      // UI adapter responsibility
3      // to schedule incoming events on UI thread
4      runOnUiThread(new Runnable() {
5          @Override public void run() {
6              final TextView tvS =
7                  (TextView) findViewById(R.id.seconds);
8              tvS.setText(Integer.toString(time / 10) +
9                  Integer.toString(time % 10));
10         }
11     });
12 }
```

Alternatively, you may wonder whether we can stay true to the single-threaded rule and reschedule all events on the UI thread at their sources. This is possible using mechanisms such as the `runOnUiThread` method and has the advantage that the other model components such as the state machine no longer have to be thread-safe. The event sources, however, would now depend on the adapter; to keep this dependency manageable and our event sources testable, we can express it in terms of a small interface (to be implemented by the adapter) and inject it into the event sources.

```
1 public interface UIThreadScheduler {
2   void runOnUiThread(Runnable r);
3 }
```

Some GUI frameworks, such as Java Swing, provide non-view components for scheduling tasks or events on the UI thread, such as javax.swing.Timer. This avoids the need for an explicit dependency on the adapter but retains the implicit dependency on the UI layer.

Meanwhile, Android developers are being encouraged to use Scheduled-ThreadPoolExecutor instead of java.util.Timer, though the thread-safety concerns remain the same as before.

Implementing Time-Based Autonomous Behavior

While the entirely passive bounded counter behavior from the previous section was straightforward to implement, the countdown timer includes autonomous timer-based behaviors that give rise to another level of complexity.

There are different ways to deal with this behavioral complexity. Given that we have already expressed the behavior as a state machine, we can use the *State design pattern* [8] to separate state-dependent behavior from overarching handling of external and internal triggers and actions.

We start by defining a state abstraction. Besides the same common methods and reference to its surrounding state machine, each state has a unique identifier.

```
1  abstract class TimerState
2  implements TimerUIListener, ClockListener {
3
4    public TimerState(final TimerStateMachine sm) {
5      this.sm = sm;
6    }
7
8    protected final TimerStateMachine sm;
9
10   @Override public final void onStart() { onEntry(); }
11   public void onEntry() { }
12   public void onExit() { }
13   public void onButtonPress() { }
14   public void onTick() { }
15   public void onTimeout() { }
16   public abstract int getId();
17 }
```

In addition, a state receives UI events and clock ticks. Accordingly, it implements the corresponding interfaces, which are defined as follows:

```
1 public interface TimerUIListener {
2   void onStart();
3   void onButtonPress();
4 }
5
6 public interface ClockListener {
7   void onTick();
8   void onTimeout();
9 }
```

As we discussed in section "Understanding User Interaction as Events", Android follows an event source/listener naming idiom. Our examples illustrate that it is straightforward to define custom app-specific events that follow this same convention. Our ClockListener, for example, combines two kinds of events within a single interface.

Concrete state classes implement the abstract TimerState class. The key parts of the state machine implementation follow:

```
1 // intial pseudo-state
2 private TimerState state = new TimerState(this) {
3   @Override public int getId() {
4     throw new IllegalStateException();
5   }
6 };
7
8 protected void setState(final TimerState nextState) {
9   state.onExit();
10  state = nextState;
11  uiUpdateListener.updateState(state.getId());
12  state.onEntry();
13 }
```

Let's focus on the stopped state first. In this state, neither is the clock ticking, nor is the alarm ringing. On every button press, the remaining running time goes up by one second and the one-shot three-second idle timeout starts from zero. If three seconds elapse before another button press, we transition to the running state.

```
1 private final TimerState STOPPED = new TimerState(this) {
2   @Override public void onEntry() {
3     timeModel.reset(); updateUIRuntime();
4   }
5   @Override public void onButtonPress() {
6     clockModel.restartTimeout(3 /* seconds */);
7     timeModel.inc(); updateUIRuntime();
8   }
9   @Override public void onTimeout() { setState(RUNNING); }
10  @Override public int getId() { return R.string.STOPPED; }
11 };
```

Let's now take a look at the running state. In this state, the clock is ticking but the alarm is not ringing. With every recurring clock tick, the remaining running time goes down by one second. If it reaches zero, we transition to the ringing state. If a button press occurs, we stop the clock and transition to the stopped state.

```
1  private final TimerState RUNNING = new TimerState(this) {
2    @Override public void onEntry() {
3      clockModel.startTick(1 /* second */);
4    }
5    @Override public void onExit() { clockModel.stopTick(); }
6    @Override public void onButtonPress() { setState(STOPPED); }
7    @Override public void onTick() {
8      timeModel.dec(); updateUIRuntime();
9      if (timeModel.get() == 0) { setState(RINGING); }
10   }
11   @Override public int getId() { return R.string.RUNNING; }
12 };
```

Finally, in the ringing state, nothing is happening other than the alarm ringing. If a button press occurs, we stop the alarm and transition to the stopped state.

```
1  private final TimerState RINGING = new TimerState(this) {
2    @Override public void onEntry() {
3      uiUpdateListener.ringAlarm(true);
4    }
5    @Override public void onExit() {
6      uiUpdateListener.ringAlarm(false);
7    }
8    @Override public void onButtonPress() { setState(STOPPED); }
9    @Override public int getId() { return R.string.RINGING; }
10 };
```

Managing Structural Complexity

We can again describe the architecture of the countdown timer Android app as an instance of the Model-View-Adapter (MVA) architectural pattern. In Fig. 8, solid arrows represent (synchronous) method invocation, and dashed arrows represent (asynchronous) events. Here, both the view components and the model's autonomous timer send events to the adapter.

The user input scenario in Fig. 9 illustrates the system's end-to-end response to a button press. The internal timeout gets set in response to a button press. When the timeout event actually occurs, corresponding to an invocation of the `onTimeout` method, the system responds by transitioning to the running state.

By contrast, the autonomous scenario in Fig. 10 shows the system's end-to-end response to a recurring internal clock tick, corresponding to an invocation of the `onTick` method. When the remaining time reaches zero, the system responds by transitioning to the alarm-ringing state.

Fig. 8 The countdown
timer's Model-View-Adapter
(MVA) architecture with
additional event flow from
model to view

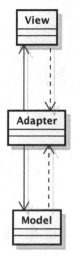

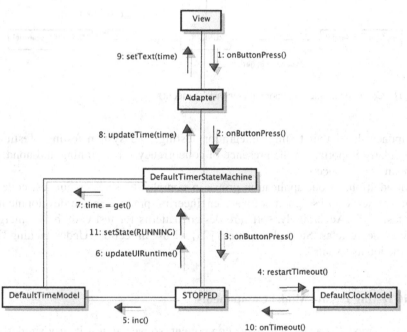

Fig. 9 Countdown timer: user input scenario (button press)

Testing GUI Applications with Complex Behavior and Structure

As we develop more complex applications, we increasingly benefit from thorough automated testing. In particular, there are different structural levels of testing:

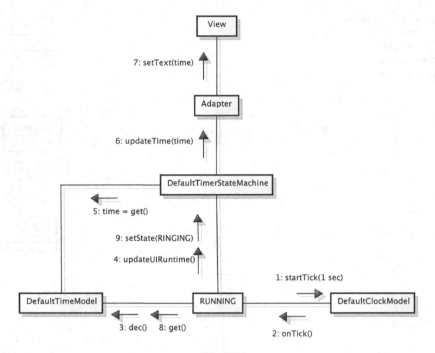

Fig. 10 Countdown timer: autonomous scenario (timeout)

component-level unit testing, integration testing, and system testing. Testing is particularly important in the presence of concurrency, where timing and nondeterminism are of concern.

In addition, as our application grows in complexity, so does our test code, so it makes sense to use good software engineering practice in the development of our test code. Accordingly, software design patterns for test code have emerged, such as the Testclass Superclass pattern [25] we use in section "Understanding User Interaction as Events".

Unit-Testing Passive Model Components

The time model is a simple passive component, so we can test it very similarly as the bounded counter model in section "Understanding User Interaction as Events".

Unit-Testing Components with Autonomous Behavior

Testing components with autonomous behavior is more challenging because we have to attach some kind of probe to observe the behavior while taking into account the presence of additional threads.

Let's try this on our clock model. The following test verifies that a stopped clock does not emit any tick events.

```
1  @Test
2  public void testStopped() throws InterruptedException {
3    final AtomicInteger i = new AtomicInteger(0);
4    model.setClockListener(new ClockListener() {
5      @Override public void onTick() { i.incrementAndGet(); }
6      @Override public void onTimeout() { }
7    });
8    Thread.sleep(5500);
9    assertEquals(0, i.get());
10 }
```

And this one verifies that a running clock emits roughly one tick event per second.

```
1  @Test
2  public void testRunning() throws InterruptedException {
3    final AtomicInteger i = new AtomicInteger(0);
4    model.setClockListener(new ClockListener() {
5      @Override public void onTick() { i.incrementAndGet(); }
6      @Override public void onTimeout() { }
7    });
8    model.startTick(1 /* second */);
9    Thread.sleep(5500);
10   model.stopTick();
11   assertEquals(5, i.get());
12 }
```

Because the clock model has its own timer thread, separate from the main thread executing the tests, we need to use a thread-safe `AtomicInteger` to keep track of the number of clock ticks across the two threads.

Unit-Testing Components with Autonomous Behavior and Complex Dependencies

Some model components have complex dependencies that pose additional difficulties with respect to testing. Our timer's state machine model, e.g., expects implementations of the interfaces `TimeModel`, `ClockModel`, and `Timer-UIUpdateListener` to be present. We can achieve this by manually implementing a so-called *mock object*[8] that unifies these three dependencies of the timer state machine model, corresponding to the three interfaces this mock object implements.

---

[8]There are also various mocking frameworks, such as Mockito and JMockit, which can automatically generate mock objects that represent component dependencies from interfaces and provide APIs or domain-specific languages for specifying test expectations.

```
 1  class UnifiedMockDependency
 2  implements TimeModel, ClockModel, TimerUIUpdateListener {
 3
 4      private int timeValue = -1, stateId = -1;
 5      private int runningTime = -1, idleTime = -1;
 6      private boolean started = false, ringing = false;
 7
 8      public int     getTime()   { return timeValue; }
 9      public int     getState()  { return stateId; }
10      public boolean isStarted() { return started; }
11      public boolean isRinging() { return ringing; }
12
13      @Override public void updateTime(final int tv) {
14        this.timeValue = tv;
15      }
16      @Override public void updateState(final int stateId) {
17         this.stateId = stateId;
18      }
19      @Override public void ringAlarm(final boolean b) {
20        ringing = b;
21      }
22
23      @Override public void setClockListener(
24          final ClockListener listener) {
25        throw new UnsupportedOperationException();
26      }
27      @Override public void startTick(final int period) {
28        started = true;
29      }
30      @Override public void stopTick() { started = false; }
31      @Override public void restartTimeout(final int period) { }
32
33      @Override public void reset() { runningTime = 0; }
34      @Override public void inc() {
35        if (runningTime != 99) { runningTime++; }
36      }
37      @Override public void dec() {
38        if (runningTime != 0) { runningTime--; }
39      }
40      @Override public int  get() { return runningTime; }
41  }
```

The instance variables and corresponding getter methods enable us to test whether the SUT produced the expected state changes in the mock object. The three remaining blocks of methods correspond to the three implemented interfaces, respectively.

Now we can write tests to verify actual scenarios. In the following scenario, we start with time 0, press the button once, expect time 1, press the button 198 times (the max time is 99), expect time 99, produce a timeout event, check if running, wait 50 s, expect time 49 (99–50), wait 49 s, expect time 0, check if ringing, wait 3 more seconds (just in case), check if still ringing, press the button to stop the ringing, and make sure the ringing has stopped and we are in the stopped state.

```
 1 @Test
 2 public void testScenarioRun2() {
 3   assertEquals(R.string.STOPPED, dependency.getState());
 4   model.onButtonPress();
 5   assertTimeEquals(1);
 6   assertEquals(R.string.STOPPED, dependency.getState());
 7   onButtonRepeat(MAX_TIME * 2);
 8   assertTimeEquals(MAX_TIME);
 9   model.onTimeout();
10   assertEquals(R.string.RUNNING, dependency.getState());
11   onTickRepeat(50);
12   assertTimeEquals(MAX_TIME - 50);
13   onTickRepeat(49);
14   assertTimeEquals(0);
15   assertEquals(R.string.RINGING, dependency.getState());
16   assertTrue(dependency.isRinging());
17   onTickRepeat(3);
18   assertEquals(R.string.RINGING, dependency.getState());
19   assertTrue(dependency.isRinging());
20   model.onButtonPress();
21   assertFalse(dependency.isRinging());
22   assertEquals(R.string.STOPPED, dependency.getState());
23 }
```

Note that this happens in *"fake time"* (fast-forward mode) because we can make the rate of the clock ticks as fast as the state machine can keep up.

Programmatic System Testing of the App

The following is a system test of the application with all of its real component present. It verifies the following scenario in *real time*: time is 0, press button five times, expect time 5, wait 3 s, expect time 5, wait 3 more seconds, expect time 2, press *stopTick* button to reset time, and expect time 0. This test also includes the effect of all state transitions as assertions.

```
 1 @Test
 2 public void testScenarioRun2() throws Throwable {
 3   getActivity().runOnUiThread(new Runnable() {
 4     @Override public void run() {
 5       assertEquals(STOPPED, getStateValue());
 6       assertEquals(0, getDisplayedValue());
 7       for (int i = 0; i < 5; i++) {
 8         assertTrue(getButton().performClick());
 9       }
10     }
11   });
12   runUiThreadTasks();
13   getActivity().runOnUiThread(new Runnable() {
14     @Override public void run() {
15       assertEquals(5, getDisplayedValue());
```

```
16      }
17    });
18    Thread.sleep(3200); // <-- do not run this in the UI thread!
19    runUiThreadTasks();
20    getActivity().runOnUiThread(new Runnable() {
21      @Override public void run() {
22        assertEquals(RUNNING, getStateValue());
23        assertEquals(5, getDisplayedValue());
24      }
25    });
26    Thread.sleep(3200);
27    runUiThreadTasks();
28    getActivity().runOnUiThread(new Runnable() {
29      @Override public void run() {
30        assertEquals(RUNNING, getStateValue());
31        assertEquals(2, getDisplayedValue());
32        assertTrue(getButton().performClick());
33      }
34    });
35    runUiThreadTasks();
36    getActivity().runOnUiThread(new Runnable() {
37      @Override public void run() {
38        assertEquals(STOPPED, getStateValue());
39      }
40    });
41  }
```

As in section "Understanding User Interaction as Events", we can run this test as an in-container instrumentation test or out-of-container using a simulated environment such as Robolectric.

During testing, our use of threading should mirror that of the SUT: The button press events we simulate using the `performClick` method have to run on the UI thread of the simulated environment. While the UI thread handles these events, we use `Thread.sleep` in the main thread of the test runner to wait in pseudo-real-time, much like the user would wait and watch the screen update.

Robolectric queues tasks scheduled on the UI thread until it is told to perform these. Therefore, we must invoke the `runUiThreadTasks` method *before* attempting our assertions on the UI components.

Keeping the User Interface Responsive with Asynchronous Activities

In this section, we explore the issues that arise when we use a GUI to control long-running, processor-bound activities. In particular, we'll want to make sure the GUI stays responsive even in such scenarios and the activity supports progress reporting and cancelation. Our running example will be a simple app for checking whether a number is prime. The complete code for this example is available online [19].

Fig. 11 Screenshot of an
Android app for checking
prime numbers

The Functional Requirements for the Prime Checker App

The functional requirements for this app are as follows:

- The app allows us to enter a number in a text field.
- When we press the *check* button, the app checks whether the number we entered is prime.
- If we press the *cancel* button, any ongoing check(s) are discontinued.

Figure 11 shows a possible UI for this app.

To check whether a number is prime, we can use this iterative brute-force algorithm.

```
1 protected boolean isPrime(final long i) {
2   if (i < 2) return false;
3   final long half = i / 2;
4   for (long k = 2; k <= half; k += 1) {
5     if (isCancelled() || i % k == 0) return false;
6     publishProgress((int) (k * 100 / half));
7   }
8   return true;
9 }
```

For now, let's ignore the `isCancelled` and `updateProgress` methods and agree to discuss their significance later in this section.

While this is not an efficient prime checker implementation, this app will allow us to explore and discuss different ways to run one or more such checks. In particular, the fact that the algorithm is heavily processor-bound makes it an effective running example for discussing whether to move such activities to the background (or remote servers).

The Problem with Foreground Tasks

As a first attempt, we now can run the `isPrime` method from within our event listener in the current thread of execution (the main GUI thread).

```
1   final PrimeCheckerTask t =
2     new PrimeCheckerTask(progressBars[0], input);
3   localTasks.add(t);
4   t.onPreExecute();
5   final boolean result = t.isPrime(number);
6   t.onPostExecute(result);
7   localTasks.clear();
```

The methods `onPreExecute` and `onPostExecute` are for resetting the user interface and displaying the result.

As shown in Table 1 below, response times (in seconds) are negligible for very small numbers but increase roughly linearly. "≪1" means no noticeable delay, and "∗" means that the test was canceled before it completed.

The actual execution targets for the app or `isPrime` implementation are

- Samsung Galaxy Nexus I9250 phone (2012 model): dual-core 1.2 GHz Cortex-A9 ARM processor with 1 GB of RAM (using one core)
- Genymotion x86 Android emulator with 1 GB of RAM and one processor running on a MacBook Air
- MacBook Air (mid-2013) with 1.7 GHz Intel Core i7 and 8 GB of RAM
- Heroku free plan with one web dyno with 512 MB of RAM

For larger numbers, the user interface on the device freezes noticeably while the prime number check is going on, so it does not respond to pressing the cancel button. There is no progress reporting either: The progress bar jumps from zero to 100 when the check finishes. In the UX (user experience) world, any freezing for more than a fraction of a second is considered unacceptable, especially without progress reporting.

Table 1 Response times for checking different prime numbers on representative execution targets

Execution target prime	Phone	Emulator	Computer	Web service
1013	$\ll 1$	$\ll 1$	$\ll 1$	$\ll 1$
10007	1	$\ll 1$	$\ll 1$	$\ll 1$
100003	3	1	$\ll 1$	$\ll 1$
1000003	27	6	$\ll 1$	1
10000169	*	60	2	2
100000007	*	*	8	8

Reenter the Single-Threaded User Interface Model

The behavior we are observing is a consequence of the single-threaded execution model underlying Android and similar GUI frameworks. As discussed in section "Thread Safety in GUI Applications: The Single-Threaded Rule", in this design, all UI events, including user inputs such as button presses and mouse moves, outputs such as changes to text fields, progress bar updates, and other component repaints, and internal timers, are processed sequentially by a single thread, known in Android as the *main thread* (or UI thread). We will continue to say *UI thread* for clarity.

To process an event completely, the UI thread needs to dispatch the event to any event listener(s) attached to the event source component. Accordingly, single-threaded UI designs typically come with two rules:

1. To ensure responsiveness, code running on the UI thread must never block.
2. To ensure thread-safety, only code running on the UI thread is allowed to access the UI components.

In interactive applications, running for a long time is almost as bad as blocking indefinitely on, say, user input. To understand exactly what is happening, let's focus on the point that events are processed sequentially in our scenario of entering a number and attempting to cancel the ongoing check.

- The user enters the number to be checked.
- The user presses the check button.
- To process this event, the UI thread runs the attached listener, which checks whether the number is prime.
- While the UI thread running the listener, all other incoming UI events—pressing the cancel button, updating the progress bar, changing the background color of the input field, etc.—are *queued* sequentially.
- Once the UI thread is done running the listener, it will process the remaining events on the queue. At this point, the cancel button has no effect anymore, and we will instantly see the progress bar jump to 100% and the background color of the input field change according to the result of the check.

So why doesn't Android simply handle incoming events concurrently, say, each in its own thread? The main reason not to do this is that it greatly complicates the design while at the same time sending us back to square one in most scenarios: Because the UI components are a shared resource, to ensure thread safety in the presence of race conditions to access the UI, we would now have to use mutual exclusion in every event listener that accesses a UI component. Because that is what event listeners typically do, in practice, mutual exclusion would amount to bringing back a sequential order. So we would have greatly complicated the whole model without effectively increasing the extent of concurrency in our system (see also section "Thread Safety in GUI Applications: The Single-Threaded Rule" above).

There are two main approaches to keeping the UI from freezing while a long-running activity is going on.

Breaking Up an Activity Into Small Units of Work

The first approach is still single-threaded: We break up the long-running activity into very small units of work to be executed directly by the UI thread. When the current chunk is about to finish, it schedules the next unit of work for execution on the UI thread. Once the next unit of work runs, it first checks whether a cancelation request has come in. If so, it simply will not continue, otherwise it will do its work and then schedule its successor. This approach allows other events, such as reporting progress or pressing the cancel button, to get in between two consecutive units of work and will keep the UI responsive as long as each unit executes fast enough.

Now, in the same scenario as above—entering a number and attempting to cancel the ongoing check—the behavior will be much more responsive:

- The user enters the number to be checked.
- The user presses the check button.
- To process this event, the UI thread runs the attached listener, which makes a little bit of progress toward checking whether the number is prime and then schedules the next unit of work on the event queue.
- Meanwhile, the user has pressed the cancel button, so this event is on the event queue *before* the next unit of work toward checking the number.
- Once the UI thread is done running the first (very short) unit of work, it will run the event listener attached to the cancel button, which will prevent further units of work from running.

Asynchronous Tasks to the Rescue

The second approach is typically multi-threaded: We represent the entire activity as a separate asynchronous task. Because this is such a common scenario, Android provides the abstract class `AsyncTask` for this purpose.

```
1  public abstract class AsyncTask<Params, Progress, Result> {
2      protected void onPreExecute() { }
3      protected abstract Result doInBackground(Params... params);
4      protected void onProgressUpdate(Progress... values) { }
5      protected void onPostExecute(Result result) { }
6      protected void onCancelled(Result result) { }
7      protected final void publishProgress(Progress... values) {
8          ...
9      }
10     public final boolean isCancelled() { ... }
11     public final AsyncTask<...> executeOnExecutor(
12         Executor exec, Params... ps) {
13         ...
14     }
15     public final boolean cancel(boolean mayInterruptIfRunning) {
16         ...
17     }
18 }
```

The three generic type parameters are Params, the type of the arguments of the
activity; Progress, they type of the progress values reported while the activity
runs in the background, and Result, the result type of the background activity.
Not all three type parameters have to be used, and we can use the type Void to
mark a type parameter as unused.

When an asynchronous task is executed, the task goes through the following
lifecycle:

- onPreExecute runs on the UI thread and is used to set up the task in a thread-
  safe manner.
- doInBackground(Params...) is an abstract template method that we
  override to perform the desired task. Within this method, we can report progress
  using publishProgress(Progress...) and check for cancelation
  attempts using isCancelled().
- onProgressUpdate(Progress...) is scheduled on the UI thread whenever the background task reports progress and runs whenever the UI thread gets
  to this event. Typically, we use this method to advance the progress bar or display
  progress to the user in some other form.
- onPostExecute(Result) receives the result of the background task as an
  argument and runs on the UI thread after the background task finishes.

Using AsyncTask in the Prime Number Checker

We set up the corresponding asynchronous task with an input of type Long,
progress of type Integer, and result of type Boolean. In addition, the task has
access to the progress bar and input text field in the Android GUI for reporting
progress and results, respectively.

The centerpiece of our solution is to invoke the `isPrime` method from the main method of the task, `doInBackground`. The auxiliary methods `isCancelled` and `publishProgress` we saw earlier in the implementation of `isPrime` are for checking for requests to cancel the current task and updating the progress bar, respectively. `doInBackground` and the other lifecycle methods are implemented here:

```
1  @Override protected void onPreExecute() {
2    progressBar.setMax(100);
3    input.setBackgroundColor(Color.YELLOW);
4  }
5
6  @Override protected Boolean doInBackground(
7      final Long... params) {
8    if (params.length != 1)
9      throw new IllegalArgumentException(
10        "exactly one argument expected");
11   return isPrime(params[0]);
12  }
13
14  @Override protected void onProgressUpdate(
15      final Integer... values) {
16   progressBar.setProgress(values[0]);
17  }
18
19  @Override protected void onPostExecute(final Boolean result) {
20   input.setBackgroundColor(result ? Color.GREEN : Color.RED);
21  }
22
23  @Override protected void onCancelled(final Boolean result) {
24   input.setBackgroundColor(Color.WHITE);
25  }
```

When the user presses the cancel button in the UI, any currently running tasks are canceled using the control method `cancel(boolean)`, and subsequent invocations of `isCancelled` return `false`; as a result, the `isPrime` method returns on the next iteration.

How often to check for cancelation attempts is a matter of experimentation: Typically, it is sufficient to check only every so many iterations to ensure that the task can make progress on the actual computation. Note how this design decision is closely related to the granularity of the units of work in the single-threaded design discussed in section "Thread Safety in GUI Applications: The Single-Threaded Rule" above.

Execution of Asynchronous Tasks in the Background

So far, we have seen how to define background tasks as subclasses of the abstract framework class `AsyncTask`. Actually executing background tasks arises as an orthogonal concern with the following strategies to choose from for assigning tasks to worker threads:

- *Serial executor:* Tasks are queued and executed by a single background thread.
- *Thread pool executor:* Tasks are executed concurrently by a pool of background worker threads. The default thread pool size depend on the available hardware resources; a typical pool size even for a single-core Android device is two.

In our example, we can schedule `PrimeCheckerTask` instances on a thread pool executor:

```
1    final PrimeCheckerTask t =
2      new PrimeCheckerTask(progressBars[i], input);
3    localTasks.add(t);
4    t.executeOnExecutor(AsyncTask.THREAD_POOL_EXECUTOR, number);
```

This completes the picture of moving processor-bound, potentially long-running activities out of the UI thread but in a way that they can still be controlled by the UI thread.

Additional considerations apply when targeting symmetric multi-core hardware (SMP), which is increasingly common among mobile devices. While the application-level, coarse-grained concurrency techniques discussed in this chapter still apply to multi-core execution, SMP gives rise to more complicated low-level memory consistency issues than those discussed above in section "Fundamentals of Thread Safety". An in-depth discussion of Android app development for SMP hardware is available here [12].

Summary

In this chapter, we have studied various parallel and distributed computing topics from a user-centric software development perspective. Specifically, in the context of mobile application development, we have studied the basic building blocks of interactive applications in the form of events, timers, and asynchronous activities, along with related software modeling, architecture, and design topics.

The complete source code for the examples from this chapter, along with instructions for building and running these examples, is available from [17]. For further reading on designing concurrent object-oriented software, please have a look at [7, 22, 23, 28].

Acknowledgements We are grateful to our former graduate students Michael Dotson and Audrey Redovan for having contributed their countdown timer implementation, and to our colleague Dr. Robert Yacobellis for providing feedback on this chapter and trying these ideas in the classroom. We are also grateful to the anonymous CDER reviewers for their helpful suggestions.

References

1. Kevin Ashton. That 'Internet of Things' thing. RFID Journal, http://www.rfidjournal.com/articles/view?4986, July 2009. Accessed: 2016-12-09.
2. Kent Beck. *Test Driven Development: By Example*. Addison-Wesley Professional, 2002.
3. Google Webmaster Central Blog. Rolling out the mobile-friendly update. https://webmasters.googleblog.com/2015/04/rolling-out-mobile-friendly-update.html, April 2015. Accessed: 2016-12-12.
4. Stefano Borini. Understanding model-view-controller. https://www.gitbook.com/book/stefanoborini/modelviewcontroller, 2016. Accessed: 2016-12-09.
5. Jemma Brackebush. How mobile is overtaking desktop for global media consumption, in 5 charts. Digiday, http://digiday.com/publishers/mobile-overtaking-desktops-around-world-5-charts/, June 2016. Accessed: 2016-12-10.
6. Jason H. Christensen. Using restful web-services and cloud computing to create next generation mobile applications. In *Proceedings of the 24th ACM SIGPLAN Conference Companion on Object Oriented Programming Systems Languages and Applications*, OOPSLA '09, pages 627–634, New York, NY, USA, 2009. ACM.
7. Thomas W. Christopher and George K. Thiruvathukal. *High Performance Java Platform Computing*. Prentice Hall PTR, Upper Saddle Ridge, NJ, 2000.
8. Erich Gamma, Richard Helm, Ralph Johnson, and John Vlissides. *Design Patterns: Elements of Reusable Object-oriented Software*. Addison-Wesley Longman Publishing Co., Inc., Boston, MA, USA, 1995.
9. Google. Android developer reference. http://developer.android.com/develop/, 2009–2018. Accessed: 2016-12-09.
10. Google. Memory consistency properties. https://developer.android.com/reference/java/util/concurrent/package-summary.html#MemoryVisibility, 2009–2018. Accessed: 2016-12-09.
11. Google. Processes and threads. http://developer.android.com/guide/components/processes-and-threads.html, 2009–2018. Accessed: 2016-12-09.
12. Google. SMP primer for Android. http://developer.android.com/training/articles/smp.html, 2009–2018. Accessed: 2016-12-09.
13. Google. Android Studio: Use Java 8 language features. https://developer.android.com/studio/write/java8-support.html#supported_features, 2017. Accessed: 2018-02-05.
14. The Guardian. Mobile web browsing overtakes desktop for the first time. https://www.theguardian.com/technology/2016/nov/02/mobile-web-browsing-desktop-smartphones-tablets/, November 2016. Accessed: 2016-12-10.
15. Graham Hamilton. JavaBeans specification. Technical report, Sun Microsystems, inc, 1997.
16. Per Brinch Hansen. *Operating System Principles*. Prentice-Hall, Inc., Upper Saddle River, NJ, USA, 1973.
17. Konstantin Läufer, George K. Thiruvathukal, and Robert H. Yacobellis. Loyola University Chicago Computer Science COMP 313/413 course examples. https://github.com/lucoodevcourse/, 2012–2018.
18. Konstantin Läufer, George K. Thiruvathukal, and Robert H. Yacobellis. Loyola University Chicago Computer Science COMP 313/413 course examples: Click counter. https://github.com/lucoodevcourse/clickcounter-android-java, 2012–2018.
19. Konstantin Läufer, George K. Thiruvathukal, and Robert H. Yacobellis. Loyola University Chicago Computer Science COMP 313/413 course examples: Prime number checker. https://github.com/lucoodevcourse/primenumbers-android-java, 2012–2018.
20. Konstantin Läufer, George K. Thiruvathukal, and Robert H. Yacobellis. Loyola University Chicago Computer Science COMP 313/413 course examples: Simple threads. https://github.com/lucoodevcourse/simplethreads-java, 2012–2018.
21. Konstantin Läufer, George K. Thiruvathukal, and Robert H. Yacobellis. Loyola University Chicago Computer Science COMP 313/413 course examples: Stopwatch. https://github.com/lucoodevcourse/stopwatch-android-java, 2012–2018.

22. Doug Lea. *Concurrent Programming in Java. Second Edition: Design Principles and Patterns*. Addison-Wesley Longman Publishing Co., Inc., Boston, MA, USA, 2nd edition, 1999.
23. Jeff Magee and Jeff Kramer. *Concurrency: State Models and Java Programs*. John Wiley & Sons, Inc., New York, NY, USA, 1999.
24. Robert C. Martin and Micah Martin. *Agile Principles, Patterns, and Practices in C# (Robert C. Martin)*. Prentice Hall PTR, Upper Saddle River, NJ, USA, 2006.
25. Gerard Meszaros. *XUnit Test Patterns: Refactoring Test Code*. Prentice Hall PTR, Upper Saddle River, NJ, USA, 2006.
26. Brad A. Myers. A brief history of human-computer interaction technology. *interactions*, 5(2):44–54, March 1998.
27. Oracle. Java platform, standard ed. 8 API specification: Class EventQueue. http://docs.oracle.com/javase/8/docs/api/java/awt/EventQueue.html, 1993–2018. Accessed: 2016-12-09.
28. Tim Peierls, Brian Goetz, Joshua Bloch, Joseph Bowbeer, Doug Lea, and David Holmes. *Java Concurrency in Practice*. Addison-Wesley Professional, 2005.
29. James Rumbaugh, Ivar Jacobson, and Grady Booch. *Unified Modeling Language Reference Manual, The (2nd Edition)*. Pearson Higher Education, 2004.
30. Douglas C. Schmidt, Michael Stal, Hans Rohnert, and Frank Buschmann. *Pattern-Oriented Software Architecture: Patterns for Concurrent and Networked Objects*. John Wiley & Sons, Inc., New York, NY, USA, 2nd edition, 2000.
31. Hong Zhu, Patrick A. V. Hall, and John H. R. May. Software unit test coverage and adequacy. *ACM Comput. Surv.*, 29(4):366–427, December 1997.

Parallel Programming for Interactive GUI Applications

Nasser Giacaman and Oliver Sinnen

Abstract This chapter will help you understand the rules that you must adhere to when developing a concurrent application with a graphical user interface (GUI). Regardless of the technology you use (for example, developing an Android mobile app or a desktop application using the .NET Framework), the concepts presented here are standard for the GUI toolkits you will use. The most important aspect includes ensuring the application does not freeze or become unresponsive, by employing background threads. This in turn leads to the other important consideration, which relates to ensuring access to any GUI components does not introduce potential race conditions. Collectively, the concepts presented here relate to the single-thread rule that governs almost all GUI toolkits you will likely come across.

Relevant core courses: GUI Concurrency is a topic suitable for any CS2-equivalent course. The material covered in this chapter would be typically covered in around 3–4 h of class time (about a week's worth of lectures). Rather than focusing on parallel programming, the focus is on thread-safety issues pertaining to GUI applications (and does not include general introductory threading). The topic is also suitable for any course that incorporates GUI development, as well as PDC courses at any level.

Relevant PDC topics: There is no specific categorization for GUI concurrency in the NSF/IEEE-TCPP Curriculum Initiative on Parallel and Distributed Computing [1]. However, the topic presented here is essential for CS2 courses that involve a GUI module. More specifically, this topic is vital for any software developer creating *interactive GUI applications*, and not just for the parallel programming enthusiast. Since GUI concurrency has much of its essence based on standard parallelization/concurrency concepts, we will be touching on the

N. Giacaman (✉) · O. Sinnen
Parallel and Reconfigurable Computing Lab, Department of Electrical and Computer Engineering, The University of Auckland, Auckland, New Zealand
e-mail: n.giacaman@auckland.ac.nz; o.sinnen@auckland.ac.nz

following subtopics – but directly in the context of GUI applications. According to Bloom's classification, students are expected to *Apply* these subtopics:

- Programming

  - Parallel programming paradigms (shared memory, task/thread spawning).
  - Semantics and correctness issues.

Learning outcomes:

- Students will understand the importance of concurrency and apply it in the context of GUI applications.
- Students will be able to discuss the main issues associated with GUI applications. This includes two fundamental primary themes:

  - Maintaining a responsive GUI application by introducing concurrency for event handlers with human-perceived delays (i.e. allowing the GUI-thread/EDT/UI-thread to promptly return to the event loop to avoid the backlog of events).
  - Ensuring thread-safety of GUI components (i.e. background threads must not access GUI components, only the GUI-thread/EDT/UI-thread may do this).

- In addition to the essential correctness themes above, students will also be able to supply intermittent updates (from the background threads) to the GUI-thread to support improved user-perceived performance (e.g. regular updates to a progress bar).
- Students will be able to understand the different ways in which a GUI application becomes unresponsive, and apply the correct techniques to overcome this:

  - The GUI-thread is never to invoke blocking functions, even when waiting for asynchronous tasks.
  - The GUI-thread is never to process any events that involve perceived delays.

Context for use: It is intended that this chapter be used directly by students to help them understand the underlying concepts of GUI concurrency. Rather than just declare the rules of GUI concurrency, it is important to explain the bigger picture why those rules are in place. The analogy presented has been used in lectures for the above courses with positive feedback from students.

Essential Concurrency Definitions

Before we start discussing GUI concurrency, we will briefly mention the most relevant definitions. Most of these you would have come across already. A **thread** is a programming entity that allows a stream of instructions to be executed independent of (and at the same time as) other instructions. A **task** (or more specifically a **Runnable** in Java) is a packaged entity of code to be executed by a thread. **Locks** are a protection mechanism that ensures only one thread executes a piece of code at any one point in time. As this chapter progresses, more definitions (especially in the context of GUI applications) are introduced. See section "Here Comes the Auditor" for a summary of these concepts.

The Cash Balance Problem

To help us understand concurrency in a graphical user interface (GUI) application, we are going to develop some storylines to explain it in non-technical terms. The first storyline introduced in this section is a rather classical example that helps explain the major problem that concurrency introduces in general. As simple as it may seem, this problem is the fundamental issue underpinning GUI concurrency, so it is important we have a clear appreciation of the inherent problem.

Figure 1 illustrates a company's policy in maintaining the cash balance by using a book. The policy includes three primitive steps to be followed whenever an employee needs to update the balance. First, the employee must observe the balance on the open page of the book (for simplicity, we assume only one balance is written on each page of the book). Once the employee has taken a mental note of the current balance, the employee momentarily performs a simple calculation on their own calculator. With this new balance in mind, the employee returns to the book, flips the page (without taking notice of what page was open) and writes the new balance on the next empty page. This page becomes the new balance.

While it may seem like a rather straightforward and harmless set of steps, the obvious situation is when an employee performs steps 1–3 at the same time as another employee. For example, if Anne and Bob both observe the book balance (1a and 1b respectively), then they both enter $520 into their calculators. Anne adds 20 to her calculator (2a), which then reports 540. In the meantime, Bob is adding 30 to his calculator (2b), which then reports 550. Anne deposits the $20 into the pile of cash, flips the page and writes $540 on a new page (3a). Just as she finishes, Bob is also depositing his $30 into the pile of cash, flips the page and writes $550 on a new page. The book balance is inconsistent with the actual amount of cash, which is actually $570! Neither Anne or Bob is to blame – they were simply following company procedure!

As naive as the three steps seem in updating something as simple as the cash balance, these are the exact same steps involved in updating an integer in our program! A statement as simple as "`count++;`" expands to three instructions that the processor must execute:

1. Read the value from memory and into a register (a small amount of fast storage located on the processor), much in the same way Anne glanced the value in the book and recorded it into her calculator.
2. Perform the increment in the register, much in the same way Anne performed the addition on the calculator.
3. Write the result from the register back to memory (with disregard to the current value in memory), much in the same way Anne copied the result from her

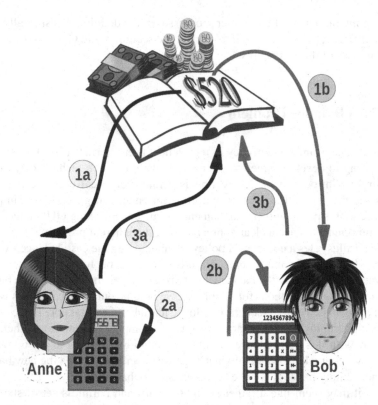

Fig. 1 In this company, a record of the cash balance is maintained in a book. The company policy to maintain the book balance is rather primitive, and involves three simple steps that each employee follows. (1) An employee observes the balance on the currently open page in the book. (2) The employee turns away and calculates the new balance using a calculator. (3) The employee returns to the book, flips the page (without noticing if it had changed since being observed in Step 1) and writes the new balance on the next page

calculator onto a new page in the book (with disregard to the current value written in the book).

We can even simulate the cash balance problem in a simple program[1]:

```
int currentBookBalance = 520;
...
// Anne
int observedAmount = currentBookBalance;        // 1a
blink();
int calculatedAmount = observedAmount + 20;  // 2a
blink();
```

---

[1]The first code example (cash balance problem) is included in the Appendix, and all the example codes are downloadable from http://parallel.auckland.ac.nz/files/gui-chapter-examples.zip

```
currentBookBalance = calculatedAmount;          // 3a
...
// Bob
int observedAmount = currentBookBalance;         // 1b
blink();
int calculatedAmount = observedAmount + 30;   // 2b
blink();
currentBookBalance = calculatedAmount;          // 3b
...
System.out.println("Final balance = $" + currentBookBalance);
```

When we run this program (eg01.CashBalanceProblem.java), we experience what is known as a **race condition** (a programming bug where the output depends on the uncontrollable timing and intertwining of the steps since multiple threads are writing to the same memory location). In our example, we notice that sometimes the final result is $540, while at other times it is $550. It is never the expected $570. The blink() function is rather an exaggeration to help illustrate the point by forcing the intertwining of the three steps between Anne and Bob by introducing a time delay between steps 1, 2 and 3.

Solving the Cash Balance Problem: Without Locks?

You probably anticipated this section to solve the cash balance problem by protecting access to the cash balance using fancy concepts such as mutual exclusion and locks, right? Well, sorry to disappoint you, but we're not going to do that here.[2] *If* we were going to take this approach, then we would be talking about how we put the cash balance inside a room that has a lock on the inside of the door. The rule is that only one person is allowed inside the room, in which case they have full access to the books while everyone else waits outside. When either of Anne or Bob wants to deposit money into the cash balance, they enter the room, perform the three steps, then exit to allow another person to perform the steps.

While using locks seems like a reasonable solution to avoid corrupting the cash balance, the complexity of managing this approach quickly escalates as we introduce more and more items that need protecting. Imagine we have multiple account books, that somewhat relate to each other. We would need to protect each and every one of these books in the same manner. If each book was placed in a separate room with its own lock, how do we ensure we do not deadlock as Anne accesses a book then also wants another book locked by Bob (who in turn wants the book already locked by Anne)?

Devising a set of policies to manage all these books in a correct (let alone efficient) manner is very complicated. So, instead of allowing all the employees to have access to the books, we say that *none* of them is allowed direct access to

---

[2]This isn't to say that locks cannot be used, but rather that we are going to solve this scenario without locks.

Fig. 2 A new policy is put into place, not only to manage the book regarding the cash balance, but all the account books. The new rule states that only Gemma should touch the books, and that any access to them must be through her. This means that Anne and Bob must now write memos for Gemma to take action on the cash balance. If multiple employees wish to access any of the books at the same time, these requests (i.e. the memos) are queued up for Gemma to process one at a time

the books! What we do is employ a new person, Gemma, to be solely responsible for any direct access to the books, including our original cash balance. If Gemma is the only employee that accesses the cash balance, then this will naturally ensure the balance remains correct at all times. Figure 2 illustrates the new policy in place, where the same three steps to modify the cash balance exist, only this time it is always performed by Gemma. We can also see how Gemma is responsible for the other books.

So, what about Anne and Bob when they want to modify the cash balance? It would not be such a good idea if they directly talked to Gemma, since she might be busy performing some other tasks. Instead, it would make more sense if they wrote their request on a memo and placed that memo in the pile next to Gemma. When she gets a chance, Gemma will pick up one of the memos from the pile and complete the instructions requested on it. This is known as the *single thread rule*, where a dedicated thread is assigned the sole responsibility of accessing unprotected data. The single thread rule is implemented in `eg01.CashBalanceWithMemos.java`:

```
BlockingQueue<Memo> pileOfMemos = new LinkedBlockingQueue<Memo>();
...
// Anne creates a Memo requesting $20 to be added
pileOfMemos.add(new Memo(20));
...
// Bob creates a Memo requesting $30 to be added
pileOfMemos.add(new Memo(30));
```

This program differs from the first one, in that Anne and Bob never directly access currentBookBalance. Instead, they each create a Memo and place it on the pileOfMemos. The Memo class is a Runnable instance, defining the three steps necessary to modify the currentBookBalance:

```
class Memo implements Runnable {
    private int amountToAdd;
    Memo(int a) {
        this.amountToAdd = a;
    }
    public void run() {
        int observedAmount = currentBookBalance;// 1
        blink();
        int calculatedAmount = observedAmount +amountToAdd; // 2
        blink();
        currentBookBalance = calculatedAmount;// 3
    }
}
```

Gemma then polls the pileOfMemos, taking one Memo at a time and completing the instructions on it:

```
// Gemma
Memo nextMemo = null;
while ((nextMemo = (Memo)pileOfMemos.poll(1,TimeUnit.SECONDS)) != null) {
    nextMemo.run();
}
```

If Gemma waits longer than 1 s, she assumes no more Memos will arrive and ends her work. Notice that locks were not necessary to protect the currentBookBalance, since Gemma is the only one that has direct access to it. Because she executes one Memo at a time, there is never any intertwining of the three instructions within a Memo. When we execute this program, we will always get the correct result of $570.

Here Comes the Auditor

We think of Gemma's role in the company as being the accountant; to ensure the correctness of the company books, she is the only one within the company allowed to access the books directly. At some stage, a tax auditor may contact the company and inquire about the state of the company's financial records (Fig. 3). Naturally, the auditor has authority (and skills) to inspect the company books without corrupting them (it is his job, after all). In this regards, the books become a medium of communication where the outside world sees the state of the company. If the auditor requires specific jobs from the company, he will telephone the company and explain what he needs. Gemma would answer the phone and take note of the request. Hopefully, Gemma will be able to fulfill the auditor's request in a short amount of time. Since there is only one telephone, and Gemma is the only company representative allowed to operate it, then prolonged handling of any request will mean that other (external) people trying to contact the company will be frustrated as they encounter a busy dial tone. Not only will this occur when Gemma is busy on the telephone, but it will also occur during the time she is completing instructions on the memos given to her from Anne and Bob. Any attempt to call the company at this time will again frustrate the outside world, as the phone rings and rings without being answered.

So, how does this all correspond to a GUI application? In a GUI application, we have the same policies and interactions in place. Here is how the analogy relates to a GUI application:

- The *company* represents the **application**.
- The company *books* represent the **GUI components** that reflect the state of the application. Much like how there are many forms of books a company may maintain, there are many forms of GUI components an application may maintain. Some are forms of input (e.g. text fields and buttons), while others are forms of output (e.g. progress bars and message dialogs). Regardless, they are all GUI components and it is not safe for multiple access.
- The *auditor* (or anyone outside the company) represents the **users** of the application.
- A phone *call* represents an **event** from a user that requires attention. The *arrival of a new memo* to the pile also represents an event.
- The *employees* within the company represent the **threads** within the application. More specifically, *Gemma's* role as *accountant/receptionist* represents the **GUI thread**'s role of sole responsibility for the GUI components. *Anne and Bob* represent the **background threads**, and they should never access the GUI components.
- A *memo* represents a **Runnable (set of instructions)**.
- A *busy dial tone* experienced by the outside world represents an **unresponsive application** or **"frozen" GUI**. In fact, any time Gemma is doing any form of processing (e.g. on the phone, or executing a memo), this corresponds to the GUI thread being busy handling an event. Such processing should be kept to a minimum, ensuring Gemma is kept as free as possible. In other words, the GUI thread should be as idle as possible so that it can respond immediately to any new events without noticeable lag.
- The *pile of memos*[3] (and list of phone messages) represents the **event queue**, containing events yet to be handled by the GUI thread.

---

[3] Although "pile" is used in this analogy, the memos will be processed in a first in first out (FIFO) manner.

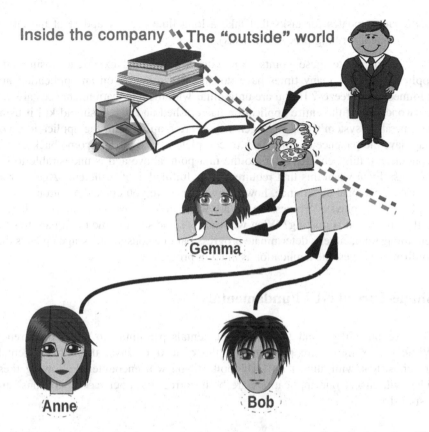

Fig. 3 As the accountant for the company, Gemma is responsible for all the book keeping. The auditor represents an external entity, or client, interested in interacting with the company (i.e. the **users** of the application). The auditor is able to interpret the company's state from the book records and is able to communicate with the company by making phone calls. Gemma is also the only person within the company that responds to the phone calls. If Gemma is busy on the phone, then the outside world gets a busy dial tone. This will inevitably frustrate anyone from outside trying to communicate with the company. Anne and Bob do not interact with the phone, much in the same way they should not access the books

- When Gemma is idle and "*on the lookout*" for memos and messages to arrive, this corresponds to the GUI thread being in the **event loop**. This is the ideal situation, meaning the GUI thread is ready and waiting to respond instantaneously as soon as an event arrives.

So, what does it mean having the "outside" world interacting with the application? In the analogy, these represent customers or auditors that will interact with the company through the phone. We need to ensure this is all responsive. In terms of the application, this represents the user interaction. To be truly responsive, the outside world demands:

1. Continuous responsiveness that never results in a frozen GUI, and

2. Frequent updates for tasks that take a long time (i.e. an update at the end is insufficient for long-processing tasks).

We can appreciate these points from our own personal experience using GUI applications. How many times have you pressed a button on an application and it immediately freezes? If you are unfamiliar with the application, chances are you are wondering if the entire application has crashed and if you should kill it using the operating system's task manager. If you are familiar with the application, you may have the patience to wait for it to complete its actions and come back to life. Nonetheless, this behavior refers to the first point above and is undesirable to say the least. Even when this first requirement is fulfilled, is it sufficient? Again, from your own personal experience, how many times have you clicked a button and the application displays a "Processing, please wait" message but gives no other hint as to the progress it is making? What we want is some sort of clue that quantifies the remaining time, either a determinate progress bar or a constant message updates that confirm to us "yes, the application is making progress".

Single-Thread GUI Fundamentals

This section will present the two fundamentals pertaining to GUI concurrency. While the examples are presented in the context of Java, these fundamentals are consistent with almost all GUI toolkits you will encounter. Following these rules will ensure our applications are both correct (without race conditions) and responsive.

Fundamental 1: Correctness

So, what is the relationship between the single-thread rule of section "Solving the Cash Balance Problem: Without Locks?" and the GUI aspects discussed in section "Here Comes the Auditor"? As hinted in section "Here Comes the Auditor", the GUI components of an application must be protected from possible corruption due to potential race conditions. The easiest way to protect these components is to use the single-thread rule. Almost all the popular GUI toolkits you will come across follow this rule [2–4], where they dedicate a specific thread to access the GUI components (just like in our analogy where Gemma was dedicated to access the books). This thread is most commonly called the **UI Thread**, the **GUI Thread**, or the **Event Dispatch Thread** (EDT) as in Java [5].

To simulate the race condition using a real GUI application, we create our own `ProgressBar` class. This class represents an actual GUI component (it extends Java Swing), which means we can add it to any GUI application. The purpose of this class is to represent the functionality of a real progress bar (for example, `javax.swing.JProgressBar`), but also to illustrate the potential race condition that may arise in using such a GUI component. Only a snip-

pet of this class is shown, but you can have a look at the complete code in
eg02.ProgressBar.java:

```
public class ProgressBar extends JLabel {
    private int value = 0;
    private double max = 100;
    ...
    public void increment(int delta) {
        int oldValue = value;        // read from memory to CPU register
        minorCPUstall();
        oldValue = oldValue + delta; // update value in register
        minorCPUstall();
        value = oldValue;            // write to memory from CPU register
        setText(toString());         // update GUI
    }
    public int getPercent() {
        return (int)(100*value/max);
    }
    public String toString() {
        return getPercent()+"%";
    }
}
```

Figure 4 shows a simple GUI application (eg02.BadGUI.java) that makes
use of this progress bar. There are two buttons below the progress bar:

- **Anne +2**: create a new thread that does some work then increments the progress
  bar by 2.
- **Bob +3**: create a new thread that does some work then increments the progress
  bar by 3.

If we have a look at how the code is implemented to achieve this seemingly innocent
behavior, we see that both threads have direct access to the progress bar instance:

```
public class BadGUI extends JFrame implements
ActionListener {
    private JButton btnAnne = new JButton("Anne +2");
    private JButton btnBob = new JButton("Bob +3");
    private ProgressBar progressBar = new ProgressBar();
    ...
    public void actionPerformed(ActionEvent e) {
        if (e.getSource() == btnAnne) {
            btnAnne.setEnabled(false);

            Thread anne = new Thread() {
                public void run() {
                    doWork(); // This is correctly performed by non-GUI-thread
                    progressBar.increment(2);  // Bad! Accessed by non-GUI thread
                    btnAnne.setEnabled(true);  // Also bad!
                }
            };
            anne.start();

        } else if (e.getSource() == btnBob) {
            ... // equivalent code for Bob's thread
        }
    }
}
```

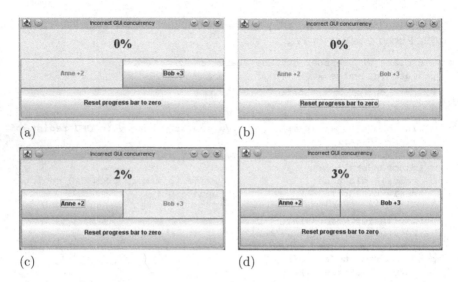

(a)   (b)

(c)   (d)

Fig. 4 Bad practice: a race condition when updating a GUI component (eg02.`ProgressBar`) from multiple threads. Since both Anne's and Bob's threads update the progress bar directly themselves, there is the likelihood that the incorrect value results in the progress bar. Instead of showing 5%, it will either show 2% or 3%, depending on which thread was last. Full example in eg02.`BadGUI.java`. (**a**) Anne's thread starts to add 2 to the current progress bar value. (**b**) Close behind, Bob's thread starts to add 3 to the current progress bar value. (**c**) Anne's thread finishes updating the progress bar. (**d**) Bob's thread finishes, and overrides the update made by Anne's thread

If pressed one at a time, with sufficient time between the completion of each action (i.e. wait for the button to be enabled again), then there is no problem; the value of the progress bar increases to 5%. However, if we were to quickly press the two buttons (as in Fig. 4a–d), then one of the threads will override the value of the progress bar that the other thread has written (rather than incrementing onto the updated value). This is because both threads have access to the same GUI component, and the 3 steps of updating a value might with some (bad) luck be interleaved. Not only is this program incorrect since the threads access our custom-made `ProgressBar`, but they also perform the re-enabling on the buttons!

To overcome this problem, we must conform to the single-thread rule discussed in section "Solving the Cash Balance Problem: Without Locks?". The same program is repeated again (eg02.`GoodGUI.java`), only this time using the correct approach by conforming to the single-thread rule:

```
public class GoodGUI extends JFrame implements ActionListener {
    ...
    public void actionPerformed(ActionEvent e) {
        if (e.getSource() == btnAnne) {
            btnAnne.setEnabled(false);

            Thread anne = new Thread() {
                public void run() {
                    // This is correctly performed by non-GUI-thread
```

```
            doWork();

            // GUI-related work moved to a "memo" for the GUI thread
            SwingUtilities.invokeLater(new Runnable() {
                public void run() {
                    progressBar.increment(2);
                    btnAnne.setEnabled(true);
                }
            });
        }
    };
    anne.start();

    } else if (e.getSource() == btnBob) {
        ... // equivalent code for Bob's thread
    }
  }
}
```

The difference here is that the updating of the progress bar and button are no longer performed by Anne's and Bob's threads. Instead, the instructions to update the GUI are wrapped inside a `Runnable` instance (representing the memo in our analogy) and passed to the GUI toolkit, requesting that the GUI thread invoke these instructions. This memo is submitted by the respective background thread (e.g. anne) after the `doWork()` computation is completed. Regardless of which memo will be picked up by the GUI thread, they will always be executed one at a time since they get piled up in the GUI thread's event queue. Go ahead and modify the `minorCPUstall()` function inside the `ProgressBar` class to increase this stall amount. The correct result is always achieved!

Fundamental 2: Responsiveness

Section "Fundamental 1: Correctness" demonstrated the single-thread rule in the context of GUI applications in order to protect the GUI components. In other words, the purpose was to ensure program *correctness*. This section will now demonstrate the single-thread rule with another purpose in mind: ensuring a *responsive application*. While this is not required in contributing towards the correctness or functionality of the application, it is essential in contributing towards a positive user experience and therefore overall user satisfaction. In fact, you could even consider an unresponsive application as dysfunctional!

Purpose of Concurrency

Based on section "Here Comes the Auditor", we already understand that we need to allow the GUI thread to be idle as much as possible in order for it to patrol the event loop and therefore react to new events without delay. To achieve this, the GUI thread employs background threads that perform all the long-lasting processing that would otherwise preoccupy the GUI thread for an unacceptable amount of time. Ultimately, this allows the GUI thread to immediately return to the event loop in anticipation of new events arriving, while the background threads are doing the real work. This is the concept of *GUI concurrency*, where the background thread (executing time-consuming computation) is working concurrently with (i.e. at the same time as) the GUI thread (patrolling the event loop). You will also hear terminology such as "the time-consuming computation is executed *asynchronously*", which is a fancy way of saying the time-consuming computation is progressing independently of the patrolling of the event loop (i.e. on its own time and in "its own little world").

Classifying a GUI's Streams of Instructions

As already hinted in the previous section, the underlying concept behind concurrency is that of a thread. By having multiple threads, we can logically perform multiple streams of instructions at the same time. In the context of a GUI application, we are interested in separating the instructions into two particular streams of instructions. Each of these streams of instructions will be executed by a thread in order for the streams to progress concurrently:

- **Event management mechanism**: this refers to the instructions that define the administration relevant to the event loop, including the enqueuing, dequeuing and handling of events on the event queue. Fortunately, GUI toolkits typically provide an implementation for this event handling mechanism so that programmers do not need to manage it (or even see it!). This also includes the nomination of the GUI thread to manage all of this communication in the event loop. All that programmers need to do is specify the stream (or block) of instructions for the handling of those events (i.e. the response to a particular event).
- **Event handling logic**: this refers to the programmer-defined stream of instructions that depict what should happen when an event is encountered. The GUI thread will initially commence handling the event, but it is the responsibility of the programmer to determine if the computation will be time-consuming. If so, then the programmer needs to "free the GUI thread" by creating a new thread to take over. The GUI thread therefore classifies the event as being "sorted out", and immediately returns to the event loop. This will avoid any "freezing" of the application.

In Java, the most primitive approach to achieve this is using `Threads` and the `SwingUtilities` class to hook into the event management system whenever necessary. This is shown by the program of Fig. 5, which demonstrates both a responsive and unresponsive handling of events within the same application. The code snippet below refers to the event handling logic when any of the 3 buttons are pressed:

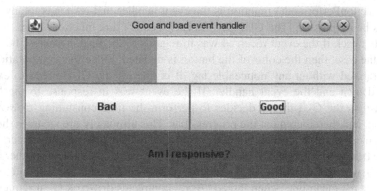

Fig. 5 Good and Bad application. This application contains a standard progress bar (top row), 2 buttons with time-consuming tasks, and a third button to test responsiveness of the application (it changes color as soon as it is pressed). When either of the "Good" or "Bad" buttons is pressed, the progress bar is incremented. The only difference is that the "Bad" button freezes the entire application until the action is completed, whereas the "Good" button maintains application responsiveness, allowing other buttons to be pressed. The full code is found in eg03.GoodAndBadGUI.java

```
public void actionPerformed(ActionEvent e) {

    // the GUI thread can quickly create a new color
    if (e.getSource() == btnResponsive) {
        btnResponsive.setBackground(createRandomColour());
        return;
    }

    // the other buttons involve some time-consuming work being performed
    if (e.getSource() == btnBad) {
        // The current thread (the GUI thread) does the work itself...
        doWork();
        // ... and then updates the progress bar
        progressBar.setValue(progressBar.getValue()+1);
    } else {
        // The GUI thread asks a background thread to take over...
        Thread bob = new Thread("Bob") {
            public void run() {
                // the work is performed by the background thread...
                doWork();
                // ... and the background thread asks the GUI thread to
                //                           update the progress bar
                SwingUtilities.invokeLater(new Runnable() {
                    public void run() {
                        progressBar.setValue(progressBar.getValue()+1);
                    }
                });
            }
        };
        bob.start();
    }
}
```

The actionPerformed() function is the event handler that responds to any of the buttons being pressed. The GUI thread always enters this function. Here, we first query to check if the event received was in regards to the responsiveness button. If this is the case, then the color of the button is updated. Since this computation can be performed without any noticeable lag, it is fine for the GUI thread to execute this and then end the event handler. If the event was in response to the "Bad" button, then the GUI thread decides to perform the time-consuming doWork() function, and then increment the progress bar. This ultimately preoccupies the GUI thread, meaning other events cannot be responded to. The final situation refers to the desired behavior, where the doWork() function is being assigned to another thread ("Bob"), which allows the GUI thread to end the event handler and respond to other events. In the meantime, when thread "Bob" completes doWork(), it requests the GUI thread to update the progress bar (since only the GUI thread should access GUI components). When you run the examples, notice the output printed that state the name of the thread executing the respective sections of code.

More Elegant Library Support: SwingWorker

The code snippets of section "Single-Thread GUI Fundamentals" demonstrated how we can achieve a responsive and thread-safe application by resorting to using primitive libraries existing in the Java library (in this case the Thread and SwingUtilities classes). While this approach got the job done and met our requirements, it does pose some disadvantages:

- It contained a large amount of boilerplate code to create background threads and send memos back to the GUI thread. This problem will be exacerbated should we need to send intermittent memos to the GUI thread (i.e. not just at the very end).
- Notice how we are creating a new thread every time the "Good" button is pressed. In most cases, this will not be an issue if we are not expecting to have too many tasks. However, if we end up having lots of threads that perform a large amount of computation, then we risk the chance of reducing performance of the application since a lot of time will be dedicated to managing the threads rather than executing the work. A smarter solution would create a fixed number of threads, and instead queue the work to be executed as a thread frees up.

To solve the points above, yet to retain respect to the single-thread GUI model, Java introduced the SwingWorker class. This provides a more elegant solution by dealing with the creation and management of a team of background threads, while also reducing the boilerplate code required. The code snippet below demonstrates how the event handler is modified:

```
public void actionPerformed(ActionEvent e) {

    // ... same as before
```

```
if (e.getSource() == btnBad) {
    // ... same as before
} else {
    Memo memo = new Memo();
    memo.execute();
}
}
```

What has changed? The code remains essentially identical to that of section "Classifying a GUI's Streams of Instructions", except now we create a Memo instance and tell it to execute(). It definitely looks more elegant than the code we had before! All the logic in regards to doing the work and updating the progress bar we define in Memo:

```
class Memo extends SwingWorker {
    protected Void doInBackground() {
        doWork();
        return null;
    }
    protected void done() {
        progressBar.setValue(progressBar.getValue()+1);
    }
}
```

You will notice that this class is not too complicated at all. In fact, SwingWorker helps guide us by specifying which functions we should be implementing. In our simple example, doInBackground() is the place we specify any time-consuming computation that will be passed on to the background thread. The done() function refers to any computation that must be performed by the GUI thread when doInBackground() is completed.

You are probably wondering, where is the background thread? This is the other elegance to this solution, in that the programmer does not need to create or manage the background threads that will execute the SwingWorker instances. This is all managed automatically by the library using a pool of threads that are dedicated to processing doInBackground() functions. If you execute the example code provided (eg04.GoodAndBadSwingWorker.java), the only difference you will notice is in the output printed. Notice how the names of the threads are now something like "SwingWorker-pool-1-thread-5" or "SwingWorker-pool-1-thread-2", which refers to the threads that are being automatically managed to execute background work. In order to see the multiple background threads being managed by the SwingWorker class, try pressing the "Good" button as fast as you can 15 times. Every time you press the button, it enqueues a memo that will eventually be processed by one of the threads. By reading the thread names, we notice that the same 10 threads are being recycled (the exact number might be slightly different for you).

You will also notice that it is still the GUI thread that is executing the done() function. Figure 6 illustrates how we can visualize SwingWorker in terms of our company analogy. The newly submitted jobs refer to the execute() function being performed on a newly created SwingWorker instance. When one of the SW-threads is idle, it picks up the next memo from the pile and completes the

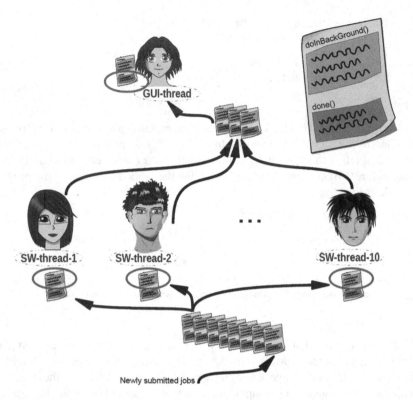

Fig. 6 SwingWorker is designed to meet both the correctness and responsiveness fundamentals of GUI concurrency. The top right corner shows how we can visualize a SwingWorker instance as a memo containing 2 sections. The blue section (`doInBackground()`) is reserved for one of the SW threads, while the orange section (`done()`) is reserved for the GUI thread. There are 2 queues: the first is when the job is submitted and it waits for a SW thread, while the second is when the background portion is completed and the memo is passed on to the GUI thread to execute the GUI-related portion

top blue section. Upon completion, that memo is passed on to the GUI thread to complete the bottom orange section. Concurrency is achieved by having multiple background threads that are available to execute the long-processing computations. By separating the GUI-related computation in a different section, this also achieves responsiveness since the GUI thread is not unnecessarily occupied.

Improving User Experience with Intermittent Results/Updates

A big part of GUI applications is ensuring a positive user experience. In this regards, responsiveness not only means avoiding a freezing user interface, but also providing *regular* updates to the user. This is especially important for background jobs that take a long amount of time. Examples include displaying search results (e.g. searching through emails) as they are found, or progressively rendering thumbnails of images in a folder. Bear in mind that we want to still conform to the GUI concurrency fundamental of correctness. This means that it must be the GUI thread updating the user interface with the information – but how does the GUI thread know about the background thread's progress on a given task? The general idea is simple:

- The background thread, as it processes the doInBackground() section, decides that it has accomplished a significant amount of work that warrants celebration. Since it is not allowed to access the GUI components directly itself, it simply publishes this achievement and resumes processing the remainder of the doInBackground() section.
- The GUI thread, upon hearing the update request, takes the published data and displays it on the GUI.

How is this achieved using SwingWorker? Before seeing the code, lets have a look at the general concept with the help of Fig. 7. As before, we have the doInBackground() and done() sections that are executed by the background SW-threads and the GUI thread respectively. There is a new section, process(List), which is executed by the GUI thread whenever an "attachment" is added on the SwingWorker "memo". How do these attachments get there? This is the job of the background thread, to publish() these items whenever it feels it has made substantial progress in the background processing. Rather than waiting for the GUI-thread to acknowledge receipt of the attachment, the background thread continues processing the remainder of the doInBackground(). This is how the attachments potentially "pile up" for the GUI thread to process() (hence a List of intermittent results to process).

One of the most common cases of publishing intermittent results is when a progress bar is used. Figure 8a shows such an example, while below is the code snippet (full code in eg03.ManyUpdates.java) that demonstrates the correct way to frequently update the status of a progress bar:

```
class Memo extends SwingWorker<Void, Integer> {
    protected Void doInBackground() {
        for (int i = 1; i <= 10; i++) {
            doWork();
            publish(i); // create a new "attachment"
        }
        return null;
    }
    protected void process(List<Integer> attachments) {
        // process all the attachments that have piled up
        for (int attachment : attachments) {
            progressBar.setValue(10*attachment);
        }
    }
    protected void done() {
        // doInBackground() ended, so re-enable start button
        btnStart.setEnabled(true);
    }
}
```

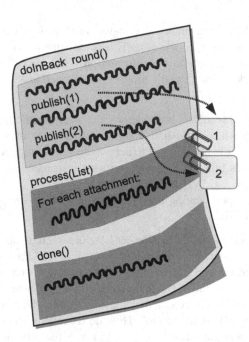

Fig. 7 Our final visualization of Swingworker to include how intermittent results/updates are propagated from a background thread to the GUI thread. As the background thread processes the doInBackground(), it frequently decides to publish intermittent data. We imagine this as an attachment to the SwingWorker instance (i.e. the memo). The GUI thread sees these attachments, and executes the process() section for the list of attachments as they come along. As before, the orange sections are GUI-related (to be executed by the GUI thread), while the blue section refers to non-GUI and time-consuming work

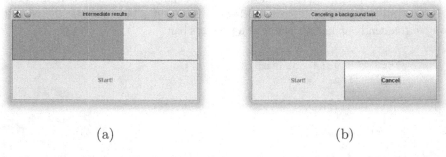

(a)   (b)

Fig. 8 (a) An example application demonstrating a background computation with intermittent updates via the progress bar. (b) The same application is extended to allow canceling of the background computation

Canceling Background Tasks

If you ran the example of Fig. 8a, you probably eventually felt like something was not quite complete. Did you notice how we had no way to cancel the task? Clearly, there was no cancel button on the GUI. If we were to introduce such a cancel button (as in Fig. 8b), what does this mean in the context of a background task that was executed concurrently? Well, first of all, whoever wishes to cancel() the background task obviously needs access to the very same SwingWorker instance that was initially told to execute(). In other words, we need to declare our SwingWorker memo at a scope such that it is still accessible to the event handler:

```
public class ManyUpdatesWithCancel extends JFrame implements ActionListener {
    private JButton btnStart = new JButton("Start!");
    private JButton btnCancel = new JButton("Cancel");

    private Memo memo;   // instance at a scope accessible to all handlers
    ...
    public void actionPerformed(ActionEvent e) {
        if (e.getSource() == btnCancel) {
            memo.cancel(true);
        } else if (e.getSource() == btnStart) {
            memo = new Memo();
            memo.execute();
        }
    }
}
```

Figure 9 shows how canceling is implemented for background tasks. Assume the SwingWorker memo instance is being executed by one of the background SW-threads. In the meantime, the cancel button was pressed, so we invoke cancel() on that same memo instance. Abruptly killing the background SW-thread as it is executing doInBackground() is an unsafe practice. Instead, what happens is the memo is stamped with a cancel request. However, for this to really take effect, the background SW-thread needs to frequently check the memo status just in case it has been stamped with cancel. This is achieved by calling isCancelled(), which checks for the status. If this returns true, then the doInBackground() method is ended with an early return statement.

Since the doInBackground() method has essentially ended (regardless of whether it was canceled or not), the GUI-thread takes over the memo by executing the done() method. It even executes any remaining attachments in the process() method. For this reason, we sometimes might want to take alternative decisions inside done() and process() depending on whether the memo had been canceled. In this case, the GUI-thread may also use the isCancelled() method to check if the memo had been canceled. The example of eg03.ManyUpdatesWithCancel.java demonstrates this by discarding attachments if the memo was canceled, and only displaying a final message if done() is processed without having received a cancel request.

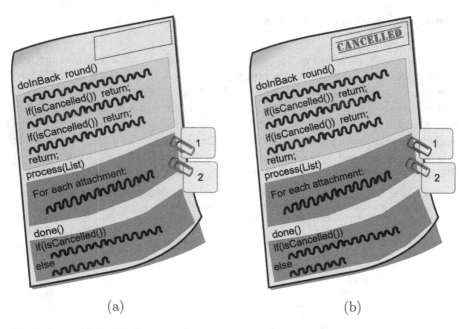

Fig. 9 Once a SwingWorker memo has started executing, it needs to periodically check if it has been requested to cancel. This is to allow the background SW-thread to tidy up and end the `doInBackground()` in a clean manner. (**a**) Before a cancel request has been made. (**b**) After a SwingWorker memo instance has been instructed to `cancel()`. The act of canceling a memo essentially means it is stamped; however, this has no effect unless it is checked for, and acted upon

Wrapping Up

In this chapter, we appreciate the necessity of multi-threading in the context of applications that possess a graphical user interface (GUI). This is especially important as multi-core processors have become the norm for desktop and mobile devices, since the sorts of applications running on these systems will interact with users via the GUI. The limitations of the GUI toolkits available for these paradigms means that programmers need to adhere to the single-thread rule. This ultimately means two things. First, the event handlers that the GUI thread responds to must be kept minimal without noticeable lag to ensure a responsive application. This is achieved by off-loading the long processing computation to a background thread. Second, the background thread must never directly access any GUI component during that time, and should instead request the GUI thread to do so. Ultimately, when developing your next GUI application, you need to keep in mind the user-perceived performance of responsiveness, by implementing the concurrency features discussed in this chapter. This includes intermittent progress updates from background threads, as well as canceling the background tasks. But, as you do so, always remember the single-thread rule.

Appendix

eg01.CashBalanceProblem.java

```java
package eg01;

public class CashBalanceProblem {

  private static int currentBookBalance = 520;

  public static void main(String[] args) {
    // Anne will execute the three steps, in order to add $20
    Thread anne = new Thread() {
      public void run() {
        int observedAmount = currentBookBalance;      // 1a
        blink();
        int calculatedAmount = observedAmount + 20;  // 2a
        blink();
        currentBookBalance = calculatedAmount;        // 3a
      }
    };
    anne.start();

    // Bob will execute the three steps, in order to add $30
    Thread bob = new Thread() {
      public void run() {
        int observedAmount = currentBookBalance;       // 1b
        blink();
        int calculatedAmount = observedAmount + 30;   // 2b
        blink();
        currentBookBalance = calculatedAmount;         // 3b
      }
    };
    bob.start();

    // Wait for both Anne and Bob to finish the three steps
    try {
      anne.join();
      bob.join();
    } catch (InterruptedException e) {
      e.printStackTrace();
    }

    System.out.println("Final_balance_=_$"+currentBookBalance);
  }

  // Simulate blinking
  public static void blink() {
    try {
      Thread.sleep(100);
    } catch (InterruptedException e) {
```

```
        e.printStackTrace();
    }
  }
}
```

eg01.CashBalanceWithMemos.java

```java
package eg01;

import java.util.concurrent.BlockingQueue;
import java.util.concurrent.LinkedBlockingQueue;
import java.util.concurrent.TimeUnit;

public class CashBalanceWithMemos {

  private static int currentBookBalance = 520;

  // the queue represents a pile of memos
  private static BlockingQueue<Memo> pileOfMemos = new
      LinkedBlockingQueue<Memo>();

  // a definition on a Memo, which states what needs to be done (
      i.e. the three steps to modify the cash balance)
  static class Memo implements Runnable {
    private int amountToAdd;
    Memo(int a) {
      this.amountToAdd = a;
    }

    @Override
    public void run() {
      int observedAmount = currentBookBalance;        // 1
      blink();
      int calculatedAmount = observedAmount + amountToAdd;    //
          2
      blink();
      currentBookBalance = calculatedAmount;          // 3
    }
  }

  public static void main(String[] args) {

    // Anne creates a new Memo, requesting $20 to be added
    Thread anne = new Thread() {
      public void run() {
        pileOfMemos.add(new Memo(20));
      }
    };
    anne.start();

    // Bob creates a new Memo, requesting $30 to be added
    Thread bob = new Thread() {
      public void run() {
```

```
        pileOfMemos.add(new Memo(30));
      }
    };
    bob.start();

    // Gemma goes to the pile of Memos, takes one at a time. If
    //    the pile is empty for more than 1 second, Gemma stops.
    Thread gemma = new Thread() {
      public void run() {
        Memo nextMemo = null;
        try {
          while ((nextMemo = (Memo)pileOfMemos.poll(1,TimeUnit.
              SECONDS)) != null) {
            nextMemo.run();
          }
        } catch (InterruptedException e) {
          e.printStackTrace();
        }
      }
    };
    gemma.start();

    // wait for Anne, Bob and Gemma to finish
    try {
      anne.join();
      bob.join();
      gemma.join();
    } catch (InterruptedException e) {
      e.printStackTrace();
    }

    System.out.println("Final_balance_=_$"+currentBookBalance);
  }

  // Simulate blinking
  public static void blink() {
    try {
      Thread.sleep(100);
    } catch (InterruptedException e) {
      e.printStackTrace();
    }
  }
}
```

References

1. S. K. Prasad, A. Chtchelkanova, M. G. F. Dehne, A. Gupta, J. Jaja, K. Kant, A. L. Salle, R. LeBlanc, A. Lumsdaine, D. Padua, M. Parashar, V. Prasanna, Y. Robert, A. Rosenberg, S. Sahni, B. Shirazi, A. Sussman, C. Weems, and J. Wu, "NSF/IEEE-TCPP Curriculum Initiative on Parallel and Distributed Computing – Core Topics for Undergraduates, Version 1," http://www.cs.gsu.edu/~tcpp/curriculum, 2012.

2. D. Lea, *Concurrent programming in Java: design principles and patterns*, 2nd ed. Addison-Wesley, 1999.
3. P. Hyde, *Java Thread Programming*. Sams, 2001.
4. E. Ludwig, "Multi-threaded user interfaces in java," Ph.D. dissertation, University of Osnabrück, Germany, May 2006.
5. Oracle. (2017) Lesson: Concurrency in Swing. http://docs.oracle.com/javase/tutorial/uiswing/concurrency.

Scheduling in Parallel and Distributed Computing Systems

Srishti Srivastava and Ioana Banicescu

Abstract Recent advancements in computing technology have increased the complexity of computational systems and their ability to solve larger and more complex scientific problems. Scientific applications express solutions to complex scientific problems, which often are data-parallel and contain large loops. The execution of such applications in parallel and distributed computing (PDC) environments is computationally intensive and exhibits an irregular behavior, in general due to variations of algorithmic and systemic nature. A parallel and distributed system has a set of defined policies for the use of its computational resources. Distribution of input data onto the PDC resources is dependent on these defined policies. To reduce the overall performance degradation, mapping applications tasks onto PDC resources requires parallelism detection in the application, partitioning of the problem into tasks, distribution of tasks onto parallel and distributed processing resources, and scheduling the task execution on the allocated resources. Most scheduling policies include provisions for minimizing communication among application tasks, minimizing load imbalance, and maximizing fault tolerance. Often these techniques minimize idle time, overloading resources with jobs and control overheads. Over the years, a number of scheduling techniques have been developed and exploited to address the challenges in parallel and distributed computing. In addition, these scheduling algorithms have been classified based on a taxonomy for an understanding and comparison of the different schemes. These techniques have broadly been classified into static and dynamic techniques. The static techniques are helpful in minimizing the individual task's response time and do not have an overhead for information gathering. However, they require prior knowledge of the system and they cannot address unpredictable changes during runtime. On the other hand, the dynamic techniques have been developed to address unpredictable changes, and maximize resource utilization at the cost of information gathering overhead. Furthermore, the scheduling algorithms have also been characterized as optimal or sub-optimal,

---

S. Srivastava (✉) · I. Banicescu
University of Southern Indiana, Evansville, IN, USA

Mississippi State University, Starkville, MS, USA
e-mail: fsrishti@usi.edu; ioana@cse.msstate.edu

© Springer International Publishing AG, part of Springer Nature 2018   313
S. K. Prasad et al. (eds.), *Topics in Parallel and Distributed Computing*,
https://doi.org/10.1007/978-3-319-93109-8_11

cooperative or non-cooperative, and approximate or heuristic. This chapter provides content on scheduling in parallel and distributed computing, and a taxonomy of existing (early and recent) scheduling methodologies.

- **Relevant core courses:** DS/A, ParAlgo, DistSystems.
- **Relevant PDC topics:** shared memory (C), distributed memory (C), data parallel (C), parallel tasks and jobs (K), scheduling and mapping (C), load balancing (C), performance metrics (C), concurrency (K), dependencies (K), task graphs (K).
- **Learning outcomes:** The chapter provides an introduction of scheduling in PDC systems such that it can be easily understood by undergraduate students, who are exposed to this topic for the first time. The chapter is intended to provide learning to undergraduate students, who are beginners in the field of high performance computing. Therefore, the goal of this book chapter is to present an overview of scheduling in parallel and distributed computing. Using the knowledge from this chapter, students are expected to understand the basics and importance of scheduling in parallel and distributed computing, understand the difference between different classes of scheduling algorithms and the computational scenarios for their application, and be able to compare different scheduling strategies based on various performance metrics, such as execution time, overhead, speedup, efficiency, energy consumption, and others. In addition, a number of useful resources related to scheduling in PDC systems have been provided for instructors.
- **Context for use:** The material is designed for being incorporated into core courses such as, data structures and algorithms (DS/A), or advanced courses such as, parallel algorithms (ParAlgo), and distributed systems (DistSystems). The material is intended for students who already have an understanding of the basic concepts and terminology of parallel and distributed computing systems.

Introduction

The scheduling problem has been formulated with several definitions across many different fields of application. The problem of job sequencing in manufacturing systems forms the basis for scheduling in parallel and distributed computing systems, and is also recognized as one of the original scheduling problems. Similar to the job sequencing problem in a manufacturing process, a scheduling system is comprised of a set of consumers, a set of resources, and a scheduling policy. A basic scheduling system is illustrated in Fig. 1, where a task in a computer program, a bank customer, or a factory job are examples of consumers, and a processing element in a computer system, a bank teller, or a machine in a factory are examples of resources in a scheduling system. A scheduler acts as an intermediary between the consumers and the resources to optimally allocate resources to consumers according to the best available scheduling policy [1].

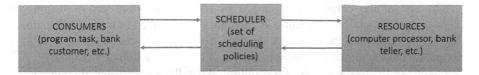

Fig. 1 A basic scheduling framework

In parallel and distributed computing, multiple computer systems are often connected to form a multi-processor system. The network formed with these multiple processing units can vary from being tightly coupled high speed shared memory systems to relatively slower loosely coupled distributed systems. Often, processors communicate with each other by exchanging information over the interconnection structure. One of the fundamental ideas behind task scheduling is the proper distribution of program tasks among multiple processors, such that the overall performance is maximized by reducing the communication cost. Various task scheduling approaches have a trade-off, between performance and scheduling overhead, associated with them for different applications in parallel and distributed computing. A solution to a scheduling problem determines both the allocation and the execution of order of each task. If there is no precedence relationship among the tasks, then the scheduling problem is known as a task allocation problem [1].

Scheduling is a feature of parallel computing that distinguishes it from sequential computing. The Von Neumann model provides generic execution instructions for a sequential program, where a processor fetches and executes instructions one at a time. As a parallel computing analogy to the sequential model, parallel random access memory (PRAM) was formulated as a shared memory abstract machine [2, 3]. However, no such practical model has yet been defined for parallel computing. Therefore, many different algorithms have been developed for executing parallel programs on different parallel architectures. Scheduling requires allocation of parallel parts of an application program onto available computational resources such that the overall execution time is minimized. In general, the scheduling problem is known to be NP-Complete [4–6]. Therefore, a large number of heuristics have been developed towards approximating an optimal schedule. Different heuristics are applicable in different computational environments depending on various factors, such as, problem size, network topology, available computational power, and others. Based on the heuristics a large number of scheduling algorithms have been developed and the performance of these algorithms also vary with the type of computational environment. One of the goals of this chapter is to clarify the differences among scheduling algorithms, and their application domains. In general, during the scheduling of program tasks on parallel and distributed computing systems, the tasks are often represented using directed graphs called task graphs and the processing elements and their interconnection network is represented using undirected graphs. A schedule is represented using a timing diagram that consists of a list of all processors and all the tasks allocated to every processor. The tasks are ordered on a processor by their starting times [1].

The rest of the chapter is organized as follows. An overview of mapping algorithms onto parallel computing architectures is described in section "Mapping Algorithms onto Architectures". A detailed taxonomy of scheduling in parallel and distributed computing is explained in section "A Scheduling Taxonomy". A discussion of the recent trends in scheduling in parallel and distributed computing systems is given in section "Examples of Recent Trends in Scheduling".

Mapping Algorithms onto Architectures

The mapping problem consists of assigning the subtasks of an application to processors, so that its execution time is minimized. The basic steps involved are: detecting parallelism, partitioning the problem into independent sub tasks, and scheduling these subtasks on processors. Performing any of these steps in isolation can lead to poor mappings, and therefore, low performance. The parallelism in a program depends on the nature of the problem and the algorithm employed by the programmer. To obtain high performance, a problem must contain sufficient parallelism. Parallelism detection is usually independent of the target machine. In contrast, partitioning and scheduling are highly dependent on architectural parameters, such as the number of processors, processor speed, communication overhead, scheduling overhead, etc. Partitioning attempts to match the granularity of the parallel subtasks to that of the target machine. Scheduling assigns subtasks to processors and orders their execution. The goals of scheduling are to spread the load as evenly as possible to processors and to minimize data communication.

Scheduling schemes can be static or dynamic. In static schemes, subtasks are assigned to processors at compile time either by the programmer or by the compiler. There is no runtime overhead. The disadvantage of static allocation is that the unpredictable runtime execution of subtasks can lead to load imbalance. Dynamic scheduling schemes assign subtasks to processors at runtime. Dynamic assignment of tasks can improve processor utilization, with a trade-off for an additional allocation overhead. Dynamic assignments can be distributed or centralized. In a centralized allocation scheme, there is a pool of tasks that is accessible by all idle processors. Accessing the central pool may be a bottleneck when the number of processors is large. In a distributed allocation scheme, tasks are allocated on the basis of processor negotiation. Distributed allocation may result in sub-optimal load balancing, as scheduling decisions are mainly based on local information.

For some applications, it may be necessary to order the execution of tasks with data dependencies. Executing data dependent tasks on different processors requires costly synchronization and communication. Therefore tasks allocated to different processors should be made as independent of each other as possible. Synchronization and communication overhead depend upon several factors, such as, the algorithm, the subdomain size, and the machine characteristics. An effective scheduling algorithm must ensure that computational tasks with dependencies are mapped onto processors that can communicate with low latency. Therefore, work allocation is not independent of work partitioning. Mapping should, thus, consider the communication topology during the partitioning step. This leads to a need for a close match between the topology of the dependency graph of the tasks and the communication topology of the machine.

Parallelism Detection

An important component for parallel and distributed computing is a technique that detects and schedules the parallelism in a sequential program, possibly by applying code transformations to effectively utilize the system resources. This process of detecting parallelism is done by examining the code for fine grain operations (such as, parallel operations in program statements) and/or coarse grain operations (such as, vector operations or loop parallelization), depending on the target architecture. Coarse grain parallelism is best detected using the program source code while the detection of fine grain parallelism usually requires an intermediate level program representation. Techniques for the detection of both coarse and fine grain parallel operations have been developed to take advantage of various parallel architectures [7].

Coarse grain parallelism found in sequential programs is mainly in the form of vectorizable computations. Considerable research attention has been devoted to the detection of vectorizable loops in Fortran programs. The techniques include the detection of coarse grain parallelism useful in generation of code for loosely coupled multiprocessor systems. Research in the detection and utilization of fine grain parallelism has also received some attention. A technique that has effectively tackled the problem of detecting fine grain parallelism across basic blocks is trace scheduling which uses a control flow graph representation of a program [8].

In general, algorithms for parallelism detection transform the code so that each statement is surrounded by the same number of loops before and after the transformation. Parallelism detection is optimal if, after transformation, each statement is surrounded by a maximal number of parallel loops. The only constraint that a parallelism detection algorithm must respect is that the partial order of operations defined by the dependencies in the program are preserved. Parallelism detection is a wide topic and has been a research topic in the area of compiler optimization [7].

Partitioning

A process or a task is the basic unit of program execution, and a parallel application is one that has multiple processes or tasks actively performing computation at one time. Partitioning is the process of decomposing a serial application into multiple concurrently executing parts. In parallel and distributed computing applications, task and data parallelism are two of the most commonly referenced parallel patterns [9]. A task parallel application is decomposed into concurrent units that execute separate instructions simultaneously. On the other hand, a data parallel application is decomposed into concurrent units that execute the same instructions on distinct data sets. Moreover, applications in parallel and distributed computing exhibit spatial and temporal patterns indicating their execution in time and space. For instance, the location of a data point in memory represents the spatial index for that application, and the order in which the data points are accessed for application execution represents the temporal index of that application. Different partitioning strategies are developed to distinguish parallel patterns in an application and further employ temporal and spatial partitioning as required. A generic procedure for determining the dimensionality of the instructions and data of an application to prepare it for partitioning, is summarized as follows [10]:

1. Determine what constitutes a single input to define the temporal dimension of the program's data. For some programs an input might be a single reading from a sensor. In other cases an input might be a file, data from a keyboard or a value internally generated by the program.
2. Determine the distinct components of an input to define the spatial dimension of the program's data.
3. Determine the distinct functions required to process an input to define the spatial dimension of the program's instructions.
4. Determine the partial ordering of functions using topological sort on the program dependence graph to define the temporal dimension of the program's instructions.

The problem of building a partition with the smallest partitioning cost is known to be intractable [11]. Therefore, research in this area has been focused on developing approximation algorithms to provide a solution to the partitioning problem.

Task Allocation and Scheduling

Task allocation is a relevant concept in distributed systems. Given a distributed system made up of a number of processing elements connected together using an interconnection network and a distributed application consisting of communicating tasks, allocation techniques assign tasks to processing elements, to optimize the execution of the application as a whole. Task allocation is considered when there is no precedence among the tasks forming a program or an application [1]. Scheduling is an ordering of the execution of the application tasks on the available processing elements. Often, task allocation and scheduling are used interchangeably and are considered to be performed together. Moreover, scheduling is considered to encompass the previous steps of parallelism detection, partitioning, and task allocation.

There are four components in any scheduling system: the target machines, the parallel tasks (defined as a set of sequential tasks, where different tasks can be executed in parallel if there are no dependencies), the generated schedule, and a performance criterion. The following mathematical description, of these four components of a scheduling system, has been adopted from [1].

Target Machine

The target machine is assumed to be made up of m heterogeneous processing elements connected using an arbitrary interconnection network. Each processing element can run one task at a time and all tasks can be processed by any processing element. Formally, the target machine characteristics can be described as a system $(P, [P_{ij}], [S_i], [I_i], [B_i], [R_{ij}])$ as follows:

1. $P = \{P_1, \cdots, P_m\}$ is a set of processors forming the parallel architecture. P_{ij} is an $m \times m$ interconnection topology matrix of processors as its rows and columns, and each matrix element represents a link between corresponding processors.
2. S_i, $1 \leq i \leq m$, is the speed of processor P_i.
3. I_i, $1 \leq i \leq m$, is the startup cost of initiating a message on processor P_i.
4. B_i, $1 \leq i \leq m$, is the startup cost of initiating a process on processor P_i.
5. R_{ij} is the transmission rate over the link connecting two adjacent processors P_i and P_j.

The connectivity of the processing elements can be represented using an undirected graph called the target machine graph as illustrated in Fig. 2.

Fig. 2 An example of a target machine with eight processors (m = 8) forming a three dimensional hypercube network. The nodes are labeled with integers indicating the processor numbers

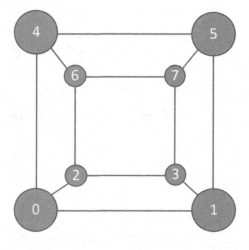

Parallel Application Tasks

A parallel program is modeled as a partially ordered set (poset) $(T, <)$, where T is a set of tasks. The relation $u < v$ implies that the computation of task v depends on the results of the computation of task U, and therefore, task u must be computed for delivering the result to the processor computing the task v. The characteristics of a parallel program can be defined as the system $(T, <, [D_{ij}], [A_i])$ as follows [1]:

1. $T = \{t_1, \cdots, t_n\}$ is a set of application tasks to be executed.
2. $<$ is a partial order defined on T, which specifies the operational precedence constraints.
3. $[D_{ij}]$ is an $n \times n$ communication data matrix, where $D_{ij} \geq 0$ is the amount of data required to be transmitted from task t_i to task t_j.
4. $[A_i]$ is an n-length vector specifying the computational requirements of a task t_i in terms of number of instructions.

The ordered tasks are represented using a directed acyclic graph, which is called a task graph. A directed edge, (i, j), between two tasks t_i and t_j indicates that t_i must be completed before a processor starts executing t_j. An example of a task graph is illustrated in Fig. 3.

Given a parallel program model in the form of a task graph and a description of the target machine, task execution time (T_{ij}) and communication delay $(C(i_1, i_2, j_1, j_2))$, between two processors j_1, j_2 executing tasks i_1, i_2 respectively, can be calculated as follows [1]:

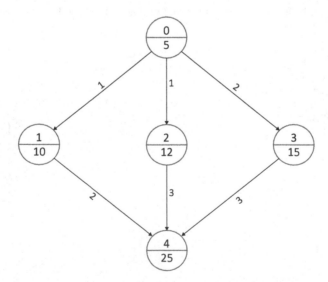

Fig. 3 A task graph with five tasks represented as nodes showing task numbers and task execution times (for example, milliseconds), and directed edges, indicating the order of execution of tasks, labeled with communication costs

$$T_{ij} = \frac{A_i}{S_j} + B_j \tag{1}$$

$$C(i_1, i_2, j_1, j_2) = \frac{D_{i_1 i_2}}{R_{j_1 j_2}} + I_{j_1} \tag{2}$$

The Schedule

Given a task graph $G = (T, A)$ for a target machine consisting of m processors, a schedule is a function f that maps each task to a processor at a specific starting time. A schedule $f(v) = (i, t)$, indicates that a task $v \in T$ is scheduled to be processed by processor p_i starting at time $= t$ units. No two tasks can have equal scheduling function. If $v < u$, where $v, u \in T$ and $f(v) = (i, t_1)$, $f(u) = (j, t_2)$, then $t_1 < t_2$. A schedule is considered feasible if it preserves all task precedence relations and communication restrictions. A Gantt chart is used to represent a schedule with task start and finishing times [1]. An example of a system that takes as input the task graph and the target machine representation, and gives out a Gantt chart representing the schedule as an output is illustrated in Fig. 4.

Performance Measures

The primary goal for scheduling in parallel and distributed computing systems is to achieve load balancing and to minimize the overall application execution time. The performance measure used to achieve this goal is the parallel execution time. The scheduling objective then is to minimize the parallel execution time for minimizing the overall completion time of an application. This, in turn, requires the minimization of the overall schedule length. Given a task graph $G = (T, A)$, the length of a schedule is the maximum finishing time of any task belonging to G. Formally [1],

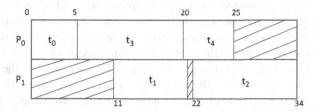

Fig. 4 A Gantt chart representing a schedule for the task graph shown in Fig. 3 on a machine with two processors P_0 and P_1. The shaded area represents the waiting time for each processor based on the task communication delays, assuming the tasks are initially located at processor P_1

$$length(f) = t_{max}, \text{ where } t_{max} = max\{t + T_{ij}\} \text{ and } f(i) = (j, t)$$
$$\forall i \in T, 1 \leq j \leq m \tag{3}$$

A Scheduling Taxonomy

Parallel and distributed computing has increasingly gained capacity to include a large range of applications. However, the power of a parallel and distributed computation can only be exploited to its full potential with efficient management and allocation of system resources relative to the computational load of the system. This motivation led to a large number of research, which focused on proposing solutions, in the form of scheduling techniques, to solve the problem of resource management in parallel and distributed computing systems. However, this has resulted in the development of various scheduling methodologies leading to the use of inconsistent terminology, problem formulations, and assumptions. Different techniques have been developed for optimizing different performance goals that used different performance metrics. Therefore, to unify the vast number of available scheduling methodologies for parallel and distributed computing, under a common, uniform set of terminology, Casavant and Kuhl [12] proposed a taxonomy that allows the classification of distributed scheduling algorithms according to a common and manageable set of salient features. This section details upon the proposed taxonomy along with a discussion on scheduling at global or system level, and at local or operating system level.

As already described in the previous section, the scheduling problem consists of three main components: (i) consumer(s), (ii) resource(s), and (iii) scheduling policy. Often, there is an assumption in parallel and distributed computing that considers a slight difference in the terms scheduling and allocation. Allocation is viewed in terms of resource allocation from the perspective of a resource, and scheduling is viewed from the perspective of a consumer in a computing system. Therefore, it is often assumed that allocation and scheduling are terms that exhibit a similar general mechanism from different viewpoints. Considering the three components, a scheduling system is evaluated via (1) performance, and (2) efficiency. Performance in a scheduling system is directly related to consumer satisfaction, which depends on how the scheduler allocates resources to process the consumer demands. Efficiency is measured in terms of the overhead and the cost to access the required resource.

There are two kinds of classification schemes for categorizing the scheduling algorithms: (i) hierarchical classification, and (ii) flat classification. The taxonomy presented in [12] is based on a hierarchical classification. However, a hierarchical classification does not capture all the issues in a scheduling system. Therefore, a flat classification that covers a number of scheduling parameters, which are not considered in a hierarchical scheme.

Hierarchical Classification

A tree based hierarchical classification of the taxonomy in [12] is illustrated in Fig. 5.

(a) *Local and global scheduling*: Local scheduling is performed at the operating system (OS) level and manages the assignment of tasks or processes to the time-slices of a single processor. Global scheduling is done at system level and provides a mechanism for allocating application tasks onto available processing elements. The classification discussed below has been developed for global scheduling techniques. Local scheduling will be discussed in more detail later in this section.

(b) *Static versus dynamic*: a choice between static and dynamic scheduling indicates the time at which the scheduling or allocation decisions are to be determined. Static scheduling algorithms are based on the assumption that the information regarding the application tasks, processes within these tasks, and the characteristics of the processing elements are available before the scheduling decision is made. Hence, each application task has a static assignment to a specific processor. Moreover, every time the scheduler encounters the same task, it assigns the task to that specific processor. Therefore, static scheduling algorithms are developed for a particular system configuration. Further, the scheduler may generate a new static assignment of tasks to processors, if the system topology or the task configurations change over a period of time. Static scheduling algorithms are also referred to as deterministic scheduling algorithms. Dynamic scheduling algorithms are based on a more realistic assumption that little or no a priori knowledge is available about the resource requirements of an application task, or about the computational environment in which the application will execute during its lifetime. In dynamic scheduling, an allocation decision is not made until the application tasks begin execution in the dynamic computational environment.

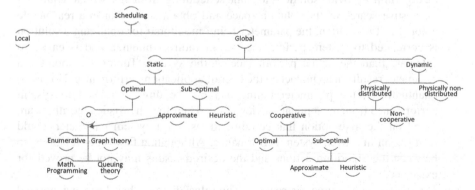

Fig. 5 Hierarchical classification based taxonomy for distributed scheduling algorithms [12]

(c) *Optimal versus sub-optimal*: In static scheduling, where complete information regarding the state of the computational system, and the resource requirements of application tasks are known a priori, optimal scheduling can be achieved based on some optimization function, such as, a function for minimizing the parallel execution time, a function for maximizing resource utilization, or a function for maximizing system throughput. However, for a different case of static scheduling, where some system parameters are computationally infeasible, suboptimal scheduling algorithms are more useful. Suboptimal scheduling algorithms are further categorized as approximate and heuristic algorithms, which are discussed next. Further, static optimal and static suboptimal-approximate scheduling is further categorized to employ the following techniques:

- Solution space enumeration and search.
- Graph theory
- Mathematical programming
- Queuing theory

(d) *Approximate versus heuristic*: Approximate solutions settle for a "good enough" solution as soon as it can be obtained, instead of searching the entire solution space for an optimal solution. Such solutions are often based on the assumption that a good solution can be recognized with minimal overhead. Moreover, in cases, where a metric is available for evaluating a solution that is obtained using approximate algorithms, result in decreased overhead time that is required to obtain the first acceptable schedule. The factors determining when an approximate algorithm should be used are: (i) availability of a function to evaluate a solution, (ii) time required to evaluate a solution using the function, (iii) availability of a metric to calculate the value of a solution, and (iv) availability of a mechanism for efficiently reducing the search space. The other suboptimal category belongs to heuristic-based algorithms. These are static algorithms, which are based on realistic assumptions regarding prior knowledge about the application and system characteristics. Unlike approximate algorithms, heuristic algorithms provide solutions to static scheduling problems, which require an exhaustive search of the solution space and obtain a solution in a reasonable amount of time. Often, the parameter being monitored for obtaining a solution is correlated to system performance in an indirect manner, and is easier to calculate than the actual performance of the system. Tuning the monitored parameter results in an impact on the overall application performance. However, quantitatively, the parameter tuning can not be directly related to system performance from an application viewpoint. Therefore, heuristic algorithms are based on the assumption that certain actions, on a system parameter, could result to an improved system performance. Although, a first-order relationship between the algorithm actions and the desired results may not be proved for existence.

(e) *Distributed versus non-distributed*: This classification has been categorized under dynamic scheduling algorithms. In dynamic scheduling, the decision for assigning tasks to processors is made during runtime. This classification categorizes dynamic scheduling techniques that either distribute the responsibility of assignment decisions among several processors (physically distributed approach), or that use a single processor for the work involved in making scheduling decisions (physically non-distributed approach). Therefore, this classification distinguishes between dynamic scheduling techniques, based on the logical authority of the decision-making process for task allocation.

(f) *Cooperative versus non-cooperative*: this classification distinguish between dynamic scheduling techniques, which target cooperation between the distributed components (cooperative), or the techniques that are developed for systems, where individual processors make decisions independent of the actions of the other processors (non-cooperative). In the non-cooperative case, individual processors are autonomous entities that make decisions for the use of their resources independently, disregarding the effect of their decision on the other processors in the system. In the cooperative case, every processor, in addition to delivering its own scheduling task, is responsible for working with the other processors to achieve a common system-wide goal.

In addition to the attributes that have been categorized using the hierarchical classification, there are a number of other distinguishing characteristics of scheduling in parallel and distributed systems that are not captured under any branch of the tree-structured taxonomy [12]. These attributes of a scheduling system could be sub categorized under several nodes of the hierarchical structure. Therefore, for the sake of clarity, these characteristics of a scheduling system are represented as a flat classification providing an extension to the existing hierarchical taxonomy.

Flat Classification

(a) *Adaptive versus nonadaptive*: An adaptive scheduling algorithm provides a solution for mapping application tasks to processing elements in the presence of runtime variations in application, algorithm, and system parameters. Such an adaptive scheduler is capable of taking multiple parameters into consideration while formulating a scheduling decisions. An adaptive scheduler modifies the value of a parameter in response to the behavior of the system. Often, such a system is known as a reward based system, where the scheduler receives reward, in the form of system performance, upon an action that it executes in the form of a specific resource assignment. Based on the reward, the scheduler may reformulate its allocation policy by tuning certain system parameters, if those parameters are inconsistent with the desired execution performance. On the other hand, a nonadaptive scheduler does not modify its basic scheduling mechanism due to variations in system activity. A non-adaptive scheduler manipulates the input parameters in the same way regardless of the system behavior.

(b) *Load balancing*: Runtime variations in application, algorithm, or system characteristics, along with poor scheduling decisions, lead to load imbalance among the executing processors in a parallel and distributed computing system. Often, load imbalance is one of the major reasons for performance degradation causing poor resource utilization, increased execution time and decreased system throughput. Recently, scheduling algorithms, focusing on load balancing, have received a great deal of attention. The goal of such scheduling algorithms is to allow processes on all nodes to finish execution at the same rate. A homogeneous system configuration facilitates this approach due to similar characteristics of all the processing elements. A load balancing scheduling system can further be categorized as a centralized system, or a distributed system. In a centralized system, a single master node is responsible for maintaining the information about the workload on the other processing elements. Further, in case of a load imbalance, the central node is responsible for transferring work from a heavily loaded processor to an idle or lightly loaded processor. However, in case of a highly imbalanced environment, the centralized node can become a bottleneck generating a large overhead leading to performance degradation. In a distributed scheduling system, each processor is responsible for maintaining the current state of information about the workload of other processors. In such a system, workload information is circulated over the network at regular time intervals, or as demanded by a processor. The processors are responsible for cooperating such that work can be transferred from a heavily loaded processor to a lightly loaded processor. However, with an increase in the skewed distribution of heavily loaded and idle processors, a distributed approach can generate large communication overhead where processors spend more time transferring work over the network than performing any useful work leading to a degraded performance. Often, load balancing scheduling algorithms are based on the assumption that the workload information, available for making load balancing decisions, is always accurate.

(c) *Bidding*: Scheduling techniques that utilize a bidding approach for assigning tasks to processors, deliver a cooperative scheduler such that enough information is exchanged between task nodes and processor nodes to facilitate an efficient allocation to optimize the overall performance of the system. As a basic mechanism of bidding, each processor node behaves as a manager and a contractor. The manager represents a task in a waiting state which is waiting to be allocated some computational resources. The contractor represents a processor node that is waiting to be allocated to a task node for execution. The manager announces the state of the task waiting for a computational resource. Further, the manager node receives bids from the potential contractor nodes. The amount and type of information exchanged, between the manager and the contractor, are the major factors in determining the efficiency of the bidding-based scheduler. Such a scheduling system is based on the notion of a fully autonomous collection of nodes, such that the manager has the freedom to select autonomously from a collection of bidding computational nodes, and the contractors are allowed to reject any assigned work if it leads to violation of local performance goals. Cloud brokers are an example of a bidding based scheduling system in cloud computing environments [13].

(d) *Probabilistic*: Probabilistic scheduling algorithms employ random selection of task to processor mapping from a large number of permutations of the available mappings, to reduce the prohibitive amount of time that would otherwise be required for analytically examining the entire solution space. The methodology generates a large number of different schedules via iteratively using the random selection process. Further, the generated set of randomly selected schedules is analyzed for selecting the best schedule from this set. Probabilistic scheduling is based on the assumption that enough variation is introduced by the random selection (using a certain probability distribution) to allow at least one good solution to enter into the randomly chosen set.

(e) *One-time assignment versus dynamic reassignment*: Scheduling methodologies that use one-time assignment technique are often used for jobs in the traditional batch processing environment in a parallel and distributed system. Such techniques generate a fixed schedule at a single point in time. Although many dynamic scheduling techniques use one-time assignment approach, they are considered static such that once a schedule has been generated for task allocation at runtime, no further changes can be made to that schedule. The scheduler generates a mapping of tasks to resources based on the information (in the form of estimated execution times or other system resource demands) provided by the application user. However, the variations that occur in the application and the system parameters at runtime are not considered by the generated schedule. Moreover, a user that understands the characteristics of the underlying computational system and the application, may provide false information to the system for manipulating the system to achieve better results.

Scheduling techniques that employ dynamic reassignment iteratively improve on earlier scheduling decisions. Dynamic reassignment is based on information on smaller computation units that are monitored over a time interval. Such techniques use dynamically created information, available from monitoring resources, to adapt to variations in application and system parameters. Therefore, dynamic reassignment can also be viewed as an adaptive approach for scheduling. Often, such an approach requires migrating tasks among processors generating an overhead. Thus, the use of such techniques should be weighed for trade-off between the generated overhead and the performance gain.

Operating System Scheduling

The classification of the scheduling strategies that have been discussed so far have been designed for global scheduling at system level. However, once the tasks are mapped to a processor, there is a need for a local scheduling mechanism that manages the execution of processes mapped to that processor. Scheduling at operating system level, also known as process scheduling, is an activity of a process manager that manages process selection, mapping, and removal of a process for a processor, according to a particular local scheduling methodology. Process scheduling is an integral part of operating systems running in the processing elements of parallel and distributed computing systems. A good process scheduling scheme allows multiple processes to be loaded simultaneously into the executable memory and share the CPU using time multiplexing.

During the lifetime of a process, it spends some time executing instructions (computing) and then makes some I/O request, for example, to read or write data to a file or to get input from a user. The period of computation between I/O requests is called a CPU burst. *Interactive processes* spend more time waiting for I/O and generally experience short CPU bursts. A text editor is an example of an interactive process with short CPU bursts. *Compute-intensive processes*, conversely, spend more time running instructions and less time on I/O. They exhibit long CPU bursts. A video transcoder is an example of a process with long CPU bursts. Even though it reads and writes data, it spends most of its time processing that data. A comparative example of an interactive process and a compute-intensive process switching between I/O and CPU burst cycles is shown in Fig. 6.

Almost all programs have some alternating cycle of CPU number crunching and waiting for I/O of some kind. In a simple system running a single process, the time spent waiting for I/O is wasted, and those CPU cycles are lost forever. A scheduling system allows one process to use the CPU while another is waiting for I/O, thereby making full use of otherwise lost CPU cycles. The challenge is to optimize the overall system performance and efficiency, subject to dynamically varying conditions. When the process enters into the system, then this process is put into a job queue. This queue consists of all processes in the system. The operating system also maintains other queues such as device queues. A device queue contains multiple processes waiting for a particular I/O device. Each device has its own device queue. A newly arrived process is put in the ready queue. Processes wait in ready queue for allocating the CPU. Once the CPU is assigned to a process, then that process will execute. To provide good time-sharing performance, the scheduler preempts a running process to let another one run. When an I/O request for a process is complete, the process moves from the waiting state to the ready state and gets placed on the ready queue. The process scheduler is the component of the operating system that is responsible for deciding whether the currently running process should continue running and, if not, which process should run next. There are four events that may occur where the scheduler needs to step in and make this decision:

(a)   (b)

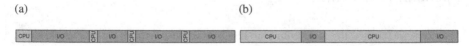

Fig. 6 A comparative example of differences between the I/O and CPU burst cycles of an interactive process versus a compute-intensive process. (**a**) Interactive process. (**b**) Compute-intensive process

1. The current process goes from the running to the waiting state because it issues an I/O request or some operating system request that cannot be satisfied immediately.
2. The current process terminates.
3. A timer interrupt causes the scheduler to run and decide that a process has run for its allotted interval of time and it is time to move it from the running to the ready state.
4. An I/O operation is complete for a process that requested it and the process now moves from the waiting to the ready state. The scheduler may then decide to preempt the currently-running process and move this newly-ready process into the running state.

A scheduler is a preemptive scheduler if it has the ability to get invoked by an interrupt and move a process out of a running state to let another process run. The last two events in the above list may cause this to happen. If a scheduler cannot take the CPU away from a process then it is a cooperative, or non-preemptive scheduler. Older operating systems, such as Microsoft Windows 3.1 or Apple MacOS prior to OS X, are examples of cooperative schedulers.

A number of local scheduling algorithms are being widely used by different operating systems. There are several performance metrics that form the optimization criteria for selecting the most appropriate scheduling algorithm for a specific computing environment. Following is a list of these performance metrics that play an important role in the selection of a particular process scheduling algorithm:

- CPU utilization – percentage of CPU being used for computational work.
- Throughput – number of processes completed per unit time.
- Turnaround time – time required for a particular process to complete, from submission time to completion.
- Waiting time – time spent by a process in the ready queue.
- Response time – The time taken in an interactive program from the issuance of a command to completion a response to that command.

First come first serve (FCFS) is the most straightforward approach to scheduling processes that are stored in a first-in, first-out (FIFO) ready queue. When the scheduler needs to run a process, it picks the process that is at the head of the queue. This scheduler is non-preemptive. *Round robin (RR)* scheduling is a preemptive version of FCFS scheduling. Processes are dispatched in a FIFO sequence, such that each process is allowed to run for a limited amount of time. This time interval is known as a time-slice or quantum. If a process does not complete within the time slice, the process is preempted and placed at the back of the ready queue. The *shortest remaining time first (SRTF)* scheduling algorithm is a preemptive version of an older non-preemptive algorithm known as *shortest job first (SJF)* scheduling. In SJF, the queue of jobs is sorted by estimated job length so that the smaller processes get to run first. This minimizes average response time. In SRTF, the algorithm sorts the ready queue by the estimated CPU burst time of a process. In *priority scheduling*, each process is assigned a priority based on a pre-defined criteria. A process, in the ready queue, with the highest priority gets to run next (UNIX-derived systems tend to use smaller numbers for high priorities while Microsoft systems tend to use higher numbers for high priorities). If the system uses preemptive scheduling, a process is preempted whenever a higher priority process is available in the ready queue. For a more detailed study on operating system process scheduling, the reader is referred to the literature in [14].

Examples of Recent Trends in Scheduling

With the evolution of the complexity of parallel and distributed computing, there has been a wide range of development of various scheduling algorithms and methodologies that can cater to the growing needs of the modern computing systems. A few examples of the recent trends in the development of scheduling in parallel and distributed computing will be discussed in this section. The examples have been selected such that they cover multiple classification categories of scheduling from the taxonomy described in the previous section. The examples begin with a description of work that have proposed and compared static, dynamic-nonadaptive, and dynamic-adaptive scheduling techniques employed in traditional high performance computing systems for scientific applications, followed by a discussion of a number of heuristic-based scheduling techniques employed in grid computing systems. Further, an example of scheduling strategies for cloud computing systems, which are defined as one of the modern parallel and distributed computing systems, will be discussed.

Dynamic Load Balancing Via Loop Scheduling in High Performance Computing

High performance computing was developed to serve the interests in the accurate modeling and simulation of various complex phenomena from various scientific areas. The scientific applications are often routines that perform varying number of repetitive computations (in the form of DO/FOR loops) over very large data sets. Moreover, these applications may exhibit irregular behavior leading to differing execution times of each iteration. In scientific applications, a loop iteration (or a chunk of loop iterations) with variable execution time is considered to be a task with varying execution time.

Dynamic loop scheduling (DLS) algorithms provide application-level load balancing of loop iterations, with the goal of maximizing application performance on the underlying system. Many DLS methods are based on probabilistic analyses, and therefore possess the capability to be inherently robust against unpredictable variations in application and system characteristics. A number of DLS algorithms

(a) (b)

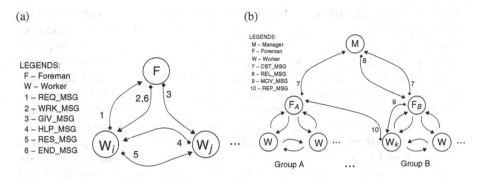

Fig. 7 Dynamic loop scheduling management approaches. (**a**) Centralized management. (**b**) Hierarchical management system

have been proposed in the last decade and have been integrated into several scientific applications, yielding significant performance improvements [15]. The DLS methods are further categorized as *non-adaptive* and *adaptive*. The non-adaptive DLS techniques have been described in a survey presented in [16]. However, the dynamic non-adaptive techniques did not address the unpredictable changes in the computational environment at runtime. Therefore, adaptive DLS techniques were developed to address this problem [17, 18]. Most of the above adaptive methods use a combination of runtime information about the application and the system, to estimate the time the remaining tasks will require to finish execution, in order to achieve the best allocation possible for optimizing application performance via load balancing.

Most loop scheduling methods are developed assuming a central ready work queue of tasks (central management approach), where idle processors obtain chunks of tasks to execute. The scheduling decisions are centralized in the master node, which is also known as the foreman node. However, accessing the foreman may become a bottleneck when a large number of workers attempt to simultaneously communicate with it. To address this bottleneck, a two-level (hierarchical) management strategy is employed, which uses multiple-foremen and partitioned disjoint processor groups of worker nodes. Each processor group executes concurrently independent parts of the problem. Forming processor groups dynamically assists the DLS methods to leverage the best possible application performance on the large-scale platform [19]. Figure 7 illustrates the centralized, and the distributed management approach used in dynamic loop scheduling methods.

Heuristic Scheduling for Grid Computing

Grids computing is a new trend in parallel and distributed computing. Computational grids are distributed systems with independent, and non-interactive compute intensive workloads. Unlike conventional high performance computing systems such as cluster computing, grid computing is more loosely coupled, heterogeneous, and geographically dispersed. Moreover, scheduling in a grid computing environment is different from scheduling in a traditional computing system, where a scheduler only manages a single local cluster and has control over the cluster resources, whereas a grid scheduler has no control over the distributed resources, and its availability of information about the system state is limited. Scheduling and resource allocation decisions in grid computing systems are approached differently for computational grid versus data grid. The scheduling techniques implemented in a compute grid focuses on managing computational resources, such as, processor compute cycles. In a data grid, the scheduler focuses on managing the distributed data and the related communication over the grid network connecting the distributed geographical locations. The scheduling problem in a grid computing system can be viewed as an optimization problem which is known to be NP-Complete [20]. Therefore, recent research has shown that heuristic techniques are increasingly being used for solving the scheduling optimization problem.

Ant Colony Optimization (ACO) is a heuristic algorithm that employs local search for combinatorial problems. ACO has been used to solve several NP-hard problems such as the traveling salesman problem, graph coloring problem, vehicle routing problem, and others. As a recent study, a modified version of the ACO algorithm, called the Balanced ACO (BACO) algorithm, has been used for grid scheduling to optimize the system makespan [21]. Using this algorithm, the grid scheduler selects a resource for mapping to the job request by finding the largest entry in the Pheromone Indicator (PI) matrix among the available jobs to be executed, where jobs are independent of each other. Another framework that combines the Fuzzy C-Mean clustering ACO algorithm to improve the scheduling in a heterogeneous grid is presented in [22]. Herein, the Fuzzy C-Mean algorithm is used for classification of the jobs into separate classes, and the ACO algorithm maps the jobs to the appropriate resources that are relevant to those classes. A scheduling algorithm for task scheduling using particle swarm optimization (PSO) heuristic for an improved job classification is given in [23]. The heuristic approach is used to map jobs to grid resources based on the calculated task length of a job and the calculated processing power of a grid resource. This method has been developed to optimize resource utilization in a grid environment. Tabu Search (TS) heuristic has also been used in a scheduling technique in grid computing using the GridSim tool in [24]. The basic principle of TS is employ local search techniques after reaching a local optimum and prevent cycling back to previously visited solutions by the use of a storage data structure called Tabu list. Further, TS can be used in conjunction with other heuristic approaches such as genetic algorithm, constraint programming, and integer programming technique, for improved performance results.

Scheduling Advances for Cloud Computing

The advent of cloud computing has revolutionized the concept of parallel and distributed computing. Cloud computing enables the access to computational resources, information, and technology to users as services over the Internet. The services that are provided in a cloud computing environment have been categorized into three main classes: (i) Infrastructure as a Service (IaaS), (ii) Platform as a Service (PaaS), and (iii) Software as a Service (SaaS). These services are provided on demand in a pay-per-use manner via the Internet. Cloud computing differs from traditional computing environments, such as cluster computing and grid computing, as it uses virtualization for resource management. This allows cloud computing resources to be scheduled as cloud services, and are provided to the end-user as a utility [25]. Recently, the concept of a cloud broker has evolved and cloud computing environments are being considered as federated systems that consist of a large number of resources as a federation [13]. However, cloud computing provides a finite pool of virtual on-demand resources, therefore, requiring efficient scheduling and resource allocation techniques that can manage the dynamic and competitive computing environment.

Cloud computing is seen as a three-layered framework consisting of an infrastructure layer, a platform layer, and a software layer. Thus, scheduling methodologies have been proposed for resource management in and between all these layers. A taxonomy of scheduling in the three cloud resource layers has been defined in [26]. The architecture consisting of the three layers, the IaaS, PaaS, and SaaS stacks, and a classification of the scheduling requirements for each of the layers is illustrated in Fig. 8. Scheduling in the software service layer requires delivering software in the form of user applications, tasks, workflows, and others, while optimizing the efficiency and maintaining the QoS requirements. Scheduling in the platform service layer requires mapping virtual resources to physical resources such that there is minimal load balance, and minimized power consumption. Scheduling in the infrastructure service layer requires delivery of physical computational and communication resources to the above two layers for efficient application to resource mapping, with minimal application or virtual machine migration, in a federated cloud computing environment.

Given that cloud computing is still an emerging technology, solutions for scheduling and resource management are fairly recent developments in the field. Some of the solutions to the scheduling problem for different aspects of cloud computing have been proposed as combinatorial solutions in [27–29], and as heuristic approaches in [30–32].

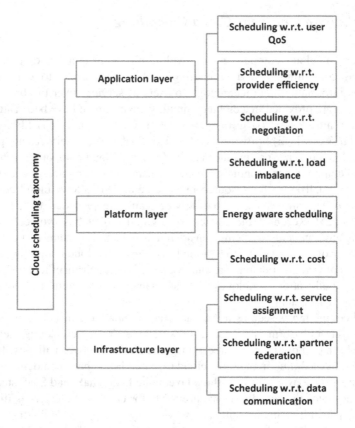

Fig. 8 Taxonomy of the cloud resource scheduling at different service layers with a focus on different scheduling challenges and objectives [26]

Chapter Review

This chapter provides a fundamental description of scheduling in parallel and distributed computing systems. The knowledge presented in here is a result of a survey and collection of information from a number of state-of-the-art work (provided as references) done in this field. Scheduling has been defined as a collective task consisting of the following sub-tasks: detecting parallelism, partitioning the problem into independent sub-tasks, and scheduling these sub-tasks on processors. Often, when scheduling is referred, it is assumed to encompass the afore mentioned sub-tasks. A generic scheduling system is comprised of four components: the target machines, the parallel tasks, the generated schedule, and a performance criterion. Over the years, a number of scheduling techniques have been developed to define the mapping policy for executing applications or tasks in a parallel and distributed computing environment. A taxonomy, proposed in [12], for the classification of various scheduling techniques has been described in section "A

Scheduling Taxonomy". Further, a distinction between application level scheduling and process scheduling at OS level is given via a description of scheduling at global and local level, respectively. A few examples of scheduling in traditional parallel and distributed computing systems, such as clusters and grid, and modern computing systems, such as clouds, have also been discussed to explain the differences in the scheduling approaches and objectives for such systems.

Exercises

1. Conduct a comparison between static and dynamic approaches. Exemplify with some cases, where one approach might be better than the other.
2. Suggest a performance metric that would be most appropriate for each of the following scenario:

   - job scheduling in a manufacturing plant
   - management for an aircraft waiting for landing clearance
   - customers waiting for a teller in a banking system

3. Show an example of a case, where load balancing is more important than minimizing the finishing times of every machine.
4. Discuss the differences between scheduling at global and local levels. How does a poor scheduling decision at one of these levels affect the performance at the other level?
5. The Ready queue of an operating system at a particular time instance is given in Table 1. The behavior of each process (if it were to use the CPU exclusively) is as follows. A process runs for the CPU burst given, then requests an I/O operation that takes 10 ms, then runs for another CPU burst of equal duration to its first CPU burst and then terminates. However, the four processes must share the CPU. Assume that the I/O operations can proceed in parallel. Draw a chart showing the execution of these processes under the round robin policy, with time quantum = 2.
6. Discuss the differences in the objectives and the challenges for scheduling in a cluster computing environment, a grid computing environment, and a cloud computing environment.

Table 1 Ready queue of an operating system with process CPU burst in milliseconds

Process	Next CPU burst
P1	2
P2	3
P3	7
P4	18

References

1. H. El-Rewini, T. G. Lewis, and H. H. Ali, *Task Scheduling in Parallel and Distributed Systems.* Upper Saddle River, NJ, USA: Prentice-Hall, Inc., 1994.
2. N. Immerman, "Expressibility and parallel complexity," *SIAM Journal on Computing*, vol. 18, no. 3, pp. 625–638, 1989.
3. J. C. Wyllie, "The complexity of parallel computations," Cornell University, Tech. Rep., 1979.
4. E. G. Coffman and J. L. Bruno, *Computer and job-shop scheduling theory.* John Wiley & Sons, 1976.
5. O. H. Ibarra and C. E. Kim, "Heuristic algorithms for scheduling independent tasks on nonidentical processors," *J. ACM*, vol. 24, no. 2, pp. 280–289, Apr. 1977. [Online]. Available: http://doi.acm.org/10.1145/322003.322011
6. D. Fernandez-Baca, "Allocating modules to processors in a distributed system," *IEEE Transactions on Software Engineering*, vol. 15, no. 11, pp. 1427–1436, Nov 1989.
7. R. Gupta and M. L. Soffa, "Region scheduling: An approach for detecting and redistributing parallelism," *Software Engineering, IEEE Transactions on*, vol. 16, no. 4, pp. 421–431, 1990.
8. J. A. Fisher, "Trace scheduling: A technique for global microcode compaction," *IEEE Transactions on Computers*, vol. C-30, no. 7, pp. 478–490, July 1981.
9. M. D. McCool, A. D. Robison, and J. Reinders, *Structured parallel programming: patterns for efficient computation.* Elsevier, 2012.
10. H. Hoffmann, A. Agarwal, and S. Devadas, "Partitioning strategies for concurrent programming," *MIT Open Access Articles*, 2009.
11. V. Sarkar, *Partitioning and Scheduling Parallel Programs for Multiprocessors.* Cambridge, MA, USA: MIT Press, 1989.
12. T. L. Casavant and J. G. Kuhl, "A taxonomy of scheduling in general-purpose distributed computing systems," *IEEE Transactions on Software Engineering*, vol. 14, no. 2, pp. 141–154, Feb 1988.
13. R. Mehrotra, S. Srivastava, I. Banicescu, and S. Abdelwahed, "Towards an autonomic performance management approach for a cloud broker environment using a decomposition-coordination based methodology," *Future Generation Comp. Syst.*, vol. 54, pp. 195–205, 2016. [Online]. Available: http://dx.doi.org/10.1016/j.future.2015.03.020
14. A. Silberschatz, P. B. Galvin, and G. Gagne, *Operating System Concepts*, 9th ed. Wiley Publishing, 2009.
15. S. Srivastava, I. Banicescu, F. M. Ciorba, and W. E. Nagel, "Enhancing the functionality of a gridsim-based scheduler for effective use with large-scale scientific applications," in *2011 10th International Symposium on Parallel and Distributed Computing*, July 2011, pp. 86–93.
16. A. R. Hurson, J. T. Lim, K. M. Kavi, and B. Lee, "Parallelization of doall and doacross loops - a survey," *Advances in computers*, vol. 45, pp. 53–103, 1997.
17. I. Banicescu and V. Velusamy, "Load balancing highly irregular computations with the adaptive factoring," in *Parallel and Distributed Processing Symposium., Proceedings International, IPDPS 2002, Abstracts and CD-ROM*, April 2002, pp. 12 pp–.
18. I. Banicescu, V. Velusamy, and J. Devaprasad, "On the scalability of dynamic scheduling scientific applications with adaptive weighted factoring," *Cluster Computing*, vol. 6, no. 3, pp. 215–226, 2003.
19. R. Cariño, I. Banicescu, T. Rauber, and G. Rünger, "Dynamic loop scheduling with processor groups." in *ISCA PDCS*, 2004, pp. 78–84.
20. J. D. Ullman, "Np-complete scheduling problems," *J. Comput. Syst. Sci.*, vol. 10, no. 3, pp. 384–393, Jun. 1975. [Online]. Available: http://dx.doi.org/10.1016/S0022-0000(75)80008-0
21. R.-S. Chang, J.-S. Chang, and P.-S. Lin, "An ant algorithm for balanced job scheduling in grids," *Future Generation Computer Systems*, vol. 25, no. 1, pp. 20–27, 2009. [Online]. Available: http://www.sciencedirect.com/science/article/pii/S0167739X08000848

22. T. Helmy and Z. Rasheed, "Independent job scheduling by fuzzy c-mean clustering and an ant optimization algorithm in a computation grid." *IAENG International Journal of Computer Science*, vol. 37, no. 2, 2010.

23. S. Selvarani and G. S. Sadhasivam, "Improved job-grouping based pso algorithm for task scheduling in grid computing," *International Journal of Engineering Science and Technology*, vol. 2, no. 9, pp. 4687–4695, 2010.

24. M. Yusof, K. Badak, and M. Stapa, "Achieving of tabu search algorithm for scheduling technique in grid computing using gridsim simulation tool: multiple jobs on limited resource," *Int J Grid Distributed Comput*, vol. 3, no. 4, pp. 19–32, 2010.

25. R. Buyya, C. S. Yeo, S. Venugopal, J. Broberg, and I. Brandic, "Cloud computing and emerging it platforms: Vision, hype, and reality for delivering computing as the 5th utility," *Future Generation computer systems*, vol. 25, no. 6, pp. 599–616, 2009.

26. Z.-H. Zhan, X.-F. Liu, Y.-J. Gong, J. Zhang, H. S.-H. Chung, and Y. Li, "Cloud computing resource scheduling and a survey of its evolutionary approaches," *ACM Computing Surveys (CSUR)*, vol. 47, no. 4, p. 63, 2015.

27. B. Speitkamp and M. Bichler, "A mathematical programming approach for server consolidation problems in virtualized data centers," *IEEE Transactions on Services Computing*, vol. 3, no. 4, pp. 266–278, Oct 2010.

28. H. N. Van, F. D. Tran, and J. M. Menaud, "Performance and power management for cloud infrastructures," in *2010 IEEE 3rd International Conference on Cloud Computing*, July 2010, pp. 329–336.

29. T. A. L. Genez, L. F. Bittencourt, and E. R. M. Madeira, "Workflow scheduling for saas / paas cloud providers considering two sla levels," in *2012 IEEE Network Operations and Management Symposium*, April 2012, pp. 906–912.

30. V. Roberge, M. Tarbouchi, and G. Labonte, "Comparison of parallel genetic algorithm and particle swarm optimization for real-time uav path planning," *IEEE Transactions on Industrial Informatics*, vol. 9, no. 1, pp. 132–141, Feb 2013.

31. M. A. Rodriguez and R. Buyya, "Deadline based resource provisioning and scheduling algorithm for scientific workflows on clouds," *IEEE Transactions on Cloud Computing*, vol. 2, no. 2, pp. 222–235, April 2014.

32. Y. L. Li, Z. H. Zhan, Y. J. Gong, J. Zhang, Y. Li, and Q. Li, "Fast micro-differential evolution for topological active net optimization," *IEEE Transactions on Cybernetics*, vol. 46, no. 6, pp. 1411–1423, June 2016.

Printed in the United States
By Bookmasters